Growth and Lactogenic Hormones

Neuroimmune Biology, Volume 2

Neuroimmune Biology

Series Editors

I. Berczi, A. Szentivanyi

Growth and Lactogenic Hormones

Neuroimmune Biology, Volume 2

Volume Editors

Lina Matera
Robert Rapaport

University of Turin, Turin, Italy
and
Mount Sinai Hospital, New York, USA

2002
ELSEVIER
AMSTERDAM – LONDON – NEW YORK – OXFORD – PARIS – SINGAPORE – TOKYO

ELSEVIER SCIENCE B.V.
Sara Burgerhartstraat 25
P.O. Box 211, 1000 AE Amsterdam, The Netherlands

First edition 2002

Library of Congress Cataloging in Publication Data
A catalog record from the Library of Congress has been applied for.

British Library Cataloguing in Publication Data
A catalogue record from the British Library has been applied for.

ISBN: 0 444 51051 6
Series ISSN: 1567 7443

∞ The paper used in this publication meets the requirements of ANSI/NISO Z39.48-1992 (Permanence of Paper).
Printed in The Netherlands.

Foreword: The Neuroimmune Biology of Growth and Lactogenic Hormones

Growth hormone (GH) has long been shown in animals to stimulate immune and inflammatory reactions. However, clinicians did not find immune abnormalities in pituitary dwarf individuals, which raised serious doubts about the role of GH in immune function [1,2]. To this date it is difficult to demonstrate immune alterations in children after GH therapy, although transient responses can be demonstrated. In contrast, *in vitro* observations with human lymphocytes indicate the role of GH in immunoregulation [3].

For the immunoregulatory role of prolactin (PRL) the first decisive evidence was obtained in hypophysectomized (Hypox) rats, which are immunodeficient [4]. Replacement doses of PRL or GH completely restored the immune reactivity of Hypox animals. Moreover, treatment with the dopaminergic drug, bromocriptine, which inhibits pituitary PRL secretion, was as immunosuppressive as was Hypox. Again the immune response could be restored with either GH or PRL treatment [5–12]. Subsequently numerous observations confirmed the immunoregulatory potential of PRL and GH, as attested for in this volume.

Although much less studied, the evidence available clearly indicates that placental lactogenic hormones (PL) also have the potential of regulating the immune system [7,13–15].

SOME TIMELY QUESTIONS AND ANSWERS

Why pituitary dwarf individuals are immunocompetent?

Pituitary dwarfs have normal serum PRL levels [16]. Animal experiments showed that PRL is able to maintain immune function in the absence of GH [6–8,11]. On this basis it is reasonable to suggest that PRL is responsible for the maintenance of immunocompetence in dwarf people as well.

Snell dwarf mice are deficient in both PRL and GH, yet show immunocompetence. Why?

These mice are deficient of the pituitary transcription factor, Pit-1, which controls the production of GH, PRL and of thyroid stimulating hormone (TSH) secretion. In the recent literature these animals have often been presented as lacking completely pituitary GH and PRL secretion. However, low serum level of GH and PRL was detectable in these animals by radioimmunoassay [17–19]. Similarly, humans with Pit-1 mutations have subnormal levels of GH, PRL and TSH, and are not negative [20]. It was also shown that Snell dwarf mice produced less lymphocyte-derived PRL (LPRL) than did their normal littermates. LPRL could be restored to normal by thyroxin treatment of lymphocytes *in vitro*. The production of placental lactogen was normal [21]. Therefore, it appears that Pit-1 deficient mice and humans do in fact, have sufficient pituitary hormone levels, which permit survival and immune function. Clearly, the joint and complete deficiency of pituitary GH and PRL has not been demonstrated to date in man or in animals. This point is further illustrated by the observation that Hypox rats are able to survive for 6–8 months because of the presence of residual PRL in their serum. If this residual PRL is neutralized by antibodies, the animals will perish within a few weeks time [22].

Mice that lack PRL or IGF-I function survive and are immunocompetent. Why?

Knockout mice, lacking either PRL or its receptor (PRLR), or IGF-I are immunocompetent. It was interpreted, therefore, that these hormones are not obligate immunoregulators, but rather, affect immune reactions as anabolic and stress modulating agents [23–25]. In actual fact the data obtained in knockout mice is a powerful confirmation of the original observations that growth and lactogenic hormones (GLH) show redundancy in the maintenance of immunocompetence [7–11]. Today a compelling body of experimental evidence, which is presented in this volume, indicates that indeed this is the case. Clearly, immune function, as many other functions in the body, are maintained by multiple genes that show redundancy [26].

Growth hormone and PRL belong to the type-I cytokine family [27]. Functional overlap and redundancy is the rule for type I cytokines (and for other cytokines as well) in the immune system. The receptor for type I cytokines consists of a ligand specific chain and of a shared signal transducing chain. For instance in the first group, where IL-2, -4, -7, -9 and -15 belong, there is a common gamma chain (γ_c), for the second group (IL3, -5 and GM-CSF) it is called the common beta chain (β_c) and for the third group (Il-6, -11, oncostatin M, leukemia inhibitory factor, ciliary neurotropic factor and cardiotrophin-1) the common chain is glycoprotein 130. Signal transduction is possible only if the ligand binding and the signal transducing chains are crosslinked by the specific cytokine. Knockout experiments in this system showed that the elimination of specific cytokines or their specific receptor chains produced minimal if any abnormalities. However, knocking out the shared signal transducing γ_c chain resulted in severe combined immunodeficiency [27,28]. These observations collectively indicate that type I cytokines are indispensable as a group for normal immune function. Apparently there is enough redundancy in this group to compensate for the lack of any particular cytokine.

Prolactin and growth hormone do not share receptor chains with any of the above cytokines. However, human GH and other primate GH are known to act on PRL receptors and to exert lactogenic activity in many species [29]. Similarly, IGF-I, IGF-II and insulin show functional overlap [13]. These facts indicate that functional redundancy exists within GLH hormones, which explains why the disabling of single genes is of no consequence for immune function.

The major signal transduction pathway, which involves the Janus kinases (JAK) and signal transducers and activators of transcription (STAT) nuclear regulatory factors, is shared between cytokines and growth and lactogenic hormones. STAT knockout mice show severe developmental and immune deficiencies [14,27,28,30]. This emphasizes the significance of this signal transduction pathway in immune development and function [31,32].

The evidence, that has accumulated to date, indicates that GLH are indispensable as a group for normal development and bodily functions, including immune function [14,15,22,24,31,33, 54] . Because the JAK-STAT transcription pathway of PRL and GH are shared with interleukins and hemopoietic growth factors [14,27,29], some regard PRL and GH as members of the hemopoietic cytokine family. However, GLH have a much wider spectrum of biological activity than any of the type I cytokines. A functional overlap with these cytokines could simply indicate the capacity of GLH hormones to maintain the hemopoietic and immune systems at times when cytokines are in short supply as well as to boost immune activity in situations of emergency.

Female mice that lack PRL or do not respond to it, do not reproduce [23–25]. In this context, one must not forget that without normal immune function reproduction is not possible. The immune system is involved in the function of the gonads, in conception, in the normal development of the fetus and it plays a role in the normal function of the mammary gland. Milk plays a very important role in the transfer of maternal antibodies and of other immune factors as well as PRL itself to the fetus. There is evidence to indicate that PRL is important for the immunological function of the mammary gland [15,33–37]. Therefore, the immunoregulatory function of PRL may be of special importance in the female reproductive compartment.

Seriously ill patients got worse after treatment with GH, why?

In patients with acute phase response (APR) the GH-IGF-I axis is suppressed. This observation prompted several clinical trials with GH, which were aimed at restoring this axis in the hope of preventing the severe catabolic state and to improve immunocompetence in the interest of increased survival. However, so far this hope did not materialize. In fact a controlled clinical trial showed that GH treatment of severely ill patients significantly elevated the proportion that did not survive [38]. Deaths attributed to "septic shock or uncontrolled infection" occurred nearly four times more commonly in GH treated patients compared to placebo receiving patients. Although no data were given regarding immune parameters, the authors suggested that alterations in immune functions may have contributed to these fatalities.

Critical illness elicits a highly coordinated and powerful acute phase reaction, whereby the immune system is switched from the adaptive mode of response to the amplification of natural immune mechanisms. The acute phase response is characterized by profound elevations of interleukin-1, interleukin-6 and tumor necrosis factor-α (TNF-α), which induce complex neuroendocrine and metabolic alterations. The hypothalamic-pituitary-adrenal axis is activated, whereas the serum levels of growth hormone, insulin-like growth factor and prolactin are suppressed. Tri-iodothyronine is also diminished (sick euthyroid syndrome). The increased serum level of cytokines and the array of neuroendocrine changes lead to fever, catabolism and to the suppression of the T lymphocyte-dependent adaptive immune system. At the same time natural immune mechanisms are amplified. There is a rapid rise in serum natural antibodies and liver-derived acute-phase proteins such as endotoxin-binding protein and C-reactive protein. These antibodies and acute phase proteins have the capacity to recognize homologous crossreactive epitopes (homotopes) on microbes and on altered self components in a polyspecific fashion and activate immune defense mechanisms after combining with the respective homotope. Host defenses against toxins and other noxious agents are also increased during the acute phase response [39–41].

The acute phase response is a massive neuroimmune and metabolic response that mobilizes all the resources of the body in the interest of host defence and survival. The findings of Takala and co-workers [38] suggest strongly that the suppression of the GH – IGF-I axis in APR is required for intense catabolism to take place. A rapid release of nutrients and of energy is necessary under these conditions in order to support maximally the defence system of the body, that includes the hypothalamus – pituitary – adrenal axis, the sympathetic nerves system, the bone marrow, CD5+ B lymphocytes, leukocytes and the liver [39–41]. The adaptive immune system is controlled by thymus-derived (T) lymphocytes and needs several days to a week for an effective response. During APR no time is available for an adaptive immune reaction, and this system is shut down, primarily by the cytokine and endocrine alterations that take place. The thymus and T cell function is heavily dependent on the GH/PRL – IGF-I axis and it is suppressed profoundly by the elevated levels of glucocorticoids and cathecolamines [39–41]. Recent observations showed that GH inhibits the production of acute phase proteins in rats with burn injury and in human hepatocytes [42,43]. These findings strongly support the above hypothesis.

One may argue that the most efficient way to fuel the intensive systemic effort for survival in APR is by the rapid breakdown of bodily tissues. GH is a powerful anabolic hormone, which supports the T lymphocyte dependent immune system, and acts as antagonist of the HPA axis that promotes APR [6,8,11,39–41,44]. The results of this controlled trial support the hypothesis that the inhibition of the HPA axis and of catabolism by GH treatment in APR hampers the bodie's defence mechanisms, which may have fatal consequences.

What is the role of GLH in the neuroimmune regulatory network?

Current evidence indicates that GLH is required for the normal growth and development of embryos as well as for the development of the immune system and the maintenance of immunocompetence. Clearly, GLH supports any adaptive immune function and natural immunity under physiological conditions. Lymphocyte precursors do not have receptors for antigen or cytokines and for this reason must rely on other physiological regulators for survival and differentiation. Even after full differentiation, naïve lymphocytes remain small and do not synthesise, neither do they respond to immune-derived cytokines. In the absence of antigenic stimulation these cells must rely on physiological systemic mediators to survive. The thymus and other lymphoid organs lose cellularity and weight in Hypox rats, which also show a profound immunosuppresion. The weight of lymphoid organs and immune reactivity can be normalized in Hypox animals by replacement doses of either PRL or GH [45]. The situation is similar in old animals to some extent, although full immune restoration by GH treatment was not possible [46]. Both GH and PRL are capable of maintaining immunocompetence, which is antagonized by the hypothalamus-pituitary adrenal axis [44]. This enables the pituitary gland to exert a true regulatory effect on the immune system. Clearly, the pituitary gland does not only maintain immunocompetence, but also, is capable of fine tuning the level of reactivity and plays a fundamental role in the induction of immunoconversion during APR [5,6,9, 41,45].

There is compelling evidence to indicate that after activation by antigen or mitogen lymphocytes produce their own PRL and/or GH. This makes the rapid proliferation required for an immune response feasible [48–53]. This situation is similar to the development of the embryo, where placental lactogenic hormones make it possible for the embryo to grow at a very rapid rate. The production of placental GLH is independent from the pituitary gland and is controlled by "placental" promoters. This allows these hormones to override the regulatory power of the maternal pituitary gland during pregnancy in the interest of assuring the proper development of the fetus. Therefore, while conception is clearly dependent on normal pituitary function, the fetus becomes independent from such influence [54].

Interestingly placental promoter was found also in association with the lymphocyte PRL gene [55]. However, Pit-1 was also detected in lymphocytes [56]. This suggests that once the lymphocyte PRL gene is activated, pituitary PRL is no longer required for lymphocyte growth or function. Once the immune response is over, most of the activated lymphocytes will undergo apoptosis, which is governed by a complex mechanism that involves the delivery of death signals, primarily by the Fas-FasL system. However, a specialized subset, called memory cells, will survive [57,58].

We observed years ago that the primary antibody response is fully pituitary dependent, whereas the secondary response shows only partial dependence. Actually, when the rats were immunized first, hypohysectomized and immunized again, they produced antibodies in response to the second stimulus, which was of similar magnitude to the primary response [6]. These results suggest that memory cells maintained their reactivity after Hypox, but the recruitment of naïve lymphocytes, which occurs in normal animals, could not take place.

The mechanism(s) for the long term survival and self-renewal capacity of memory B and T lymphocytes is not understood. It was hypothesised that memory B lymphocytes are stimulated by idiotypes, which are unique determinants of antigen receptors [59]. Major histocompatibility antigens presenting self peptides were suggested to fulfill a stimulatory role for memory T lymphocytes [60]. The role of cytokines in T cell longevity is also recognized and IL-15 was claimed to be necessary for CD8+ memory cells [61]. T lymphocyte apoptosis is inhibited by interferon- (IFNα) and IFNβ and were proposed to play a role in memory cell survival. These cytokines are able to maintain T cells without an antigenic stimulus [62]. Recently Cho and co-workers [63.] demonstrated that in recombinase deficient (RAG-1 –/–) mice, which are lymphopenic, naïve T lymphocytes undergo "homeostasis-stimulated" proliferation, which is MHC restricted, and develop into memory cells in the absence of antigenic stimulation. These cells acquire the phenotypic and functional characteristics of antigen-induced memory CD8+ T cells and lyse target cells directly and respond to lower doses of antigen than naïve cells and secrete IFNγ faster upon restimulation. Interleukin-2 or co-stimulation by CD28 are not required and effector cells are not formed during this homeostatic differentiation. These findings indicate that memory T cells may be generated and maintained under the influence of physiological immunoregulatory mechanisms, in the complete absence of immune stimulation by antigen, cytokine or adhesion signals.

Immature thymocytes of rodents are killed by glucocorticoids, whereas mature thymocytes are saved. The helper, suppressor and killer functions of T lymphocytes and the production of interleukins by them are all inhibited by glucocorticoids. In contrast, the function of memory cells and of cells mediating the graft-versus-host reaction is not inhibited by glucocorticoids [64]. Prolactin and GH antagonize the immunosuppressive effects of the ACTH-adrenal axis [6,8,11]. Taking all the evidence in consideration, one may suggest that naive T cells are maintained in the absence of antigenic stimulation by pituitary GLH, whereas memory T cells are autonomous and survive and resist glucocorticoids, most likely because they produce autocrine GLH and cytokines that enable these cells to survive and to resist adverse conditions, such as the APR.

This pattern of immune response, whereby autonomy is obtained gradually from pituitary regulation by GLH, assures maximal host defence. At the same time the maturation and selection process of lymphocytes in the thymus and bone marrow is tightly controlled by the neuroimmune regulatory system. Pituitary GLH is important for the development of lymphocytes and of the maintenance of mature naïve cells in a state of immunocompetence [22,45]. It is likely that after activation paracrine GLH gradually assumes a prominent role in the maintenance of lymphocyte function. Finally, it appears that memory cells rely on autocrine GLH for long term survival and function. This is to be substantiated further experimentally.

FROM BENCH TO BEDSIDE

The goal of this volume is to present the current evidence for the role of growth and lactogenic hormones in the neuroimmune regulatory system. The evidence presented is compelling and shows that all the requirements for proven biological significance have been fulfilled. Receptors for GLH on cells of the immune system have been characterized, signal transduction pathways have been identified and are being characterized, and the immunoregulatory activity of GLH has been demonstrated in various species, including man. It is also clear that both PRL and GH are produced within the immune system by activated cells. Placental, pituitary and tissue derived GLH hormones all play a role in neuroimmunoregulation. This redundancy serves well the adaptability and versatility of the neuroimmune regulatory network as well as of immune function.

Finally, the therapeutic use and manipulation of GLH is currently underway for the treatment/correction of various human conditions. Therefore, the ultimate criterion for the success of scientific research, i.e the application of knowledge obtained on the laboratory bench at the bedside is being fulfilled. It is very rewarding to witness one's initial research efforts to develop and reach this critical stage. No reasonable arguments can be raised any more in the face of this evidence against the fundamental role of growth and lactogenic hormones in immunoregulation. Clearly the challenge today is not to prove, but to understand, the neuroimmune regulatory role of GLH in its entire complexity.

THE FUTURE

The realization that a third systemic regulator, the immune system, is included in homeostatic and in allostatic regulation to form the Neuroimmune regulatory network, provides new foundation to Biology. This network is immensely complex and powerful and is involved in both physiological (homeostatic) and pathophysiological (allostatic) regulation. Indeed the entire biological cycle from conception till death of the individual is subject to this regulatory system. It is also clear that the defects and abnormalities of this system is the underlying cause for many diseases that include neural conditions, endocrine and metabolic diseases immune abnormalities (immunodeficiency, hypersensitivity conditions and autoimmune diseases, etc) and others [65]. A better understanding of neuroimmunoregulation is obligatory for obtaining new insights into the pathogenesis of these conditions and for the development of more rational approaches to treatment. The ultimate goal of this volume and of all the other volumes of this series is to promote the understanding of the science and to ease human suffering.

Istvan Berczi

ACKNOWLEDGEMENTS

I thank to Dr. Robert Rapaport, who has contributed significantly to the interpretation of the findings in critically ill patients after GH treatment. Many other colleagues contributed over the years to experimentation and to the development of the viewpoints expressed in this article. Notably, I owe special thanks to Drs Eva Nagy, Edris Sabbadini, Robert Shiu, Henry Friesen, Robert Matusik, Richard Warrington, Kalman Kovacs and Sylvia Asa. The experimental work discussed here was supported in part by MRC of Canada and the Arthriris Society of Canada.

REFERENCES

1. Selye H. Stress and disease. Science 1955;122:625–631.
2. Berczi I. The effect of growth hormone and related hormones on the immune sytem.

In: Berczi I, editor. Pituitary Function and Immunity. Boca Raton, FL: CRC Press, 1986;133–159.

3. Rapaport R. Growth hormone therapy and immune function. In: Berczi I, Gorczynsky R, editors. New foundation of Biology. NeuroImmune Biology Vol. 1. Elsevier, 2001;403–408.
4. Nagy E, Berczi I. Immunodeficiency in hypophysectomized rats. Acta Endocrinol 1978;89:530–537.
5. Nagy E, Berczi I. Prolactin and contact sensitivity. Allergy 1981;36:429–431.
6. Berczi I, Nagy E, Kovacs K, Horvath E. Regulation of humoral immunity in rats by pituitary hormones. Acta Endocrinol 1981;98:506–513.
7. Nagy E, Berczi I, Friesen HG. Regulation of immunity in rats by lactogenic and growth hormones. Acta Endocrinol 1983;102:351–357.
8. Berczi I, Nagy E, Asa SL, Kovacs K. Pituitary hormones and contact sensitivity in rats. Allergy 1983;38:325–330.
9. Nagy E, Berczi I, Wren GE, Asa SL, Kovacs K. Immunomodulation by bromocriptine. Immunopharmacology 1983;6:231–243.
10. Nagy E, Berczi I. Pituitary dependence of bone marrow function. Brit J Haematol 1989;71:457–462.
11. Berczi I, Nagy E, Asa SL, Kovacs K. The influence of pituitary hormones on adjuvant arthritis. Arthritis Rheum 1984;27:682–688.
12. Nagy E, Berczi I. Immunomodulation by tamoxifen and pergolide. Immunopharmacology 1986;12:145–153.
13. Berczi I, Nagy E. Placental lactogen is a hemopoietic hormone. Brit J Haematol 1991;79:355–358.
14. Berczi I, Szentivanyi A. Growth and lactogenic hormones, insulin-like growth factor, insulin. In: Szentivanyi A, Berczi I, editors. The Immune-Neuroendocrine Circuitry, Neuroimmune Biology, Vol. 3. Elsevier, 2002 (in press).
15. Soares MJ. Uteroplacental prolactin family: Immunological regulators of viviparity. In: Matera L, Rapaport R, editors. Growth and Lactogenic Hormones. Neuroimmune Biology, Vol. 2. Elsevier, 2002;187–202.
16. Kaplan SL, Grumbach MM, Friesen HG, Costom BH. Thyrotropin-releasing factor (TRF) effect on secretion of human pituitary prolactin and thyrotropin in children and in idiopathic hypopituitary dwarfism: further evidence for hypophysiotropic hormone deficiencies. J Clinical Endocrinology & Metabolism 1972;35:825–830.
17. Sinha YN, Salocks CB, Vanderlaan WP. Pituitary and serum concentrations of prolactin and GH in Snell dwarf mice. Proc Soc Exp Biol Med 1975 Oct;150(1):207–10.
18. Barkley MS, Bartke A, Gross DS, Sinha YN, Prolactin status of hereditary dwarf mice. Endocrinology 1982 Jun;110(6):2088–96.
19. Soares MJ, Bartke A, Colosi P, Talamantes F, Identification of a placental lactogen in pregnant Snell and Ames dwarf mice. Proc Soc Exp Biol Med 1984 Jan;175(1):106–8.
20. Pfaffle RW, Blankenstein O, Wuller S, Kentrup H, Combined pituitary hormone deficiency: role of Pit-1 and Prop-1. Acta Paediatr Suppl 1999 Dec;88(433):33–41.
21. Gala RR, The influence of thyroxine, growth hormone and prolactin alone and in combination on the production of prolactin-like activity by splenocytes from Snell dwarf mice. Life Sci 1995;57(2):113–22.
22. Nagy E, Berczi I. Hypophysectomized rats depend on residual prolactin for survival. Endocrinology 1991;146:2776–2784.

23. Horseman ND, Zhao W. Montecino-Rodrigez E, Tanaka M, Nakashima K, Engle SJ, Smith F, Markoff E, Dorshkind K. Defective mammopoiesis, but normal hematopoiesis, in mice with a targeted disruption of the prolactin gene. EMBO J 1997;16:6926–35.
24. Bole-Feysot C, Goffin V, Edery M, Binart N, Kelly PA. Prolactin (PRL) and its receptor: Actions, signal transduction pathways and phenotypes observed in PRL-receptor knockout mice. Endocrine Rev 1998;19:225–68.
25. Dorshkind K, Horseman ND, The roles of prolactin, growth hormone, insulin-like growth factor-I, and Thyroid hormones in lymphocyte development and function. Insights from genetic models of hormone and hormone receptor deficiency. Endocrine Rev 2000;21:292–312.
26. Berczi, I. Neuroimmune Biology – An introduction. In: Berczi I, Gorczynski R, editors. New Foundation of Biology; Neuroimmune Biology Series, Vol. 1. Elsevier, 2001;3–45.
27. Warren JL. Type I cytokines, interferons and their receptors. In: Paul WE, editor. Fundamental Immunology 4th edition. New York: Lippincott-Raven, 1999;741–74.
28. Notarangelo LD, Gilliani S, Mazza C, Mella P, Savoldi G, Rodriguez-Perez C, Mazzolari E, Fiorini M, Duse M, Plebani A, Ugazio AG, Vihinen M, Candotti F, Schumacher RF. Of genes and phenotypes: the immunological and molecular spectrum of combined immune deficiency. Defects of the gamma (c) -JAK3 signaling pathway as a model. Immunol Rev 2000;178:39–48.
29. Freeman ME, Kanyicska B, Lerant A, Nagy G. Prolactin: structure, function, and regulation of secretion. Physiol Rev 2000;80:1523–631.
30. Ivashkiv LB. Jak-STAT signaling pathways in cells of the immune system. Rev Immunogenet 2000;2:220–30.
31. Berczi I, The role of the growth and lactogenic hormone family in immune function. Neuroimmunomodulation 1994;1:201–216.
32. Berczi I, Szentivanyi A. Immunocompetence. In: Berczi I, Szentivanyi A, editors. The Immune-Neuroendocrine Circuitry. Neuroimmune Biology, Vol. 2. Elsevier, 2002.
33. Baines MG, Olfactory stimuli and allo-recognition. In: Berczi I, Gorczynski R, editors. New Foundation of Biology, Neuroimmune Biology, Vol. 1. Elsevier, 2001;313–329.
34. Choudhury SR, Knapp LA.Human reproductive failure I: immunological factors. Hum Reprod Update 2001 Mar–Apr;7(2):113–34.
35. Ellis LA, Mastro AM, Picciano MF. Do milk cytokines and hormones influence neonatal immune function? J Nutr 1997;127:985S–988S.
36. Korhonen H, Marnila P, Gill HS. Milk immunoglobulins and complement factors. Br J Nutr 2000;84 Suppl 1:S75–80.
37. Weiss Carrington P. Secretory immunology of the mammary gland. In: Berczi I, Kovacs K, editors. Hormones and Immunity. Lancaster, UK: MTP Press, 1987;172–202.
38. Takala J, Ruokonen E, Webster NR, et al. Increased mortality associated with growth hormone treatment in critically ill adults. N Engl J Med 1999;341:785–92.
39. Berczi I. Neurohormonal host defence in endotoxin shock. Ann NY Acad Sci 1998;840:787–802.
40. Haeryfar SMM, Berczi I. The thymus and the acute phase response. Cell Mol Biol 2001;47:145–156.
41. Berczi I, Bertok L, Chow D. Natural immunity and neuroimmune host defence. Ann NY Acad Sci 2001;917:248–257.
42. Jeschke MG, Herndon DN, Wolf SE, et al. Recombinant human growth hormone alters acute phase reactant proteins, cytokine expression, and liver morphology in burned rats. J

Surg Res 1999;83:122–29.
43. Derfalvi B, Igaz P, Fulop KA, Szalai C, Falus A. Interleukin-6-induced production of type II acute phase proteins and expression of junB gene are downregulated by human recombinant growth hormone in vitro. Cell Biol Int 2000;24:109–14.
44. Berczi I. The influence of pituitary-adrenal axis on the immune system. In: Berczi I, editor. Pituitary Function and Immunity. Boca Raton, FL: CRC Press, 1986;49–132.
45. Berczi I., Nagy E, Matusik RJ, Friesen HG. Pituitary hormones regulate c-myc and DNA synthesis in lymphoid tissue. J Immunol 1991;146:2201–2206.
46. Kelley KW, Brief S. Westly HJ. GH_3 pituitary adenoma cells can reverse thymic aging in rats. Proc Natl Acad Sci USA 1986;83:5663–5667.
47. Berczi I. Immunoregulation by the pituitary gland. In: Berczi I, editor. Pituitatry Function and Immunity. Boca Raton, CA: CRC Press, 1966.
48. Montgomery DW, Zukoski CF, Shah GN, Buckley AR, Pacholczyk T, Russell DH. Concanavalin A-stimulated murine splenocytes produce a factor with prolactin-like bioactivity and immunoreactivity. Biochem Biophys Res Commun 1987;145:692–698.
49. Hartmann DP, Holaday JW, Bernton DW. Inhibition of lymphocyte proliferation by antibodies to prolactin. FASEB J 1989;3:2194–2202.
50. Clevenger CV, Russell DH, Appasamy PM, Prystowsky MB. Regulation of interleukin-2 driven T-lymphocyte proliferation by prolactin. Proc Natl Acad Sci USA 1990;87:6460–6464.
51. Weigent D. Growth hormone and insulin-like growth factor-1 production by cells of the immune system. In: Matera L, Rapaport R, editors. Growth and Lactogenic Hormones. Neuroimmune Biology, Vol. 2. Elsevier, 2002; 87–100 (this volume).
52. Hooghe E. Signal transduction and modulation of gene expression by prolactin in human leukocites. In: Matera L, Rapaport R, editors. Growth and Lactogenic Hormones. Neuroimmune Biology, Vol. 2. Elsevier, 2002;123–136.
53. Sabharwal P, Glaser R, Lafuse W, Varma S, Liu Q, Arkins S, Kooijman R, Kutz L, Kelley KW, Malarkey WB. Prolactin synthesized and secreted by human peripheral blood mononuclear cells:an autocrine growth factor for lymphoproliferation. Proc Natl Acad Sci USA 1992;89:7713–6.
54. Handwerger S, Freemark M. The roles of placental growth hormone and placental lactogen in the regulation of human fetal growth and development. J Pediatr Endocrinol Metab 2000;13:343–56.
55. DiMattia GE, Gellersen B, Duckworth ML, Friesen HG. Human prolactin gene expression: The use of an alternative noncoding exon in decidua and the IM-9_P3 lymphoblastoid cell line. J Biol Chem 1990;265:16412–421.
56. Delhase M, Vergani P, Malur A, Hooghe Peters EL, Hooghe RJ. The transcription factor Pit-1/HGF-1 is expressed in hematopoietic and lymphoid tissues. Eur J Immunol 1993;23:951–955.
57. Fearon DT, Manders P, Wagner SD.Arrested differentiation, the self-renewing memory lymphocyte, and vaccination. Science 2001 Jul 13;293(5528):248–250.
58. Sprent J, Tough DF. T cell death and memory. Science 2001 Jul 13;293 (5528):245–248.
59. Nayak R, Mitra-Kaushik S, Shaila MS. Perpetuation of immunological memory: a relay hypothesis. Immunology 2001 Apr;102(4):387–395.
60. Ashton-Rickardta PG, Opferman JT. Memory T lymphocytes. Cell Mol Life Sci 1999 Oct 1;56(1–2):69–77.
61. Sprent J, Surh CD. Generation and maintenance of memory T cells. Curr Opin Immunol

2001 Apr;13(2):248–254.

62. Akbar AN, Lord JM, Salmon M. IFN-alpha and IFN-beta: a link between immune memory and chronic inflammation. Immunol Today 2000 Jul;21(7):337–342.
63. Cho BK, Rao VP, Ge Q, Eisen HN, Chen J. Homeostasis-stimulated proliferation drives naive T cells to differentiate directly into memory T cells. J Exp Med 2000 Aug 21;192(4):549–556.
64. Claman HN. Corticosteroids-immunologic and anti-inflammatory effects. In: Berczi I, Kovacs K, editors. Hormones and Immunity. Lancaster, UK: MTP Press, 1987;38–42.
65. Berczi I, Szentivanyi A. editors. The Immune-Neuroendocrine Circuitry. Neuroimmune Biology, Vol. 3. Elsevier, 2002 (in press).

Preface

For more than seventy years evidence has accumulated documenting the existence of a bi-directional communication network between growth hormone and the immune system. In the past twenty years there has been a tremendous proliferation of information detailing the workings of the growth hormone and insulin-like growth factor axis. A multitude of growth factors and binding proteins have been identified. More and more evidence supporting the important role of the growth hormone IGF network in the well functioning of the normal immune system has been documented. Some of recent developments in this area have been beautifully summarized by Professor Berczi in his introduction to this volume.

I am delighted to have been able to collaborate with my colleague Professor Matera on the production of this volume. I am pleased to have been able to call on some of my friends and colleagues for their expertise in the various areas pertaining to growth hormone IGF immune system interactions. In the first section, Professor Derek LeRoith details some of the cellular effects of growth hormone and IGF-I, which form a basis for further interactions between the growth hormone IGF axis and the immune systems. Professor Bozzola specifically addresses the effect of growth hormone and IGF-I on lymphocytes and cytokines as well as cell proliferation. Professor Cohen provides new insights into the role of not only growth hormone and IGF-I but of the IGF binding proteins on cellular function. Professor Tenore provides a comprehensive review of the expression and function of receptors for growth hormone and IGF-I in the immune systems. Professor Weingent, one of the pioneers in the field of growth hormone immune system interactions, summarizes the evidence confirming the production of growth hormone and IGF-I by cells of the immune system. Professor William Murphy focuses on the potential clinical implications of growth hormone-immune interactions. In the following section, we have summarized some of the evidence of the role of growth hormone and IGF-I in hematopoiesis. Professors Colao and Geffner describe the growth hormone immune system interactions in clinical conditions such as growth hormone excess, acromegaly and HIV disease.

I hope that through these various chapters the reader will acquire a sufficient knowledge to kindle additional interest in order to pursue new and more in-depth exploration of the fascinating and ever evolving field of Growth Hormone-IGF-Immune System interactions.

ACKNOWLEDGMENT

I would like to acknowledge first and foremost the vision, foresight, tenacity and friendship of Professor Berczi without whom the production of this volume would not have been possible. I would like to thank Professor Matera for her excellent sense of collaboration and professionalism. Finally, I would like to thank my assistant Letty Gonzalez for invaluable help in typing as well as organizing the various chapters.

Dr. Robert Rapaport

List of Corresponding Authors

Eduardo Arzt

Laboratorio de Fisiología y Biología Molecular, Universidad de Buenos Aires, Ciudad Universitaria, Pabellón II, (1428) Buenos Aires, Argentina.

Graziella Bellone

Department of Clinical Physiopathology, University of Turin, Via Genova 3, 10126 Torino, Italy

Mauro Bozzola

Dipartimento di Scienze Pediatriche, Università degli Studi di Pavia, IRCCS San Matteo, P.le Golgi 2, 27100 Pavia, Italy

Uptala Chattopadhyay

Department of Immunoregulation and Immunodiagnosis, Chittaranjan National Cancer Institute, 37 S.P. Mukherjee Road, Kolkata – 700 026, India

Annamaria Colao

Department of Endocrinology and Clinical and Molecular Oncology, "Frederico II" University, via S. Pansini 5, 80131 Naples, Italy

Carlos Diéguez

Department of Physiology, School of Medicine, University of Santiago de Compostela. Rua S. Francisco sn , 15705 Santiago de Compostela, Spain

Mitchell Geffner

Children's Hospital Los Angeles, Keck School of Medicine, University of Southern California, 4650 Sunset Blvd, Los Angeles, CA 90027, USA

Elizabeth L. Hooghe-Peters

Pharmacology Department, Medical School, Free University of Brussels, Laarbeeklaan 103, B-1090, Brussels, Belgium

Ron Kooijman

Department of Pharmacology, Medical School, Free University of Brussels, Laarbeeklaan 103, B-1090 Brussels, Belgium

Derek Le Roith

Clinical Endocrinology Branch, National Institutes of Health, Bethesda, MD 20892-1758, USA

Lina Matera
Department of Internal Medicine, University of Turin, Corso A.M. Dogliotti 14, 10126 Italy

William J. Murphy
Laboratory of Molecular Immunoregulation, National Cancer Institute-Frederick Cancer Research and Development Center, Frederick, MD, USA

Robert Rapaport
Diabetes Center, Mount Sinai Hospital, New York, NY, USA

Susan Richards
Immunology Laboratory, Cell and Protein Therapeutics R&D, Genzyme Corporation, Framingham, MA 10701, USA

Michael J. Soares
Department of Molecular & Integrative Physiology, University of Kansas Medical Center, 3901 Rainbow Boulevard, Kansas City, KS 66160, USA

Noburo Suzuki
Departments of Immunology and Medicine, St.Marianna University School of Medicine, 2-16-1, Sugao, Miyamae-ku, Kawasaki, Kanagawa 216-8511, Japan

Alfred Tenore
Department of Pediatrics (DPMSC), University of Udine School of Medicine, Udine, Italy

Li-yuan Yu-Lee
Departments of Medicine, Molecular & Cellular Biology, and ImmunologyBaylor College of Medicine, Houston, TX 77030, USA

Sara E. Walker
The University of Missouri-Columbia, Harry S. Truman Memorial Veteran's Hospital Research, 800 Hospital Drive, Columbia, Missouri 65201, USA

Douglas A. Weigent
University of Alabama at Birmingham, Department of Physiology and Biophysics, 1918 University Blvd MCLM 894, Birmingham, AL 35294-0005, USA

Stuart Alan Weinzimer
Division of Endocrinology/Diabetes, Department of Pediatrics, The Children's Hospital of Philadelphia and The University of Pennsylvania, Philadelphia, PA 19104-4399, USA

Contents

I. INTRODUCTION

Growth and Lactogenic Hormones
Edited by L. Matera and R. Rapaport

Immunoregulation by Prolactin – An Introduction

LINA MATERA

Department of Internal Medicine, University of Turin, Corso A.M. Dogliotti 14, 10126 Italy

In addition to adaptive immune reactions, host defense also relies on natural immune mechanisms. Soluble factors can finely tune the outcome of the immune response. These mediators are collectively referred to as cytokines, which are classically considered to act within the immune system in a paracrine or autocrine fashion. The immune system shows a considerable degree of autonomy and an elaborate internal regulatory network. For this reason, most scientists rejected the idea that immunoregulatory signals are also coming from the neuroendocrine system, despite the many demonstrations of communications between the neuroendocrine and the immune systems. It took a long time for papers dealing with this subject to reach high-ranked scientific journals.

A breakthrough in this direction has been the description of structural similarities between lympho-hemopoietic cytokines and the pituitary prolactin (PRL) and growth hormone (GH) [1]. The receptors for these mediators also share common features. Studies on receptor function revealed a unique intracellular signaling pathway with the participation of cytoplasmic Janus tyrosine kinases (JAK) and the transcription factors, signal transducers and activators of transcription (STAT) [2,3]. A particular cytokine receptor may mediate distinct and multiple intracellular signals, leading to the activation of various genes and functions. Conversely, different cytokines can activate the same transcription factors [4–7], that bind to common sites on the promoter of the target genes. The tissue distribution of cytokines, cytokine-receptors and transcription factors play important roles in the determination of gene activation.

Some signal transduction pathways are common to cytokines of the same family, which explains their redundant mode of action. Pleiotropism and redundancy must have evolved to ensure the integrity of a given function even in the shortage of a factor. For instance, the cooperation between PRL, cytokines and hemopoietins has been convincingly demonstrated *in vitro* [8–11]. PRL has also been shown to restore suppressed immunocompetence or hemopoiesis in animal models. On the basis of these observations PRL is now tested as a protective agent on a group of acquired immunodeficient patients receiving myeloablative anti-retroviral treatment [12]. It is also anticipated that in the future PRL could be used together with antitumoral cytokines to minimize their toxic effects [13].

A compelling body of experimental evidence indicates that PRL is a necessary factor in the cytokine/hemopoietin network. However, PRL or PRL-receptor knockout mice develop normal immune and hemopoietic function [14,15]. From these observations it may be argued that gene knock-out is an imperfect model when dealing with factors with redundant action in that alternate factors can take over the role of PRL during ontogenesis. Perhaps the additional knockout of the alternate hormone/cytokine pathways would result in abnormal lymphohemopoietic development. Another explanation for the immunocompetence of PRL and PRL-receptor knockout mice may be that PRL is not a necessary factor for normal hemopoiesis, but is required during stressful situations [16]. Thus, the increased production of PRL during stressful conditions may indicate an increased requirement. Therefore, it is conceivable that the natural role of this hormone is to counteract the necessary and still harmful effects of elevated levels of glucorticoids during stress. An abnormal increase or depletion of some cytokines can also represent a threatening stressful condition for homeostasis. Such conditions exist in patients undergoing myeloablative therapy [12], where PRL helps erythroid replacement. Similarly, the enhancing effect of PRL on antitumor responses observed *in vitro* in the absence of serum [10] may reflect a mechanism operating *in vivo* during the decrease of some serum growth factors. Therefore, the beneficial effect of PRL *in vivo* may represent a rescue mechanism from experimentally induced stress-associated immune suppression.

With all this in mind, it seems even easier to assign to PRL the role of a immuno-potentiating /inflammatory/survival factor, with a trend towards promoting the Th1 cytokine profile, in clear contrast to the well-known Th2 polarizing effect of glucocorticoid. The *in vitro* data seem all converge towards a protective role of PRL on the immune system. We are not so far from the evidence provided by the pioneer experiments of Berczi et al. As the present molecular scenario develops the involvement of PRL in diseases of the immune system will become clarified, and the clinical use of PRL for the restoration of the lympho-hemopoietic system will be placed on a rational basis.

Dr. Lina Matera

REFERENCES

1. Bazan JF. Structural design and molecular evolution of a cytokine receptor superfamily. Proc Natl Acad Sci USA 1990;87:6934–6938.
2. Rane SG, Reddy EP. Janus kinases: components of multiple signaling pathways. Oncogene 2000;19:5662–5679.
3. Heim MH. The Jak-STAT pathway: cytokine signalling from the receptor to the nucleus. J Recept Signal Transduct Res 1999;19:75–120.
4. Horseman ND, Yu-Lee L-y. Transcriptional regulation by the helix bundle peptide hormones: Growth hormone, prolactin, and hematopoietic cytokines. Endocr Rev 1994;15:627–649.
5. Pallard C, Gouilleux F, Charon M, Groner B, Gisselbrecht S, Dusanter-Fourt I. Interleukin-3, erythropoietin, and prolactin activate a STAT5-like factor in lymphoid cells. J Biol Chem 1995;270:15942–15945.
6. Gouilleux F, Pallard C, Dusanter-Fourt I, Wakao H, Haldosen LA, Norstedt G, Levy D, Groner B. Prolactin, growth hormone, erythropoietin and granulocyte-macrophage colony stimulating factor induce MGF-Stat5 DNA binding activity. EMBO J 1995;14:2005–2013.
7. Socolovsky M, Fallon AE, Lodish HF. The prolactin receptor rescues EpoR-/- erythroid progenitors and replaces EpoR in a synergistic interaction with c-kit. Blood 1998;92:1491–1496.
8. Kooijman R, Hooghe-Peters EL, Hooghe R. Adv Immunol; 1996;63:377–454.
9. Yu-Lee L-y. Molecular actions of prolactin in the immune system. Proc Soc Exp Biol Med 1997;215:35–52.

10. Matera L. Action of pituitary and lymphocyte prolactin. Neuroimmunomodulation 1997;4:171–180.
11. Clevenger CV, Rycyzyn MA, Syed F, Kline JB. Prolactin receptor signal transduction. In: Horseman ND, editor. Prolactin. Boston: Kluwer Academic Publishers, 2001;355–379.
12. Woody MA, Welniak LA, Sun R et al. Prolactin exerts hematopoietic growth-promoting effects in vivo and partially counteracts myelosuppression by azidothymidine. Exp Hematol 1999;27:811–816.
13. Richards SM, Murphy WJ. Use of human prolactin as a therapeutic protein to potentiate immunohematopoietic function. J Neuroimmunol 2000;109:56–62.
14. Bole-Feysot C, Goffin V, Edery M, Binart N, Kelly PA. Prolactin (PRL) and its receptor: actions, signal transduction pathways and phenotypes observed in PRL receptor knockout mice. Endocr Rev 1998;19:225–268.
15. Goffin V, Binart N, Clement-Lacroix P et al. From the molecular biology of prolactin and its receptor to the lessons learned from knockout mice models. Genet Anal Biomol Engin 1999;15:189–201.
16. Dorshkind K, Horseman ND. The roles of prolactin, growth hormone, insulin-like growth factor-I, and thyroid hormones in lymphocyte development and function: insights from genetic models of hormone and hormone receptor deficiency. Endocr Rev 2000:21:292–312.

II. GLH BIOLOGY, DEVELOPMENT & RECEPTORS

Growth and Lactogenic Hormones
Edited by L. Matera and R. Rapaport

The Growth Hormone – Insulin-Like Growth Factor-I Axis and Immunity

WILSON MEJIA NARANJO, MYRIAM SANCHEZ-GOMEZ and DEREK LE ROITH

National University of Colombia, Bogota, Colombia, Department of Chemistry, Laboratory of Hormones (WMN, MS-G) and Clinical Endocrinology Branch, National Institutes of Health, Bethesda MD 20892-1758 (DL), USA

ABSTRACT

There are many overlaps between the immune system and the endocrine system, and the effects of hormones on the proliferation and differentiation of immune cells is widely studied. Of particular importance is the role of the growth hormone/insulin-like growth factor-I (GH/IGF-I) axis in immune function. Whether these effects are brought about by the circulating (endocrine) hormones or via paracrine/autocrine mechanisms remains to be conclusively determined for different immune cell types and tissues. Nevertheless the importance of the GH/IGF-I axis in immune responsiveness, from the data summarized in this review, is clear.

1. INTRODUCTION

The immune and endocrine systems share a common set of ligands and receptors. An increasing body of evidence suggests that there is bi-directional communication between these two physiological systems. Cells of the immune system are known to secrete and/or have receptors for several hormones. Similarly, although cytokines like interferons are derived from immunologically competent cells such as macrophages and lymphocytes, these factors also have hormonal effects on various other organs. We propose that this observation is pivotal to a biochemical understanding of how and why there is bi-directional communication between the immune and the endocrine systems.

Increasing evidence suggests that growth hormone (GH) and the insulin-like growth factors (IGFs) and their receptors play a role in the function of the immune system [1]. The original somatomedin hypothesis stated that pituitary-derived GH stimulated the synthesis of IGF-I and triggered its export from the liver [2]. The liver is the primary source of circulating IGF-I, which is a crucial factor for postnatal growth and development. Thus, coupling of GH to IGF-I production appears optimal only with adequate dietary intake and general health [3]. More recently, it has been established that the expression of IGF-I is not limited to the liver, and the synthesis of GH is not restricted to the pituitary gland. Experimental evidence accumulated over the last decade has demonstrated that lymphoid organs such the thymus, spleen and peripheral blood produce and respond to GH by synthesizing and releasing IGF-I [4]. The presence of cell surface receptors for GH and IGF-I on different subpopulations of lymphocytes suggest that there is a local mechanism of action for these hormones in addition to the traditional endocrine mode of action [5,6].

Bi-directional communication between the endocrine and autocrine/paracrine modes of actions of these hormones appears to be a major source of immunostimulatory signals [7]. A number of experimental studies suggest that the GH-IGF-I axis plays a role in the control of lymphopoiesis and immune function. Studies performed in Snell dwarf (dw/dw) mice and in hypophysectomised rats demonstrated that humoral and cell-mediated immunity are depressed in these animals, and was ameliorated by the administration of pituitary-derived hormones [8,9]. In a recent review, Dorshkind and Horseman have pointed out some contradictions in a number of studies that make it difficult to draw definitive conclusions about the role played by the GH-IGF-I axis in immunity. However, when the available literature is reconsidered in view of the more recent results obtained from the genetic models described below, a new hypothesis emerges [10].

Two mutant mouse models show more limited endocrine defects than those observed in the Snell dwarf mouse or in hypophysectomized rodents. The spontaneous (lit/lit) mutation and the genetically engineered (Igf1-/-) mutant mouse have brought new insights into the precise effects of these hormones in lymphopoiesis [11,12]. The spontaneous genetic defect in the gene encoding GHRF results in impaired production of GH and IGF-I in (lit/lit) mice. Serum GH levels are reduced by approximately by 90% in these animals, which in turn causes a 90–95% reduction in circulating levels of IGF-I. Mice in which the gene encoding IGF-I has been disrupted by homologous recombination show growth retardation and full grown (Igf1-/-) mice are less than one third the size of wild type mice [13]. The lack of GH-induced growth in these mice supports the hypothesis that IGF-I mediates many of the effects of GH on somatic growth. This hypothesis has recently been challenged by the finding that GHR knockout mice are smaller than IGF-I knockout mice. This observation suggests that GH may have an IGF-I-independent role in growth [14]. The analysis of (lit/lit) and (Igf1-/-) mice have raised the hypothesis that neither GH nor IGF-I is an obligate immunoregulator. Instead, these factors might act as anabolic and stress-modulating hormones in cells of the immune system [10].

The importance of an autocrine/paracrine mechanism of action of the GH/IGF-I axis has become further apparent with the recent findings that the liver IGF-I specific knock-out mice (LID) do not differ from wild-type mice in body weight and length [15]. From these studies, it was concluded that hepatic IGF-I production is a major source of the circulating peptide levels, but that liver-derived IGF-I is not essential for post-natal growth and development. In addition, the study strongly suggests that local production of IGF-I may also mediate the growth-promoting effects of IGF-I. This evidence reinforces the idea that locally produced GH and IGF-I have many actions, including the regulation of whole body growth, as well as regulation of growth, maintenance, function and repair of specific tissues, such as those of the immune system [16]. The aim of this review is to discuss the possible roles of GH and IGF-I on lymphoid tissue, and the evidence for autocrine or paracrine functions of the GH/IGF-I axis. Finally, in view of the recent hypothesis, we will introduce the role of malnutrition, a stress factor, on the anabolic actions of the GH/IGF-I axis on the immune system.

2. GROWTH HORMONE

2.1. Background

Growth hormone was one of the first pituitary proteins that was shown to have profound effects on the regulation of the immune system *in vivo*. The first line of evidence came in 1967, from two different studies. In the first study, dwarf mice with very low levels of GH were found to have very depressed immunologic responses and involuted central and peripheral lymphatic tissues [17]. The second study examined the effects of antibodies to pituitary extracts [18]. Both studies led to the conclusion that GH controls the growth of lymphoid tissue. Subsequently, delayed recovery of the total leukocyte count in hypophysectomized adult rats [19] and diminished NK cell function and decreased longevity were observed in hypophysectomized middle-aged mice [20]. Clinically, hypogammaglobulinemia is found to be associated with GH deficiency [21], and GH treatment has been reported to result in a number of changes in lymphocyte subpopulations [22].

2.2. GH expression in lymphoid cells

The extra-pituitary production of GH was first established in human lymphocytes by Weigent et al. [23]. These researchers used fluorescently-labeled anti-GH antibodies to show that the number of GH-positive cells was doubled in response to mitogenic stimulation. This result was extended to show that GH mRNA is expressed in human lymphocytes [24] and translated into a GH-immunoreactive protein with a molecular weight similar to that of pituitary GH [25]. A number of cell lines and tissues have been shown to secrete GH, including the B cell lymphoma line IM-9, human and rat bone marrow, spleen, thymus, lymph nodes and rat peripheral blood lymphocytes [1]. A study of human tissues using *in situ* hybridization and RT-PCR, has also shown that GH transcripts can be detected in spleen, lymph node, tonsil and thymus [26]. The presence of the transcription factor Pit-1 in lymphoid tissues [27] strongly supports the idea that extra-pituitary GH transcription and production is regulated in lymphoid cells.

Since lymphocytes produce GHRH and somatostatin, and possess specific receptors for these peptides [28], it seems likely that GH production may be similarly regulated by GHRH and somatostatin within the immune system, as they are in the endocrine system. However, various studies have produced conflicting data. There are reports showing that GHRH induces stimulation, inhibition [29] and no effect on GH expression [30]. However, the secretion of GH by ConA-stimulated human lymphocytes was shown to be up-regulated by GH, but not affected by IGF-I [31]. These results suggest that GH is synthesized *de novo* in lymphocytes, but there may be a difference in the regulation of GH secretion between the endocrine and the immune systems.

2.3. Growth hormone receptor on lymphocytes

GH receptors have been detected in normal thymic and lymphoid cells, as well as in the transformed human lymphoid cell line IM-9 [32,33]. In fact, the first use of radioreceptor assays to detect GH receptors was reported in human cultured lymphocytes [33]. Subsequently, the GH receptor has been identified in human peripheral blood lymphocytes (PBL) [34] and in bovine and murine thymocytes [35,36]. The GH receptor has been shown to be ubiquitously expressed on the cell surface of human PBL, with the highest expression on B cells, as determined by dual fluorochrome flow cytometry [37].

The GH receptor has been shown to be widely expressed in the rat [38,39] with comparatively much lower levels of expression in lymphoid tissues. The GH receptor and GH binding protein have been found to be expressed in the same cell types [40] and tissue types [41], including thymic tissue [42]. The GH receptor has been cloned and sequenced from human and rat lymphocytes [43], and was found to be identical to the GHR cloned from liver [44].

Despite the growing knowledge of the actions of GH on cells of the immune system, further studies are needed to understand which specific subsets of lymphocytes are the primary targets for GH action. One approach to this question is the quantitative analysis of the GH receptors on subpopulations of lymphocytes. In a study performed in our laboratory, we have analyzed the distribution of the GH receptor on spleen, lymph nodes, thymus and peripheral blood lymphocytes in normal male rats (unpublished results). Flow cytometric analysis using fluorescein-coupled bovine growth hormone (bGH) as a ligand, indicated that 20% of B lymphocytes, 7% of CD4+ T lymphocytes and 6% of CD8+ respectively express the GH receptor in the spleen. Similarly in lymph nodes, 20% of B lymphocytes, 11% of CD4+ T lymphocytes and 7% of CD8+ T lymphocytes express the GH receptor.

2.4. GH signal transduction

In the late 1980's, when significant advances were being made in the understanding of how hormones and growth factors elicit their cellular responses, the mechanism of hormonal action of GH remained virtually unknown. The purification and cloning of the GH receptor, the discovery of the dimerizing stochiometry between GH and its receptor, and the crystallization of the complex, were key discoveries. These findings helped to elucidate the mechanism of action not only of GH, but also for other cytokines that involve receptor dimerization. The placing of GH in the family of hematopoietic cytokines, which includes among others, erythropoietin, granulocyte-colony stimulating factor, macrophage-colony stimulating factor, and the interleukins, has provided a theoretical basis for the finding that GH has significant activity as hematopoietic cytokine [45].

The first evidence that tyrosine phosphorylation was involved in GH signal transduction was difficult to interpret, as the GH receptor did not share sequence homology with known tyrosine kinases. This apparent contradiction was solved in 1993, when it was demonstrated that an early step in the GH signaling cascade was the physical association and activation of Janus kinase 2 (JAK2) [46]. This finding provided the basis to delineate the mechanism by which not only GH, but also other members of the family of cytokine receptors initiate cellular signaling. It is now accepted that when JAK2 is activated by GH, it phosphorylates multiple proteins on tyrosine residues, including JAK2 itself, the GH receptor, and SHC proteins [47]. These signals lead to phosphorylation and activation of the extracellular signal regulated protein kinases (ERKs) -1 and -2 [48], phosphorylation of the insulin receptor substrates that have been implicated in regulation of glucose metabolism, and phosphorylation and activation of signal transducers and activators of transcription (STATs) -1, -3, -5a and -5b, which have been implicated in the expression of a variety of GH-sensitive genes [49]. In vivo studies have shown that GH phosphorylates STAT5a and STAT5b in many tissues, including the immune system. Furthermore, it has been demonstrated that GH up-regulates the transcription of the STAT5 gene in lymphoid organs, including the thymus and peripheral blood and in cell lines transfected with the GHR cDNA [50,51].

As mentioned previously, the GHR belongs to the superfamily of cytokine receptors which appear to share several intracellular signaling mechanisms. This redundancy in signaling pathways makes it difficult to see how specificity is maintained. If the signaling pathways do overlap, then the responses of lymphoid cells to GH could be viewed as a minor phenomenon, with other ligands/receptors also mediating the same responses. However, gene-disruption studies have helped to elucidate the physiological functions of the various STAT molecules, and it seems likely that the various STATs have both essential and nonessential, or redundant, functions [52].

Much progress has been made towards the characterization of the mechanisms by which cellular signaling is initiated. However, the mechanisms by which signal transduction is turned off is equally important, but far less understood. The SOCS family of proteins has recently been identified and found to inhibit cytokine-signaling pathways, including that of GH [53,54]. This family consists of eight proteins, SOCS1-7 and CIS, each containing an SH2 domain [55]. In vitro studies have shown that GH induces the transient transcription of SOCS3 and CIS with a maximal effect on mRNA levels after 1 hour, whereas SOCS2 mRNA levels were steadily increased over time [56]. In transfected cell lines, it was shown that expression of SOCS-1 or SOCS-3 inhibited GH mediated gene transcription, while SOCS-2 and CIS expression had no effect [57]. In vivo, has been shown that GH also induces the expression of SOCS proteins in a tissue-specific manner and in response to GH administration [58].

2.5. GH actions

As described above, it is likely that GH is produced and released locally in lymphoid tissues by constitutive expression, resulting in a pattern of continuous exposure to local GH. In support of this hypothesis, the lymphoid tissues of GH-treated rats were less responsive to growth induced by GH injections than by GH infusions [1]. Similarly, it has been shown in mice that when GH is injected, 20- to 40-fold higher doses are required to reverse corticosterone-induced supression of splenic lymphocyte responses to mitogens, than when GH is administered by constant infusion through a minipump [59]. This may result from down-regulation of GH post-receptor signaling pathways induced by the bolus injections. In mice that overexpress bGH, there is an enlargement of the internal organs, particularly the spleen, which may be due to both the high levels of GH and the continuous pattern of GH exposure [60]. Some of these differences between injected and infused GH could be due to the higher serum IGF-I levels that are induced via hepatic stimulation in response to the continuous GH exposure, rather than injections of GH [61]. These findings have been confirmed in aged monkeys where rhGH was administered by continuous infusion and found to have anabolic effects on the lymphoid organs [62].

Some in vivo studies clearly suggest that GH can induce thymocytes to proliferate. Implants of GH3 pituitary cells in aging rats increases the total numbers of thymocytes and increases the percentage of CD3- bearing cells, with a parallel decrease in CD4-CD8- double-negative thymocytes, which normally accumulate in the aging rat thymus. The role of GH in thymus development was further supported by findings in GH-deficient dwarf mice. In addition to the precocious decline in thymulin serum values, these animals showed a progressive thymic hypoplasia with decreased numbers of CD4+ CD8+ double-positive thymocytes. These defects could be reversed by prolonged treatment with GH. It has also been suggested that GH may play a role in thymocyte traffic, as infusions of recombinant human GH increase the engraftment of human T cells into the thymus of mice [63].

The fact that human lymphocytes express GH receptors suggest that GH may modulate immune function in humans. However, the data on the relationship between GH and the human immune system are conflicting. In untreated GH-deficient children, immune function has been reported to produce no changes in immune function with a slight decreases in B lymphocyte number [64]. To date, no study has been made to evaluate whether these patients display abnormalities in the local GH/IGF-I axis. In GH-deficient humans, it is possible that the GH produced locally in the immune system compensates for the lack of endocrine GH. This may explain why a deficiency in pituitary GH or endocrine IGF-I, appears to have minor effects on immune function in humans, as compared to the effects of such deficiencies in rats. However, the notion that the GH/IGF-I axis enhances thymic cell proliferation is supported by a clinical case of an acromegalic patient with high circulating levels of GH and IGF-I and thymic hyperplasia [65]. Clearly, more studies are needed in order to establish a clear relationship between high levels of GH and thymic proliferation in the human.

3. INSULIN-LIKE GROWTH FACTOR-I (IGF-I)

3.1. Background

The insulin-like growth factors (IGFs) are members of the family of insulin-related peptides, which includes insulin, IGF-I and IGF-II. The IGFs are potent mitogens for many different cell types, including those of the immune system, and these factors play a central role in growth and development [66].

The biological activities of the IGF-I are controlled by various factors. First, the number of cell surface IGF-I binding sites on target cells determines the strength of the IGF-I signal. A family of cell surface receptors that includes the insulin, the type 1 IGF, and the type 2 IGF receptors mediates the biological actions of IGF-I. Most of the effects of IGF-I on growth and differentiation are elicited by the ligand-dependent activation of the type 1 receptor, a transmembrane tyrosine kinase receptor [67].

A second mechanism by which the effects of IGFs are regulated is through IGF binding proteins (IGFBPs) [66,68,69]. The IGF-I in the circulation is associated with six known IGFBPs. Most of the IGF-I is found in a 150 kD ternary complex consisting of IGF-I, IGFBP-3 (or IGFBP-5), the predominant IGFBP in plasma, (42–45 kD) and an acid-labile subunit, ALS (84–89 kD). The remaining IGF-I is either free (< 1%) or bound to the five other IGFBPs in binary complexes. The binding proteins serve to protect the IGF-I against degradation, thereby prolonging its half life, facilitating the transport to distinct tissues and both facilitating and impairing the interaction between the IGFs and their cell surface receptors. In tissues, most of the IGF-I is bound in the form of binary complexes. IGFBP-2, -4 and -5 are expressed in spleen and thymus of normal mice. The spleen also expresses IGFBP-3 and -6, whereas IGFBP-1 is not detectable in either spleen or thymus [70].

Finally, the amount of IGFs that are produced and secreted regulates the bioavailability of these growth factors. Similarly, the expression levels of the type 1 IGF receptor gene can regulate the function of IGFs. The local and circulating levels of a number of hormones and growth factors tightly control IGF-I receptor expression.

3.2. IGF-I receptor on lymphoid cells

Various approaches have been used to identify receptors for IGF-I on lymphoid tissues. Binding assays using radiolabelled IGF-I have identified binding sites for IGF-I on human lymphoid cells, such as T or B lymphomas, as well as in resting or activated peripheral lymphocytes [71].

Because IGF-I also binds to IGFBPs, which are produced by lymphoid cells and may also be localized at the cell surface, IGF-I binding assays do not necessarily give an accurate representation of the distribution of receptors in a highly heterogeneous cell populations, such as those found in the immune system.

The co-localization the human type 1 IGF-I receptor with lymphocyte markers can be monitored by two-color flow cytometry. These studies have shown that the IGF-I receptor is present on most monocytes and B lymphocytes, but on only 2% of T lymphocytes [5]. The distribution of the IGF-IR in rat splenocytes was studied using the biotinylated IGF-I analogue des(1-3) IGF-I, followed by phycoerythrin-conjugated streptavidin (PE-SA) staining [72]. Des(1-3) IGF-I is a functionally active ligand that binds well to the IGF-I receptor, but poorly to IGFBPs. The results showed that IGF-I receptors were readily detectable on a wide variety of splenocytes, including T cells, B cells and monocytes, with the highest binding capacity observed on monocytes, followed by B cells, and T cells had the lowest binding capacity. Furthermore, comparative analysis of IGF-I receptor expression on subsets of T cell showed that CD4+ cells had higher IGF-IR expression levels than did CD8+ cells.

Mutant mice for the IGF-IR have been produced. In contrast to little (lit/lit) mutants which, depending on their genetic background, some survive and reach adulthood, null mutants for the IGF-IR gene die at birth of respiratory failure and have more severe growth deficiency [73].

3.3. IGF-I peptide in lymphoid cells

Exons 1 and 2 of the IGF-I gene contain two distinct promoters that give rise to two distinct IGF-I mRNAs. IGF-I mRNA expressing exon 1 is the form found in fetal tissue, whereas the exon 2 form appears postnatally, concomitant with the acquisition of GH responsiveness. These different forms of IGF-I mRNA may supply either GH-dependent endocrine IGF-I (exon 2) or local GH-independent paracrine or autocrine IGF-I (exon 1) [74]. In myeloid cells, IGF-I transcripts are exclusively initiated within exon 1, which is characteristic of extrahepatic IGF-I mRNA. Macrophages produce high levels of IGF-I mRNA and the peptide, and lymphoid cells produce low levels [75]. In addition, bone marrow stromal cells and thymic epithelial cells release IGF-I after being stimulated with GH [76]. These results show that lymphocytes are exposed to: 1) endocrine IGF-I from the circulation; 2) their own autocrine IGF-I; and 3) possibly a third source of IGF-I (paracrine), derived from epithelial cells and stromal cells in lymphoid organs and bone marrow.

It has been shown that cytokines other that GH can also affect IGF-I synthesis in lymphoid tissue. In macrophages, tumor necrosis factor-α (TNF-α) and the colony-stimulating factors (CSFs) can induce the expression of IGF-I [77,78]. In contrast, interferon-γ (IFN-γ), which is derived from T cells, decreases the levels of IGF-I mRNA in macrophages, in a time and dose-dependent manner [79].

3.4. IGF-I and its role in the generation of humoral and cell mediated immune response

For several decades, the studies conducted in Snell dwarf (dw/dw) mice and in hypophysectomised rodents have supported the hypothesis that GH and IGF-I play a critical modulatory role in the development and function of the immune system. For instance, humoral- and cell-mediated immunity are suppressed in PRL-, GH-, IGF-I- and thyroid-hormone-deficient hypophysectomized rats and in Snell dwarf mice [17,80]. This condition is reversed upon replacement with the corresponding hormone [19]. In agreement with this observation, both GH and

IGF-I treatments lead to increases in the weight of lymphoid organs and individually enhance the differentiation and proliferation of components of the immune system [81]. In addition, IGF-I has been shown to stimulate the response to T-dependent antigens in bone marrow transplant recipients [82].

In hypophysectomized rats, rhIGF-I treatment caused an increase in the weights of the spleen and thymus [83]. This finding suggested that IGF-I plays a regulatory role in lymphoid organ growth. In adult mice that received continuous infusion of IGF-I, the increased spleen and thymus weight were associated with an increase in CD4+ T cells in both the spleen and thymus. The number of B cells was also increased specifically in the spleen. Splenocytes and lymph nodes from animals that received 2 weeks of IGF-I treatment showed an increased responsiveness to mitogens. This study suggested that the administration of rhIGF-I in aged animals increased both the number of lymphocytes and enhanced their function, and was thereby potentially beneficial to the functioning of the immune system [84].

Recombinant human IGF-I may act as a hematopoietic factor, as it has been shown to have two major effects on B cell development. As a differentiation factor, IGF-I produced by bone marrow stromal cells in the hematopoietic microenvironment plays a key role in regulating primary B lymphopoiesis [85]. As a B cell proliferation co-factor, IGF-I synergistically enhances the proliferative effect of IL-7 on pro-B cells [86].

IGF-I also affects the immune response after bone marrow transplantation. Adult mice were lethally irradiated and reconstituted with a transplant of bone marrow cells from syngeneic donors. Continuous infusion of IGF-I into these animals resulted in an increase in the total number of pre-B and mature B cells in bone marrow B lineage cells. Treatment with rhIGF-I also increased the weights of the thymus and spleen, as well as the repopulation of peripheral lymphocytes [82]. Intraperitoneal injections of rhIGF-I in normal mice also increased the number of bone marrow hematopoietic progenitor cells. Furthermore, administration of the rhIGF-I to mice with chemically induced myelosuppression increased the number of progenitor cells [87].

The role of IGF-I in thymopoiesis has been reviewed [84]. The evidence suggests that GH itself also physiologically modulates the thymus by stimulating the secretion of thymulin and the proliferation of thymic epithelial cells (TEC) in vitro. GH also induces the expression of extracellular matrix ligands and receptors and modulates the interactions of extratracellular matrix-mediated TEC and thymocytes [88]. IGF-I itself can substitute for GH in stimulating thymulin production by cultured TECs and in increasing the adhesion of TECs and thymocytes [89]. In addition, human thymocytes synthesize and secrete both GH and IGF-I [90]. Taken together, these findings led to the conclusion that GH may function as an autocrine/paracrine growth factor in the thymus via stimulating the local synthesis of IGF-I.

In peripheral lymphoid organs, administration of rhIGF-I increased T and B cell populations and elevated antibody titers following primary or secondary antigen challenge. Furthermore, when rhIGF-I was added to cultures of splenocytes from antigen-primed mice, immunoglobulin synthesis was enhanced. These studies led to the conclusion that locally produced IGF-I might stimulate B and T cell lymphopoiesis [91].

In view of the observed hematopoietic effects of IGF-I and the fact that serum concentrations of IGF-I decline during aging, the effect of rhIGF-I alone or combined with bone marrow cells from young mice on the involuted thymus was studied in aged mice. Results showed that the mice that received the combined therapy had higher cellularity compared with animals treated with hormone or bone marrow transplantation alone. This finding suggested that aging induces deficiencies in both endocrine and hematopoietic processes [92].

The effects of continuous infusion of rhGH and rhIGF-I, both individually and in combination, on the immune system of aged female monkeys have been studied [62]. Treatment with the combination of these two hormones, nearly tripled the percentage of CD4 lymphocytes and doubled the CD4/CD8 ratio in the spleen. In addition, splenic reactive follicles and splenic B cell representation were increased by rhGH, and both hormones enhanced the overall immunogenic response to tetanus toxoid. Lymph nodes also showed an increase in the percentage of CD4 cells and rhGH treatment consistently increased the surface area of lymph nodes. In peripheral blood, the CD4 and the ratio CD4/CD8 tended to decrease in response to rhIGF-I treatment, but were normalized when both rhGH and rhIGF-I were administered together. This observation led to the conclusion that these anabolic hormones might cause lymphocytes to accumulate in lymphoid organs at the expense of the numbers of lymphocytes in the circulation. In agreement with these observations, when IGF-I was administered to fetal monkey, significant effects on hematopoietic and lymphoid tissues were observed. These included an increase in the total number of fetal lymphocytes and red cell parameters and a significant elevation in the number of circulating B cells and the CD4/CD8 ratios in lymph nodes [93].

Recent studies with various genetic mouse models have given rise to new insights into the role of IGF-I in the immune system. The endocrine defects in mice that are defective in the genes encoding IGF-I are more limited and selective than those in hypophysectomized rats and in dw/dw mice. A study was conducted to assess innate, humoral, and cell mediated immune response in the GH/IGF-I deficient mice (lit/lit) [94]. When no defects were observed in humoral or cell mediated immunity in response to challenges with T- independent and T-dependent antigens and with Listeria monocytogenes, a new working hypothesis was raised. This hypothesis proposed that GH and IGF-I may act as anabolic and stress modulating hormones in most cells, including those of the immune system. It is suggested that the observed immune defects in dw/dw and lit/lit mice and the positive effects of these hormones on restoring thymopoiesis or antigen responsiveness are due to stress caused by suboptimal environmental conditions. According to this hypothesis, any positive effects of these hormones on immunity occurs primarily as an adaptation to stress.

4. THE ROLE OF THE LOCAL GH/IGF-I AXIS DURING NUTRITIONAL STRESS

The bi-directional interaction between the endocrine and the immune systems added to the possible paracrine/autocrine regulatory mechanisms involved in the GH/IGF axis play an important role in the maintenance of physiological and immunological homeostasis. A recent hypothesis has suggested that the inmunomodulatory effects of GH and IGF-I may be explained in terms of their anabolic/somatogenic actions [10]. The nutritional stress caused by protein/energy calorie restriction may be mediated by a direct effect of these somatogenic hormones and/or a partitioning of nutrient use away from skeletal muscle growth and towards tissues of higher priority, such as the immune system [95].

Nutrition is an important determinant of immune responses; deficits in protein and calories have deleterious effects on host defenses systems. Both protein and energy intake are critical in the regulation of serum levels of IGF-I, IGFBP-3, IGFBP-1, GH and GHBP, as well as the expression levels of the GHR and the IGF-IR [3]. It is known that a GH resistant state is induced during nutritional stress, depending on the specific type and length of nutrient deprivation. In humans and in several other species, when food intake is reduced, serum GH levels are increased (except in the rat) and circulating levels of IGF-I, IGFBP-3 and GH-BP are reduced [96]. In addition, the expression of GH receptor mRNA is also reduced in liver, which reduces hepatic

GH binding capacity. Despite the increased level of GH secretion, the downregulation of GH receptors is partly responsible for the GH resistance [97]. This insensitivity to GH and the protein and calorie deficits lead to a reduction in serum IGF-I, which is exacerbated by a fall in systemic IGFBP-3 levels, because of a reduced production and increased protease activity [98].

It has been well established that a low protein diet or reduced calorie intake decreases serum levels of IGF-I [3]. Approximately 75% of the circulating IGF-I is produced in the liver, which has been further demonstrated by the circulating levels of IGF-I in the liver-IGF-I specific knock-out mice [15]. However some other tissues might be producing and exporting the peptide to the circulation during caloric restriction. A 48-h fast decreased the total levels of IGF-I mRNA in lung, liver, kidney, muscle, stomach, brain and testes; in heart, IGF-I mRNA levels did not change [99]. The total IGF-I produced in skeletal muscle has been shown to be sensitive to the nutritional insult. A low protein diet decreases the amount of IGF-I mRNA and peptide in muscle [100]. There are no reports that show the effect of dietary protein or calorie intake on the IGF-I expression in lymphoid tissue. Our recent results (unpublished) show that in the spleen, IGF-I mRNA levels remain unchanged in mice fed different protein diets. This observation agrees with the idea that nutrient deficiency can block the growth-promoting properties of IGF-I, while some other properties of IGF-I, are unaffected or less affected by dietary restriction.

The anabolic actions of IGF-I are mediated by the type 1 IGF receptor. In rats, fasting as well as calorie and protein restriction increase the binding of IGF-I to certain tissues. Fasting for 48 h increased IGF-I binding to stomach, lung, testes, kidney and heart in the rat, whereas there was no change in rat brain. In this study, the number of IGF-I receptors correlated with the levels of IGF-IR mRNA in each tissue analyzed. The increase in IGF-I receptor expression on these tissues was most likely the result of the marked reduction in circulating levels of IGF-I [99]. Similarly protein or calorie restriction caused significant increases in IGF-IR mRNA in the testis and heart and muscle [101]. These results indicate that the change in IGF-I and IGF-IR mRNA levels is distinctly regulated in each tissue in response nutritional stress. The systemic concentrations of IGF-I fall, but the local concentration may either fall or remain constant in response to protein nutrition; however, binding of IGF-I by tissues is consistently increased. The expression of the IGF-I receptor gene in lymphocytes of patients with low levels of circulating IGF-I was analyzed, and the levels of IGF-IR mRNA in circulating lymphocytes from patients with LTD (Laron-type dwarfism) or IGHD (Isolated GH deficiency) were increased, as compared to normal controls. The number of IGF-I receptors on lymphocytes and erythrocytes from these patients was also higher than the observed in controls [102]. We have found similar results in lymphocytes derived from spleens of mice subjected to nutritional insult (unpublished results). Specifically, B-lymphocytes expressed the highest levels of IGF-I receptor and GH receptor in response to dietary restriction, as determined by FACS and RNase protection assays (unpublished results). These results indicate that the change in IGF-I and IGF-I receptor mRNA levels during protein or calorie deprivation is regulated in a tissue-specific, and probably cell type-specific manner. The change in receptor expression may be secondary to changes in circulating ligand levels.

Insulin-like growth factor-I is involved in the effects of GH on the thymus. The endogenous synthesis and secretion of GH and IGF-I from human thymocytes was demonstrated and the physiological role of the secreted GH suggested an autocrine/paracrine role of this hormone by stimulating the synthesis of IGF-I and thymocyte proliferation [90]. A restricted protein diet significantly decreased both GH receptor and IGF-I mRNA expression in the rat thymus, as determined by RNase protection assay. Furthermore, continuous infusion of rhGH or rhIGF-I failed to restore the expression levels of these genes, suggesting that malnutrition in rats induces a GH-resistant state in the thymus [103].

The effect of GH on regulation of STAT5 and SOCS gene transcription has been investigated in liver, thymus and peripheral blood lymphocytes from normal and malnourished female rats after stimulation with GH [51]. In liver from normal rats, the transcriptional activation depends on the GH secretory pattern. That is, pulsatile GH-stimuli up-regulate STAT5, but have no effect on CIS and decrease SOCS-3 expression. Continuous infusion of GH increases STAT5 and SOCS3 levels, while CIS remains unchanged [50]. In malnourished rats which received 50% of ad libitum fed controls over a period of 7 days, STAT5 mRNA levels were reduced in the liver, thymus and peripheral blood lymphocytes [104]. When GH was administered in pulsatile mode at 2, 4 and 8 h to rats under protein-calorie restriction, the transcription of STAT5 was increased, with an earlier response in thymus than in peripheral blood lymphocytes. However, the levels of STAT5 at 8 h were significantly lower in the three tissues analyzed as compared to the controls [105]. Taken together, these results show that SOCS3 transcription appears to be responsive to the temporal pattern of hormone stimulation, and that STAT5 may play an important role in the induction of GH resistance in malnutrition.

5. CONCLUSIONS

The effects of GH and IGF-I on the immune system are of increasing interest to both researchers and clinicians. Recombinant human GH is used to treat dwarfism, but is also been considered as a potential therapeutic agent in catabolic states, aging-related disorders, immunodeficient AIDS patients, to name a few examples. Clinical trials using rhIGF-I are also ongoing for various disease states. While these agents are used primarily for their anabolic effects, they may also have secondary effects such as potentiating cancer growth. While they have been shown to have positive effects on the immune system, one should remain vigilant especially in older patients where autoimmune antibodies maybe more prevalent.

Thus we have attempted to summarize the relevant information regarding GH and IGF-I and the immune system, realizing that this is an area of research that will require many more years of further investigation.

REFERENCES

1. Clark R. The somatogenic hormones and insulin-like growth factor-I: stimulators of lymphopoiesis and immune function. Endocrine Reviews 1997;18:157–179.
2. Salmon, WD, Daughaday WH. A hormonally controlled serum factor which stimulates sulfate incorporation by cartilage *in vitro.* J Lab Clin Med 1957;49:825–826.
3. Thissen J-P, Ketelslegers J-M, Underwood LE. Nutritional regulation of the insulin-like growth factors. Endocrine Reviews 1994;15:80–101.
4. Weigent DA, Baxter JB, Blalock JE. The production of growth hormone and insulin-like growth factor-I by the same subpopulation of rat mononuclear leukocytes. Brain, Behavior, and Immunity 1992;6:365–76.
5. Kooijman R, Willems M, De Haas CJ, et al. Expression of type I insulin-like growth factor receptors on human peripheral blood mononuclear cells. Endocrinology 1992;131:2244–50.

6. Rapaport R, Sills IN, Green L, et al. Detection of human growth hormone receptors on IM-9 cells and peripheral blood mononuclear cell subsets by flow cytometry: correlation with growth hormone-binding protein levels. Journal of Clinical Endocrinology and Metabolism 1995;80:2612–9.
7. Merchav S. The haematopoietic effects of growth hormone and insulin-like growth factor-I. J pediatr Endocrinol Metabol 1998;11:677–685.
8. Berczi I. Pituitary hormones and immune function. Acta Paediatrica Supplement 1997;423:70–5.
9. Nagy E, Berczi I, Friesen HG. Regulation of immunity in rats by lactogenic and growth hormones. Acta Endocrinologica 1983;102:351–7.
10. Dorshkind K, Horseman N. The roles of prolactin, growth hormone, insulin-like growth factor-I, and thyroid hormones in lymphocyte development and function: insights from genetic models of hormone and hormone receptor deficiency. Endocrine Reviews 2000;21:292–312.
11. Donahue LR, Beamer WG. Growth hormone deficiency in 'little' mice results in aberrant body composition, reduced insulin-like growth factor-I and insulin-like growth factor-binding protein-3 (IGFBP-3), but does not affect IGFBP-2, -1 or -4. Journal of Endocrinology 1993;136:91–104.
12. Powell-Braxton L, Hollingshead P, Giltinan D, Pitts-Meek S, Stewart T. Inactivation of the IGF-I gene in mice results in perinatal lethality. Annals of the New York Academy of Sciences 1993;692:300–1.
13. Powell-Braxton L, Hollingshead P, Warburton C, et al. IGF-I is required for normal embryonic growth in mice. Genes and Development 1993;7:2609–17.
14. Lupu F, Terwilliger J, Lee K, Segre G, Efstratiadis A. Roles of growth hormone and insulin-like growth factor-1 in mouse postnatal growth. Dev Biol 2001;229:141–162.
15. Yakar S, Liu J-L, Stannard B. Normal growth and development in the absence of hepatic insulin-like growth factor I. Proc Natl Acad Sci USA 1999;96:7324–7329.
16. LeRoith D, Bondy C, Yakar S, Liu J-L, Butler A. The somatomedin hypothesis: 2001. Endocrine Reviews 2001;22:53–74.
17. Baroni C. Thymus, peripheral lymphoid tissues and immunological responsiveness of the pituitary dwarf mouse. Experientia 1967;23:282–3.
18. PierpaoliW, Sorkin E. Cellular modifications in the hypophysis of neonatally thymectomized mice. British Journal of Experimental Pathology 1967;48:627–31.
19. Duquesnoy RJ, Mariani T, Good RA. Effect of hypophysectomy on immunological recovery after sublethal irradiation of adult rats. Proceedings of the Society For Experimental Biology and Medicine 1969;131:1176–8.
20. Harrison DE, Archer JR, Astle CM. The effect of hypophysectomy on thymic aging in mice. Journal of Immunology 1982;129:2673–7.
21. Fleisher TA, White RM, Broder S, et al. X-linked hypogammaglobulinemia and isolated growth hormone deficiency. New England Journal of Medicine 1980;302:1429–34.
22. Bozzola M, Biscaldi I, Cisternino M, et al. Long term growth hormone (GH)-releasing hormone and biosynthetic GH therapy in GH-deficient children: comparison of therapeutic effectiveness. Journal of Endocrinological Investigation 1990;13:235–9.
23. Weigent DA, Baxter JB, Wear WE, Smith LR, Bost KL, Blalock JE. Production of immunoreactive growth hormone by mononuclear leukocytes. Faseb Journal 1988;2:2812–8.
24. Weigent DA, Riley JE, Galin FS, LeBoeuf RD, Blalock JE. Detection of growth hormone

and growth hormone-releasing hormone-related messenger RNA in rat leukocytes by the polymerase chain reaction. Proceedings of the Society For Experimental Biology and Medicine 1991;198:643–8.

25. Baxter JB, Blalock JE, Weigent DA. Expression of immunoreactive growth hormone in leukocytes in vivo. Journal of Neuroimmunology 1991;33:43–54.
26. Wu D, Devi R, Malarkey W. Localization of growth hormone messenger ribonucleic acid in the human immune system – a clinical research study –. J Clin endocrinol Metab 1996;81:1278-1282.
27. Delhase M, Vergani P, Malur A, Hooghe-Peters EL, Hooghe RJ. The transcription factor Pit-1/GHF-1 is expressed in hemopoietic and lymphoid tissues. European Journal of Immunology 1993;23:951–5.
28. Guarcello V, Weigent DA, Blalock JE. Growth hormone releasing hormone receptors on thymocytes and splenocytes from rats. Cellular Immunology 1991;136:291–302.
29. Zelazowski P, Dohler KD, Stepien H, Pawlikowski M. Effect of growth hormone-releasing hormone on human peripheral blood leukocyte chemotaxis and migration in normal subjects. Neuroendocrinology 1989;50:236–9.
30. Hattori N, Ikekubo K, Ishihara T, Moridera K, Hino M, Kurahachi H. Spontaneous growth hormone (GH) secretion by unstimulated human lymphocytes and the effects of GH-releasing hormone and somatostatin. Journal of Clinical Endocrinology and Metabolism 1994;79:1678–80.
31. Hattori N, Shimomura K, Ishihara T. Growth hormone (GH) secretion from human lymphocytes is up-regulated by GH, but not affected by insulin-like growth factor-I. Journal of Clinical Endocrinology and Metabolism 1993;76:937–9.
32. Arrenbrecht S. Specific binding of growth hormone to thymocytes. Nature 1974;252:255–257.
33. Lesniak M, Gorden P, Roth GJ. Binding of 125I-human growth hormone to specific receptors in human culturedlymphocytes. Characterization of the interaction and a sensitive radioreceptor assay. J Biol Chem 1974;249:1661–1667.
34. Kiess W, Butenandt O. Specific growth hormone receptors on human peripheral mononuclear cells: reexpression, identification and characterization. J Clin Endocrinol Metabol 1985;60:740–746.
35. Landron D, Guerre-Millo M, Postel-Vinay MC, Lavau M. Relationship between increased binding and insulin-like effects of human growth hormone in adipocytes from young fa/fa rats. Endocrinology 1989;124:2305–13.
36. Ban E, Gagnerault MC, Jammes H, Postel-Vinay M-C, Haour F, Dardenne M. Specific binding sites for growth hormone in cultured mouse thymic epithelial cells. Life Sciences 1991;48:2141–2148.
37. Badolato R, Bond HM, Valerio G, et al. Differential expression of surface membrane growth hormone receptor on human peripheral blood lymphocytes detected by dual fluorochrome flow cytometry. Journal of Clinical Endocrinology and Metabolism 1994;79:984–90.
38. Mathews LS, Enberg B, Norstedt G. Regulation of rat growth hormone receptor gene expression. J Biol Chem 1989;264:9905–9910.
39. Tiong T, Herington A. Tissue distribution, characterization and regulation of the mRNA for growth hormone receptor and serum binding protein in the rat. Endocrinology 1991;129:1628–1634.
40. Lobie P, Garcia-Aragon J, Wang B, Baumbach W, Waters M. Cellular localization of the growth hormone binding protein in the rat. Endocrinology 1992;130:3057–3065.

41. Carlsson B, Billing H, Rymo L, Isaksson OGP. Expression of the growth hormone binding protein messenger RNA in the liver and extrahepatic tissues in the rat: co-expression with the growth hormone receptor. Mol Cell Endocrinol 1990;73:R1–R6.
42. Gagnerault M-C, Vinay M-CP, Dardenne M. Expression of Growth hormone receptors in murine lymphoid cells analyzed by flow cytofluorometry. Endocrinology 1996;137:1719–1726.
43. Mercado M, Da Vila N, McLeod JF, Bauman G. Distribution of growth hormone receptor messenger ribonucleic acid containig and lacking exon 3 in human tissues. J Clin Endocrinol Metabol 1994;78:731–735.
44. Leung DW, Spencer SA, Cachianes G, et al. Growth hormone receptor and serum binding protein: purification, cloning and expression. Nature 1987;330:537–543.
45. Cosman D. The hematopoeitin receptor superfamily. Cytokine 1993;5:95–106.
46. Argetsinger LS, Campbell GS, Yang X, et al. Identification of JAK2 as growth hormone receptor associated kinase. Cell 1993;74:1–20.
47. Vander Kuur J, Allevato G, Billestrup N, Norstedt G, Carter S. Growth hormone-promoted tyrosil phosphorylation of SHC proteins and SHC association with Grb2. J Biol Chem 1995;270:7587–7593.
48. Winston LA, Bertics PJ. Growth hormone stimulates the tyrosine phosphorylation of 42- and 45 kDa ERK-related proteins. J biol Chem 1992;267:4747–4751.
49. Hill CR, Treisman. Transcriptional regulation by extra-cellular signals: mechanisms and specificity. Cell 1995;80:199–210.
50. Umana-Perez A, Ortiz BL, Morales A, Norstedt G, Sanchez-Gomez M. GH regulates STAT5 and SOCS gene expression *in vivo* and *in vitro*. In: Basic and Clinical GH and IGF-I Research in the Postgenomic Era. Goteborg: Growth Hormone & IGF Research, 2000.
51. Ortiz BL, Sanchez-Gomez M, Norstedt G. Comparison of STAT5 mRNA levels in GH-treated male and female rats analysed by a solution hybridization assay. Growth Hormone & IGF Research 2000;10:236–241.
52. Liu X, Robinson G, Wagner K, Garret L, Wynshaw-Boris A, Henninghausen L. STAT5a is mandatory for addult mammary gland development and lactogenesis. Genes Dev 1997;12:179–186.
53. Endo TA, Masuhara M, Yokouchi M, et al. A new protein containig an SH2 domain that inhibits JAK kinases. Nature 1997;387:921–924.
54. Starr R, Willson TA, Murray LJ, et al. A family of cytokine-inducible inhibitors of signalling. Nature 1997;917–921.
55. Nicholson SE, Hilton DJ. The SOCS proteins: a new family of negative regulators of signal transduction. J Leukoc Biol 1998;63:665–668.
56. Adams TE, Hansen JA, Starr R, Nicola NA, Hilton DJ, Billestrup N. Growth hormone preferentially induces the rapid, transient expression of SOCS-3, a novel inhibitor of cytokine receptor signaling. J Biol Chem 1998;273:1285-1287.
57. Billestrup N, Hansen JA, Larsen SG, Lindberg K, Richter HE. SOCS and GH receptor signalling. In: 11th International Congress of Endocrinology. Sydney, 2000.
58. Tollet P, Flores-Morales A, Stavreus-Evers A, Sahlin L, Norstedt G. Growth hormone regulation of SOCS-2, SOCS-3 and CIS messenger ribonucleic acid expression in the rat. Endocrinolgy 1999;140:3693–3704.
59. Bernton E, Dave J. Prolactin, growth hormone and immune homeostasis. In: Meltzer M, Mantovani A, editors. Cellular and Cytokine Networks in Tissue Immunity. New York:

Wiley-Liss, 1991;69–75.

60. Cecim M, Bartke A, Yun J, Wagner T. Growth allometry of transgenic mice expressing the mouse metallothionein-I/bovine growth hormone gene. Transgene 1993;1:125–132.
61. Maiter D, Underwood L, Maes M, Davenport ML, Ketelslegers JM. Different effects of intermitent and continuous growth hormone (GH) administration on serum somatomedin C/Insulin-like growth factor I and liver GH receptors in hypophysectomised rats. Endocrinology 1988;123:1053–1059.
62. LeRoith D, Yanowski J, Kaldjian EP, et al. The effects of growth hormone and insulin-like growth factor I on the immune system of aged female monkeys. Endocrinology 1996;137:1071–1079.
63. Savino W, Dardenne M. Neuroendocrine Control of Thymus Physiology. Endocrine Reviews 2000;21:412–443.
64. Petersen BC, Rapaport R, Henry D, Huseman C, Moore W. Effect of treatment with biosynthetic human growth hormone (GH) on peripheral blood lymphocyte populations and function in growth hormone-deficient children. J Clin Endocrinol Metab 1990;70:1756–1760.
65. Bazzoni N, Ambrosi B, Arosio JM, Barbareschi M, Losa M, Faglia G. Acromegaly and thymic hyperplasia: a case report. J Endocrinol Invest 1990;11:931–935.
66. Jones JI, Clemmons DR. Insulin-like growth factors and their binding proteins: biological actions. Endocrine Reviews 1995;16:3–34.
67. LeRoith D, Werner H, Beitner-Johnson D, Roberts CTJ. Molecular and cellular aspects of the insulin-like growth factor I receptor. Endocrine Reviews 1995;16:143–63.
68. Bach LA, Rechler MM. Insulin-like growth factor binding proteins. Diabetes Reviews 1995;3:38–60.
69. Martin JL, Baxter RC. IGF binding proteins as modulators of IGF action. Contemporary Endocrinology: The IGF System 1999;17:227–255.
70. Domene H, Meidan R, Yakar S, et al. Role of GH and IGF-I in the regulation of IGF-I, IGF-I receptor and IGF binding protein gene expression in the rat spleen. Regul Pept 1994;52:215–226.
71. Stuart CA, Meehan RT, Neale LS, Cintron NM, Furlanetto RW. Insulin-like growth factor-I binds selectively to human peripheral blood monocytes and B-lymphocytes. Journal of Clinical Endocrinology and Metabolism 1991;72:1117–22.
72. Xu X, Mardell C, Xian CJ, Zola H, Read CL. Expression of functional insulin-like growth factor-1 receptor on lymphoid cell subsets of rats. Immunology 1995;85:394–399.
73. Liu J, Baker J, Perkins A, Robertson E, Efstrstiadis A. Mice carrying null mutations of the genes encoding insulin-like growth factor I (Igf-1) and type 1 IGF receptor (Igf1r). Cell 1993;75:59–72.
74. Rotwein P, Thomas M, Gronowski A, Bichell D, Kikuchi K. The somatomedin hypothesis revisited: early events in growth hormone action. In: Baxter R, Gluckman P, Rossenfeld R, editors. The Insulin-like Growth Factors and their Regulatory Proteins. Amsterdam: Elsevier, 1994;13–21.
75. Arkins S, Rebeiz N, Biragyn A, Reese DL, Kelley KW. Murine macrophages express abundant insulin-like growth factor-I class I Ea and Eb transcripts. Endocrinology 1993;133:2334–43.
76. Nyman T, Pekonen F. The expression of insulin-like growth factors and their binding proteins in normal human lymphocytes. Acta Endocrinologica 1993;128:168–72.
77. Fournier T, Riches D, Winston B, et al. Divergence in macrophague insulin-like growth

factor-I (IGF-I) synthesis induced by TNF-alpha and prostaglandin E2. J Immunology 1995;155:2123–2133.

78. Arkins S, Rebeiz N, Brunke-Reese DL, Minshall C, Kelley KW. The colony-stimulating factors induce expression of insulin-like growth factor-I messenger ribonucleic acid during hematopoiesis. Endocrinology 1995;136:1153–60.

79. Arkins S, Rebeiz N, Brunke-Reese DL, Biragyn A, Kelley KW. Interferon-gamma inhibits macrophage insulin-like growth factor-I synthesis at the transcriptional level. Molecular Endocrinology 1995;9:350–60.

80. Berczi I, Nagy E, Kovacs K, Horvath E. Regulation of humoral immunity in rats by pituitary hormones. Acta Endocrinologica 1981;98:506–13.

81. Kelley KW. Immunologic roles of two metabolic hormones, growth hormone and insulin-like growth factor-I, in aged animals. Nutrition Reviews 1995;53:S95–S104.

82. Jardieu P, Clark R, Mortensen D, Dorshkind K. In vivo administration of insulin-like growth factor-I stimulates primary B lymphopoiesis and enhances lymphocyte recovery after bone marrow transplantation. J Immunol 1994;152:4320–4327.

83. Guler H, Zapf J, Scheiwiller E, Froesch E. Recombinant human insulin-like growth factor 1 stimulates growth and has distinctic effects on organ size in hypophysectomized rats. Proc Natl Acad Sci USA 1988;85:4889–4893.

84. Clark R, Strasser J, McCabe S, Robbins K, Jardieu P. Insulin-like growth factor-1 stimulation of lymphopoiesis. J Clin Invest 1993;92:540–548.

85. Landreth K, Narayanan R, Dorshkind K. Insulin-like growth factor-I regulates pro-B cell differentiation. Blood 1992;80:1207–1212.

86. Gibson L, Piktel D, Landreth K. Insulin-like growth factor-1 potentiates expansion of interleukin-7-dependent pro-B cells. Blood 1993;82:3005–3011.

87. Tsarfaty G, Longo D, Murphy W. Human Insulin-like growth factor I exerts hematopoietic growth-promoting effects after in vivo administration. Exp Hematol 1994;22:1273–1277.

88. Yamada M, Hato F, Kinoshita Y, Tominaga K, Tsuji Y. The indirect participation of growth hormone in the thymocyte proliferation system. Cell Mol Biol 1994;40:111–121.

89. de Mello-Coelho V, Villa Verde D, Dardenne MSW. Pituitary hormones modulate cell-cell interactions between thymocytes and thymic epithelial cells. J Neuroimmunol 1997;76:39–49.

90. Sabharwal P, Varma S. Growth hormone synthesized and secreted by human thymocytes acts via insulin-like growth factor I as an autocrine and paracrine growth factor. J Clin Endocrinol Metab 1996;81:2663–2669.

91. Robbins K, McCabe S, Scheiner T, Strasser J, Clark R, Jardieu P. Immunological effects of insulin-like growth factor-I enhancement of immunoglobulin synthesis. Clin Exp Immunol 1994;95:337–342.

92. Montecino-Rodriguez E, Clark R, Dorshkind K. Effects of insulin-like growth factor administration and bone marrow transplantation on thymopoiesis in aged mice. Endocrinology 1998;139:4120–4126.

93. Tarantal A, Hunter M, Gargosky S. Direct administration of insulin-like growth factor to fetal rhesus monkeys. Endocrinology 1997;138:3349–3358.

94. Foster MP, Jensen ER, Montecino-Rodriguez E, Leathers H, Horseman N, Dorshkind K. Humoral and cell-mediated immunity in mice with genetic deficiencies of prolactin, growth hormone, insulin-like growth factor-I and thyroid hormone. Clinical Immunology 2000;96:140–149.

95. Davis SL. Environmental modulation of the immune system via the endocrine system.

Domestic Animal Endocrinology 1998;15:283–9.

96. Estivariz C, Ziegler T. Nutrition and the insulin-like growth factor system. Endocrine 1997;7:65–71.

97. Maiter D, Maes M, Underwood L, Fliesen T, Gerard G, Ketelslegers J. Early changes in serum concentrations of somatomedin-C induced by dietary protein deprivation in rats: contributions of growth hormone receptor and post-receptor defects. J Endocrinol 1988;118.

98. Oster MH, Fielder PJ, Levin N, Cronin MJ. Adaptation of the growth hormone and insulin-like growth factor-I axis to chronic and severe calorie or protein malnutrition. Journal of Clinical Investigation 1995;95:2258–65.

99. Lowe W, Adamo M, Werner H, Roberts C, LeRoith D. Regulation by fasting of rat insulin-like growth factor I and its receptor. Effects on gene expression and binding. J Clin Invest 1989;84:619–626.

100. Sanchez-Gomez M, Malmo K, Mejia W, et al. Insulin-like growth factor-I, but not growth hormone, is dependent on a high protein intake to increase nitrogen balance in the rat. British Journal of Nutrition 1999;81:145–152.

101. Takenaka A, Takahashi S, Noguchi T. Effect of protein nutrition on insulin-like growth factor-I (IGF-I) receptor in various tissues of rats. J Nutr Sci Vitaminol 1996;42:347–357.

102. Eshet R, Werner H, Klinger B, et al. Up-regulation of insulin-like growth factor-I (IGF-I) receptor gene expression in patients with reduced serum IGF-I levels. Journal of Molecular Endocrinology 1993;10:115–20.

103. Carrasco-Rodriguez S,Sanchez-Gomez M. Lymphopoietic effects of GH and IGF-I in rats under nutritional stress. In: 11th International Congress of Endocrinology. Sydney, 2000.

104. Ortiz BL, Sanchez-Gomez M, Norstedt G. STAT5 transcription in lymphoid tissues from nutritional restricted rats is regulated by GH. In: 11th International Congress of Endocrinology. Sydney, 2000.

105. Ortiz, B, Norstedt G, Sanchez-Gomez M. STAT5 messenger RNA in normal and malnourished rats is upregulated by growth hormone treatment. In: 1998 Growth Hormone Research Society (GRS) Conference. San Francisco CA: Growth Hormone and IGF Research, 1998.

Growth and Lactogenic Hormones
Edited by L. Matera and R. Rapaport

Reciprocal Interactions between the GH/IGF-I System and Cytokines

FABRIZIO DE BENEDETTI[1] and MAURO BOZZOLA[2]

[1]*Pediatria Generale e Reumatologia, Dipartimento di Scienze Pediatriche, Università degli Studi di Pavia, IRCCS San Matteo, P.le Golgi 2, 27100 Pavia*
[2]*Dipartimento di Scienze Pediatriche, Università degli Studi di Pavia, IRCCS San Matteo, P.le Golgi 2, 27100 Pavia*

ABSTRACT

A growing body of evidence indicates a bidirectional relationship between the neuroendocrine system and immune functions. In this chapter we discuss the interactions between the growth hormone/insulin-like growth factor system and cytokines. Much experimental evidence suggests that both GH and IGF-I have a stimulatory effect on the immune response and on the production of immune (interleukin-2 and interferon-γ) and inflammatory (tumor necrosis factor, interlukin-1, interleukin-6) cytokines. However, the *in vivo* relevance in humans of these findings has yet to be fully demonstrated. On the other hand, proinflammtory cytokines affect the GH/IGF-I axis. These effects are relevant for the understanding of the mechanisms leading to stunted growth in chronic inflammatory diseases or diseases with recurrent infections in childhood.

1. INTRODUCTION

A growing body of evidence indicates a bidirectional relationship between the neuroendocrine system and immune function. Neurohormones act on immune cells while cytokines secreted by lymphocytes and macrophages, in turn, influence neuroendocrine function. The variability of endocrine and immune responses depends on these interactions which can be facilitatory or inhibitory.

Cytokines are a large family of protein mediators including interleukins (ILs), colony-stimulating factors (CSFs), interferons (IFNs), tumor necrosis factors (TNFs), and growth factors which are produced by a wide variety of cells. Cytokines play a major role in the initiation and regulation of immune and inflammatory responses. Cytokines are glycoproteins and, similarly to hormones, are produced by one cell and act on others. In contrast to hormones, cytokines are produced at very low levels and their effects are therefore often more localized. Because many cell types produce cytokines and each cytokine interacts with multiple target cells, complicated cellular networks are formed through cytokines.

2. CAN GH MODULATE HUMORAL FUNCTION?

It is reasonable to hypothesize that if GH can influence the immune system, its absence should lead to alterations in the immune response. A number of experimental findings support this hypothesis. The dwarf Snell mouse represents a model of congenital hypopituitarism due to a mutation in the Pit-1 gene, leading to an arrest in ontogenic development of the thymus and, consequently, to a severe thymus-dependent immunodeficiency [1]. Long-term GH treatment prevents thymus involution and normalizes the immune response [1]. Experimental administration of antisera against GH leads to the suppression of antibody formation in mice suggesting a role for GH in immune function [2]. In humans, GH deficiency is not usually associated with immunodeficiency, except for the rare X linked combinations of isolated GH deficiency and agammaglobulinemia [3]. No clinical signs of immune dysfunction have been observed even in patients with severe GH deficiency due to either a mutation in the Pit-1 gene or a deletion of the GH-N gene.

In vitro studies indicate that incubation of human peripheral blood mononuclear cells with different concentrations of hGH significantly increases the number of interferon-γ-secreting cells as well as the concentration of IFN-γ [4]. Studies on GH-deficient children showed that the production of IL-1α and IL-2 by mononuclear cells requires GH. In particular, IL-1α production was normalized after 15 days of substitutive GH therapy and IL-2 after 3 months of treatment [5].

The pituitary gland is needed for the syntesis of tumor necrosis factor-α (TNF- α) by macrophages. Hypophysectomized rats had markedly depressed macrophage synthesis of TNF-α. Exogenous GH partially reversed the effect of hypophysectomy [6]. Unlike experimental data in animals, no difference in basal serum TNF-α and IL-1β concentration was observed in 15 children with GH deficiency and in 19 controls [7]. An explanation for these findings could be that only in experimental models *complete GH deficiency* can be obtained by hypophysectomy. The effect of GH or GH-dependent factors on peripheral macrophages is confirmed by the significant increase in both serum TNF- α and IL-1β concentrations after GH injection, and by their fall 24 hours later when serum GH values fell [7]. In contrast, in GH-deficient adults basal TNF-α levels are high and fall after proloned GH administration [8].

3. EFFECT OF IGF-I ON IMMUNE RESPONSES

Cells of the immune system, such as T and B lymphocytes and macrophages, express functional IGF receptors [9,10]. Moreover, since IGF-I is produced by immune cells [11], its effects on immune responses may be secondary to autocrine or paracrine mechanisms. Several observations demonstrate that IGF-I, either endogenously produced or exogenously added, affects *in vitro* immune cell replication and function. IGF-I stimulates the proliferation of T cells [12] and mediates the growth promoting effect of GH on T-lymphoblastoid cell lines [13]. In addition, it increases mitogen-induced IL-2 production [14], which might be responsible for the effect on proliferation. More recently it has been shown that costimulatory signals delivered through the CD28 molecule expressed on T lymphocytes induce expression of the IGF receptor, which in turn mediates anti-apoptotic effects protecting T cells from cell death induced by Fas activation

[15]. IGF-I has also been shown to play an important role in the in vitro regulation of natural killer cell cytotoxicity [16]. Indeed, recent data have shown that the stimulation of natural killer cell cytotoxicity by dehydroepiandrosterone is mediated by the induction of an autocrine mechanism involving generation and release of IGF-I by purified natural killer cells [17]. Similarly to GH, IGF-I induces proliferation of B lymphocytes and production of immunoglobulins [18].

The *in vivo* relevance of these *in vitro* results has yet to be demonstrated. To the best of our knowledge only two observations suggest, albeit indirectly, an *in vivo* effect of IGF-I on immune function in humans. Short-term fasting, which causes decreased levels of IGF-I, is associated with decreased *ex vivo* production of IL-2 by mitogen activated lymphocytes [19]. Prolonged starving may cause a chronic depression of IL-2 production, therefore eventually promoting the immunosuppression associated with nutrient denial. Administration of a combination of GH/IGF-I to HIV-infected patients resulted in a significant improvement in HIV-specific immune responses [20].

4. EFFECTS OF CYTOKINES ON THE GH/IGF-I SYSTEM.

In the previous paragraph we have outlined the effects of GH and IGF-I on cytokine production. A vast body of evidence shows that cytokines, and particularly the proinflammatory cytokines IL-1, IL-6 and TNF-α affect the GH/IGF-I axis. These effects are relevant in human pathological conditions for the understanding of the stunted growth associated with chronic inflammatory diseases or diseases with recurrent infections in childhood [21]. Therefore, GH/IGF-I may have a significant prospective therapeutic relevance for this common complication of chronic inflammation. In this paragraph we will describe the effects of IL-1, IL-6 and TNF-α on the GH/IGF-I axis in experimental animals and relate them with the findings in childhood chronic inflammatory diseases.

4.1. Effects on GH

IL-1, IL-6 and TNF-α appears to have different effects on the GH/IGF-I axis. Some data suggest that TNF-α may directly inhibit GH pituitary production, while IL-1 has been shown to induce GH production by pituitary cells in vitro [22]. However, data on the effect *in vivo* of the administration of TNF are still contradictory. On the contrary, chronic overexpression of IL-6 in IL-6 transgenic mice does not modify the number of pituitary GH producing cells or serum GH levels, demonstrating that IL-6 does not affect directly pituitary GH production *in vivo* [23].

4.2. Effects on IGF-I and IGF binding proteins

In mice the administration of IL-1 causes a significant decrease in circulating levels of IGF-I [24]. Moreover, in mice treated with endotoxin, which is a powerful inducer of inflammatory cytokine production, the co-administration of interleukin-1 receptor antagonist (IL-1Ra), a physiological antagonist to IL-1 produced during inflammation, inhibits in part the decrease in IGF-I induced by endotoxin [25]. A similar decrease in IGF-I levels was observed following the administration of TNF-α [26]. Both IL-1 and of TNF-α induce a decrease in hepatic IGF-I levels, thus suggesting a direct effect on liver production of IGF-I [24,26]. In agreement with this *in vivo* observation, it has been shown that IL-1 and TNF-α inhibit the expression of the IGF-I mRNA induced by GH in primary hepatocytes [27,28]. Wolf et al have also reported that in primary hepatocytes IL-1 and TNF-α inhibit the expression of mRNA for the GH receptor [28]. Taken together, these observations suggest that IL-1 and TNF-α inhibit production of IGF-I by the liver by causing a hyporesponsiveness of hepatocyte to GH.

Similarly to IL-1 and TNF-α, chronic overexpression of IL-6 in IL-6 transgenic mice and acute administration of IL-6 to normal mice cause a significant decrease in IGF-I levels [23]. However, *in vitro* IL-6 has no effect on GH-induced IGF-I mRNA expression in cultured hepatocytes [27] and normal levels of liver IGF-I protein were found in IL-6 transgenic mice [29]. Taken together these observations show that the decrease in circulating IGF-I levels induced by IL-6 *in vivo* cannot be explained by low liver IGF-I production. On the other hand, the decrease in IGF-I levels induced by IL-6 appears to be secondary to increased clearance/degradation of IGF-I. Indeed, we have recently shown that IL-6 transgenic mice have decreased levels of the ternary 150 Kd complex, comprising IGF-I – IGF binding protein-3 (IGFBP-3) and the acid labile subunit (ALS) [29]. Since the 150 Kd complex represents the major reservoir of circulating IGF-I and the association of IGF-I in this complex markedly increase the plasma half-life of IGF-I, the defective formation of this complex in IL-6 transgenic mice may be responsible for the decreased circulating IGF-I levels. This defect in 150 Kd complex is not due a decrease in ALS. Indeed, addition of exogenous IGFBP-3 and IGF-I to sera from IL-6 transgenic mice led to efficient formation of the 150 KDa ternary complex, suggesting the presence of functionally normal serum ALS [29]. In agreement with this finding, it has been reported that IL-6 does not inhibit spontaneous or GH-induced release of ALS from cultured hepatocytes [30]. On the contrary, we found that both the chronic overexpression of IL-6 in IL-6 transgenic mice and the acute administration of IL-6 to normal mice induced a marked decrease in circulating levels of IGFBP-3, suggesting that impaired ternary complex formation is secondary to decreased IGFBP-3 [29].

It is noteworthy that the effect of IL-1 and TNF on IGFBP-3 and ALS appears to be different from that of IL-6. Indeed, administration of IL-1 or TNF-a does not affect circulating IGFBP-3 levels [24,26]. On the other hand IL-1 suppresses ALS production by primary hepatocytes [30,31].

4.3. Effect of proinflammatory cytokines on somatic growth

In summary, the three proinflammatory cytokines IL-1, IL-6 and TNF-α affect the GH/IGF-I axis , albeit with different effects on the components of this system (Table I), inducing a marked decrease in circulating IGF-I. IL-6 transgenic mice with high circulating levels of IL-6 since birth show a significant decrease in postnatal growth rate, and reach an adult size, which is about 30–50% smaller than that of non-transgenic mice. This fact indicates the relevance of chronic overproduction of IL-6 and of IL-6-induced mechanisms on the IGF-I system [23]. The neutralization of IL-6 in these mice leads to a complete correction of the growth defect, thus proving that stunted growth is caused by the sustained production of IL-6 [32]. Even though IL-1 and TNF-α have an influence on the GH/IGF-I axis there does not exist, to our knowledge, an animal model that, in the same way as for IL-6, can prove a direct link between stunted growth and a chronic hyperproduction of IL-1 and/or TNF-α.

4.4. GH/IGF-I axis in chronic inflammatory diseases

As previously mentioned stunted growth is a common complication in children with chronic inflammatory diseases, such as systemic juvenile idiopathic arthritis (JIA) and Crohn's disease,

Table I In vitro and in vivo effects in experimental animals of the proinflammatory cytokines IL-1, IL-6 and TNF-α on the GH/IGF-I axis.

	IL-1	IL-6	TNF-α
In vitro GH production by pituitary cells	Increase	No effect	Decrease
In vivo GH levels	Unknown	Normal	Decreased/ Normal
In vitro IGF-I expression by hepatocytes	Decrease	No effect	Decrease
In vivo IGF-I levels	Decrease	Decrease	Decrease
In vitro ALS expression by hepatocytes	Decrease	No effect	Unknown
ALS levels in vivo	Decrease	Normal	Unknown
In vitro IGFBP-3 expression	Unknown	Unknown	Unknown
In vivo IGFBP-3 levels	Normal	Decrease	Normal

Abbreviations: IL, interleukin, GH, growth hormone; IGF-I, insulin-like growth factor-I; IGFBP-3, IGF binding protein-3; ALS, acid labile subunit.

Table II Comparison of the abnormalities of the GH/IGF-I system in the IL-6 transgenic NSE/hIL-6 mice with those present in children with systemic juvenile idiopathic arthritis (s-JIA), Crohn's disease (Crohn's), cystic fibrosis (CF) and with HIV infection (HIV-inf).

	NSE/hIL-6 mice	s-JIA	Crohn's	CF	HIV-inf
Circulating GH levels	Normal	Normal	Normal	Normal	Normal/Low
Circulating IGF-I levels	Decreased	Decreased	Decreased	Decreased	Decreased
Circulating IGFBP-3 levels	Decreased	Decreased	Decreased	Decreased	Decreased
Circulating ALS levels	Normal	Normal	Normal	Unknown	Unknown

and of diseases with recurrent infections, such as cystic fibrosis and AIDS. In these patients GH production is essentially normal [34–36], while decreased levels of IGF-I have been found [23,37–40]. While in Crohn's disease and systemic JIA ALS levels have been found to be in the normal range [29,41], low levels of IGFBP-3 have been demonstrated [29,42,43]. High levels of IL-6 have been detected in these diseases and the level of IL-6 appears to correlate directly with a variety of clinical and laboratory parameters of disease severity [44–46]. Moreover, a direct relation of IL-6 production with low IGF-I and IGFBP-3 has been observed in patients with systemic JIA and in HIV infected children with decreased growth rate [23,29,39]. Taken together, these findings and their similarities with the abnormalities in the GH/IGF-I system induced by IL-6 in experimental animals (Table II) strongly support the conclusion that chronic overproduction of IL-6 is a pivotal factor in the induction of the decrease in IGF-I and IGFBP-3 levels and of the stunted growth associated with chronic inflammation in childhood.

5. CONCLUSIONS

In this chapter we have described the bidirectional interaction between the GH/IGF-I system and cytokines, employed by the immune system to induce and modulate the inflammatory and immune responses. While a vast body of evidence indicates that GH and IGF-I affect *in vitro* immune responses and cytokine production by immune cells, the *in vivo* relevance of these effects has not yet been clarified. On the other hand, the demonstration of the effects the proinflammatory cytokines, and particularly of IL-6, on the GH/IGF-I axis provides information on the mechanisms leading to stunted growth in chronic inflammation in childhood, that may have a significant prospective therapeutic relevance, ranging from a direct inhibition of the cytokine to the administration of IGF-I and/or IGFBP-3.

REFERENCES

1. Pierpaoli W, Baroni C, Fabris N, Sorkin E. Hormones and immunological capacity. II. Reconstitution of antibody production in hormonally deficient mice by somatotropic hormone, thyrotropic hormone and thyroxin. Immunology, 1969;16:217–230.
2. Pierpaoli W, Sorkin E. Hormones and immunological capacity. I. Effect of heterologous anti-growth hormone antiserum on thymus and peripheral lymphatic tissue in mice. Induction of a wasting syndrome. J Immunol 1968;101:1036–1043.
3. Fleischer TA, White RM, Broder S, Nissley SP, Blaese RM, Mulvihill JJ, Olive G, Waldmann TA. X-linked hypogammaglobulinemia and isolated growth hormone deficiency. N Engl J Med 1980;302:1429–1434.
4. Mustafa A, Nyberg F, Mustafa M, Bakhiet M, Mustafa E, Winblad B, Adem A. Growth hormone stimulates production of interferon-α by human periferal mononuclear cells. Horm Res 1997;48:11–15.
5. Edwards CK, Lorence RM, Dunham DM Arkins S, Yunger LM, Greager JA, Walter RJ, Dantzer R, Kelley KW. Hypophysectomy inhibits the synthesis of tumor necrosis factor-α by rat macrophages: partial restoration by exogenous growth hormone or interferon-γ. Endocrinology 1998;128:989–996.
6. Casanova S, Repellin A, Schimpff RM. Production of interleukin-1α and interleukin-2 by mononuclear cells from children with growth delay in relation to the degree of growth hormone deficiency: effects of substitutive treatment. Horm Res 1990;34:209–214.
7. Bozzola M, De Amici M, Zecca M, Schimpff RM, Rapaport R. Modulating effect of human growth hormone on tumour necrosis factor-α and interleukin-1β. Eur J Endocrinol 1998;139:640–643.
8. Serri O, St-Jacques P, Sartippour M, Renier G. Alterations of monocyte function in patients with growth hormone deficiency: effect of substitutive GH therapy. J Clin Endocrinol Metab 1999;84:58–63.
9. Kooijman R, Willems M, Dettass CJC. Expression of type I insulin-like growth factor receptors on human peripheral blood mononuclear cells. Endocrinology 1992;131:2244–2250.
10. Stuart CA, Meehan RT, Neale LS, Cintro NM, Furlanetto RW. Insulin-like growth factor-I binds selectively to human peripheral blood monocytes and B lymphocytes. J Clin Endocrinol Metab 1991:72:1117–1122.
11. Baxter JB, Blalock JE, Weigent DA. Characterization of immunoreactive insulin-like

growth factor I from leukocytes and its regulation by growth hormone. Endocrinology 1991;129:1727–1734.

12. Clark R, Strasser J, McCabe S, Robbins K, Jardieu P. Insulin-like growth factor_I stimulation of lymphopoiesis. J Clin Invest 1993;922:540–548.
13. Geffner ME, Bersh N, Lippe BM, Rosenfeld RG, Hintz RL, Golde DW. Growth hormone mediates the growth of T lymphocytes cell lines via locally generated insulin-like growth factor I. J Clin Endocrinol Metab 1990;71:464–469.
14. Koojiman R, Rijkers GT, Zegers BJ. IGF-I potentiates interleukin-2 production in human peripheral blood T cells. J Endocrinol 1996;149:351–356.
15. Walsh PT, O'Connor R. The insulin-like growth factor-I receptor is regulated by CD28 and protects activated T cells from apoptosis. Eur J Immunol 2000;30:1010–1018.
16. Auernhammer CJ, Feldmeier H, Nass R, Pachmann K, Strasburger CJ. Insulin-like growth factor I is an independent coregulatory modulator of natural killer (NK) cell activity. Endocrinology 1996;137:5332–5336.
17. Solerte SB, Fioravanti M, Vignati G, Giustina A, Cravello L, Ferrari E. Dehydroepiandrosterone sulfate enhances natural killer cell cytotoxicity in humans via locally generated immunoreactive insulin-like growth factor I. J Clin Endocrinol Metab 1999;84:3260–3267.
18. Kimata H, Yoshida A. Differential effect of growth hormone, insulin-like growth factor-I, insulin-like growth factor II, and insulin on Ig production and growth on human plasma cells. Blood 1994;83:1569–1574.
19. Savendahl L, Underwood LE. Decreased interleukin-2 production form cultured peripheral blood mononuclear cells in human acute starvation. J Clin Endocrinol Metab 1997;82:1177–1180.
20. Nguyen BY, Clerici M, Venzon DJ, et al. Pilot study of the immunologic effects of recombinant human growth hormone and recombinant insulin-like growth factor in HIV-infected patients. AIDS 1998;28;12:895–904.
21. De Benedetti F, Martini A. Cytokines et retard statural au cours des maladies inflammatoires chroniques de l'enfant. Arch Pediatr 1999;6:499–502.
22. Bernton EW, Beach JE, Holaday JW, Smallridge RC, Flein HG. Release of multiple hormones by a direct action of interleukin-1 on pituitary cells. Science 1987;238:519–521.
23. De Benedetti F, Alonzi T, Moretta A, Lazzaro D, Costa P, Poli V, Martini A, Ciliberto G, Fattori E. IL-6 causes growth impairment through a decrease in insulin-like growth factor-I: a model for stunted growth in children with chronic inflammation. J Clin Invest 1997;99:643–650.
24. Fan J, Wojnar MM, Theodorakis M, Lang CH. Regulation of insulin-like growth factor (IGF)-I mRNA and peptide and IGF-binding proteins by interleukin-1. Am J Physiol 1996;270 (Regulative Integrative Comp. Physiol 39):R621-R629.
25. Lang CH, Fan J, Cooney R, Vary TC. IL-1 receptor antagonist attenuates sepsis-induced alterations in the IGF system and protein synthesis. Am J Physiol 1996;270 (Endocrinol Metab 33):E430–E437.
26. Fan J, Char D, Bagby GJ, Gelato MC, Lang CH. Regulation of insulin-like growth factor-I (IGF-I) and IGF-binding proteins by tumor necrosis factor. Am J Physiol 1996;269 (Regulative Integrative Comp. Physiol 38):R1204–R1212.
27. Wolf M, Bohm S, Brand M, Kreyman G. Proinflammatory cytokines interleukin-1β and tumor necrosis factor-α inhibit growth hormone stimulation of insulin-like growth factor-I

synthesis and growth hormone receptor mRNA levels in cultured rat liver cells. Eur J Endocrinol 1996;135:729–735.

28. Thissen JP, Verniers J. Inhibition by interleukin-1β and tumor necrosis factor-α of the insulin-like growth factor-I messenger ribonucleic acid response to growth hormone in rat hepatocyte primary culture. Endocrinol 1997;138:1078–1084.
29. De Benedetti F, Meazza C, Oliveri M, Pignatti P, Vivarelli M, Alonzi T, Fattori E, Barreca A, Martini A. Effect of interleukin-6 on IGFBP-3. A study in interleukin-6 transgenic mice and in patients with systemic juvenile idiopathic arthritis. Endocrinol 2001;142:4818–4826.
30. Barreca A, Ketelslegers JM, Arvigo A, Minuto F, Thissen JP. Decreased acid-labile subunit (ALS) levels by endotoxin *in vivo* and by interleukin-1β *in vitro*. Growth Horm IGF Res 1998;8:217–223.
31. Delhanty PJD. Interleukin-1β suppresses growth hormone-induced acid-labile subunit mRNA levels and secretion in primary hepatocytes. Biochem Biophys Res Commun 1998;243:269–272.
32. De Benedetti F, Pignatti P, Vivarelli M, Meazza C, Ciliberto G, Savino R, Martini A. In vivo neutralization of human Il-6 (hIL-6) achieved by immunization with a hIL-6 receptor antagonist. J Immunol 2001:166:4334–4340.
33. Tsatsoulis A, Siamopoulou A, Petsoukis C, Challa A, Bairaktari E, Seferiadis K. Study of growth hormone secretion and action in growth-retarded children with juvenile chronic arthritis (JCA). Growth Horm IGF Res 1999;9:143–149.
34. Allen RC, Jimenez M, Cowell CT. Insulin-like growth factor and growth hormone secretion in juvenile chronic arthritis. Ann Rheum Dis 1991;50:602–606.
35. Milunsky A, Bray GA, Londono J, Loridan L. Insulin, glucose, growth hormone and free fatty acids. Determination in patients with cystic fibrosis. Am J Dis Child 1971;121:15–19.
36. Braegger CP, Torresani T, Murch SH, Savage MO, Walker-Smith JA, MacDonald TT. Urinary growth hormone in growth impaired children with chronic inflammatory bowel disease. J Pediatr Gastroenterol Nutr 1993;16:49–52.
37. Aitman TJ, Palmer RG, Loftus J, Ansell BM, Royston JP, Teale JD, Clayton RN. Serum IGF-I levels and growth failure in juvenile chronic arthritis. Clin Exp Rheumatol 1989;7:557–561.
38. Laursen EM, Juul A, Lanng S, Hoiby N, Koch C, Muller J, Skakkebaek NE. Diminished concentrations of insulin-like growth factor-I in cystic fibrosis. Arch Dis Child 1995;72:494–497.
39. De Martino M, Galli L, Chiarelli F, Verrotti A, Rossi ME, Bindi G, Galluzzi F, Salti R, Vierucci A. Interleukin-6 release by cultured peripheral blood mononuclear cells inversely correlates with height velocity, bone age, insulin-like growth factor-I, and insulin-like growth factor binding protein-3 serum levels in children with perinatal HIV-1 infection. Clin Immunol 2000;94:212–218.
40. Johann-Liang R, O'Neill L, Cervia J, Haller I, Giunta Y, Licholai T, Noel GJ. Energy balance, viral burden, insulin-like growth factor-1, interleukin-6 and growth impairment in children infected with human immunodeficiency virus. AIDS 2000;14:683–690.
41. Bannerjee K, Croft NM, Babinska K, Edwards R, Camacho-Hubner C, Sanderson IR, Savage MO. Relationship of changes in IGF-I, IGFBP-3, ALS and leptin to inflammatory and nutritional markers during enteral feeding in children with Crohn's disease. Program of the 39th Annual Meeting of the European Society for Paediatric Endocrinology (ESPE),

Brussels, Belgium, p 63. Abstract.
42. Davies UM, Jones J, Reeve J, Camacho-Hubner C, Charlett A, Ansell BM, Preece MA, Woo PMM. Juvenile Rheumatoid Arthritis: effects of desease activity and recombinant human growth hormone on insulin-like growth factor 1, insulin-like growth factor binding proteins 1 and 3, and osteocalcin. Arthritis Rheum 1997;40:332–340.
43. Beattie RM, Camacho-Hubner C, Wacharasindhu S, Cotterill AM, Walker-Smith JA, Savage MO. Responsiveness of IGF-I and IGFBP-3 to therapeutic intervention in children and adolescents with Crohn's desease. Clin Endocrinol Oxf 1998;49:483–489.
44. De Benedetti F, Martini A. Is systemic juvenile rheumatoid arthritis an IL-6-mediated disease? J Rheumatol 1998;25:203–207.
45. Reinecker HC, Steffen M, Witthoeft T, Pflueger I, Schreiber S, MacDermott RP, Raedler A. Enhanced secretion of tumor necrosis factor-alpha, IL-6, and IL-1beta by isolated lamina propria mononuclear cells from patients with ulcerative colitis and Crohn's disease. Clin Exp Immunol 1993;94:174–181.
46. Mitsuyama K, Sata M, Tanikawa K. Significance of interleukin-6 in patients with inflammatory bowel disease. Gastroenterol Jpn 1991;26:20–28.

Growth and Lactogenic Hormones
Edited by L. Matera and R. Rapaport

Biological Significance of Insulin-Like Growth Factor Binding Proteins

STUART ALAN WEINZIMER[1] and PINCHAS COHEN[2]

[1]*Division of Endocrinology/Diabetes, Department of Pediatrics, The Children's Hospital of Philadelphia and The University of Pennsylvania, Philadelphia, Pennsylvania, USA*
[2]*Division of Endocrinology, Department of Pediatrics, Mattel UCLA Children's Hospital, Los Angeles, California, USA*

ABSTRACT

The insulin-like growth factors (IGFs), insulin-like growth factors binding proteins (IGFBPs), and the IGFBP proteases regulate somatic growth and cellular proliferation. IGFs are potent mitogenic agents whose actions are determined by the availability of free IGFs to interact with the IGF receptors. IGFBPs comprise a family of proteins that bind IGFs with high affinity and specificity and thereby regulate IGF-dependent actions. IGFBPs also mediate IGF-independent actions on cell growth and apoptosis through specific IGFBP association proteins and cleavage of IGFBPs by specific proteases. The ubiquity and complexity of the IGF/IGFBP axis have important applications to the pathophysiology of growth disorders, inflammatory disease, and cancer.

1. INTRODUCTION

The insulin-like growth factors (IGF-I and IGF-II), insulin-like growth factor binding proteins (IGFBPs), and IGFBP proteases are involved in the regulation of somatic growth and cellular proliferation, *in vitro* and *in vivo*. IGFs are potent mitogenic agents whose actions are determined by the availability of free IGFs to interact with the IGF receptors. The extent of free IGFs in a system is modulated by the rates of IGF production, binding to specific IGFBPs, and clearance. The IGFBPs comprise a superfamily of six proteins (IGFBP-1–6) that bind to IGFs with high affinity and specificity, and a family of IGFBP-related proteins (IGFBP-rPs), which are structurally similar to the IGFBPs but bind IGFs with much lower affinity. IGFBPs not only regulate IGF bioavailability and action (so-called IGF-dependent actions), but also mediate IGF-independent actions on cell growth and apoptosis. IGFBPs are produced by a variety of tissues, and each tissue has specific concentrations of IGFBPs. Cleavage of IGFBPs by proteases plays a key role in modulating concentrations and actions of IGFBPs. IGFBP association proteins, recently characterized serum, cell-surface, cytosolic, and nuclear proteins, may mediate IGF-

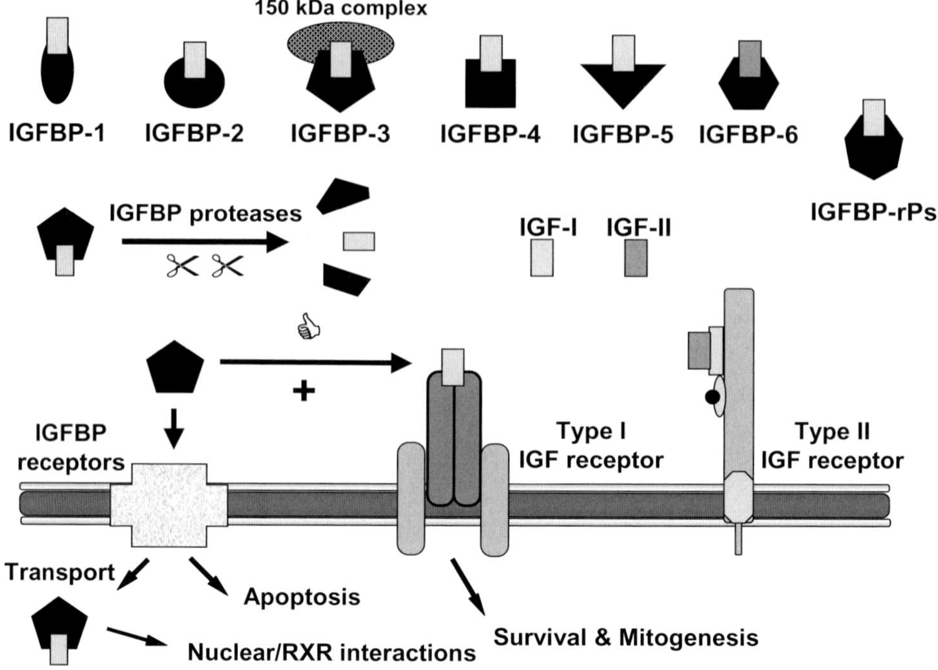

Figure 1. The IGF-IGFBP axis. See text for details.

independent actions of IGFBPs. This review aims to summarize recently gained insights about the biological actions of IGFBPs.

2. *IN VIVO* PHYSIOLOGY OF IGFBPS

Most of the IGF-I and IGF-II in serum are found in a 150-kDa ternary complex formed by a molecule of IGF, IGFBP-3, and a glycoprotein known as the acid labile subunit (ALS) [1]. IGFBP-5 also participates in a ternary complex with IGF-I and ALS [2]. Less than 1% of IGFs circulate in the free forms (Figure 1). ALS is found nearly exclusively in the intravascular space, and the ternary complex does not cross the capillary barrier [3]. The half-life of the 150 kDa complex in serum is an order of magnitude greater than either free IGF-I or free IGFBP-3 [4]. Due to the absence of ALS in tissues, most tissue IGFs are bound to IGFBPs as binary complexes leaving only small amounts of local free IGF. The liver produces most of the circulating IGFs, although physiologically important autocrine and paracrine production occurs within other tissues [5]. The liver also produces most of the circulating IGFBP-3 and ALS. Different hepatic components produce different components of the ternary complex: hepatic endothelia and Kupffer cells synthesize IGFBP-3, while hepatocytes make IGF-I and ALS [6–8]. Growth hormone stimulates hepatic production of all three components of the 150 kDa complex [9–11].

IGFBP-3 is the most abundant IGFBP in post-natal serum, existing at levels an order of magnitude higher than other IGFBPs. IGFBP-3 levels do not change acutely [12], in contrast to IGFBP-1 and -2 levels, which vary throughout the day depending on the metabolic state [11,13).

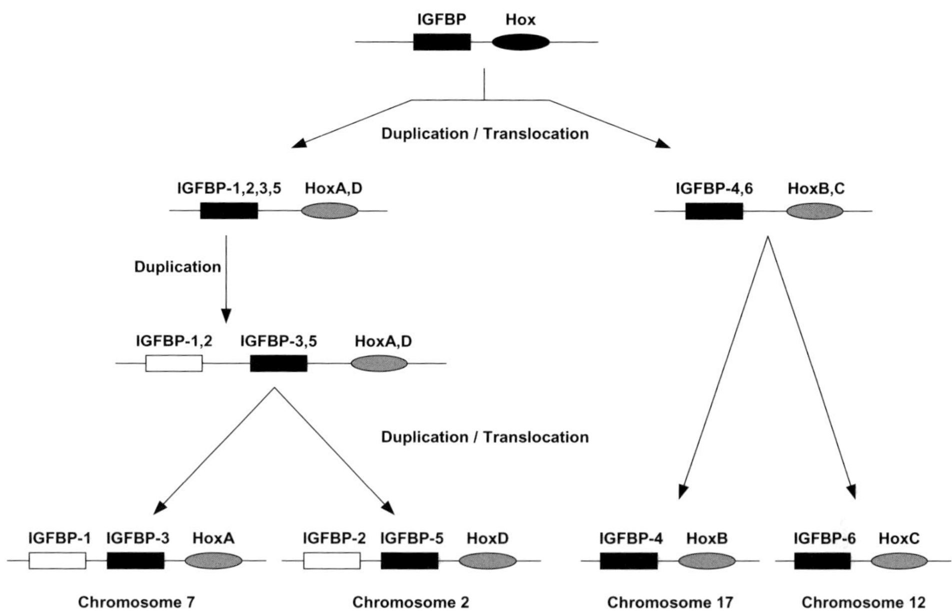

Figure 2. Co-evolution of the IGFBP and Hox genes. See text for details.

The regulators of IGFBP-4, -5, and -6 in serum are not well-characterized; however, their levels have been shown to be age-dependent [14–16], and for IGFBP-4 and IGFBP-5, GH-dependent [17].

3. EVOLUTION OF THE HIGH-AFFINITY IGFBPS

Six IGFBPs with high affinity to IGFs have been identified to date. The first five IGFBPs demonstrate high affinity for both IGF-I and IGF-II, share at least 50% homology among themselves, and share 80% homology among different species [18,19]. Most IGFBPs show greater affinity for IGF-I than IGF-II, except for IGFBP-6, which has 100-fold greater affinity for IGF-II than IGF-I [14]. Homology among IGFBPs is most conserved at the N- and C-terminal regions, while the middle region bears little similarity across different IGFBPs [1]. IGFBPs share a highly conserved set of at least 16 cysteine residues that can form disulfide bridges to stabilize their tertiary structures [12].

The evolutionary conservation of IGFBPs supports their importance in the regulatory process. IGFBP genes lie in close proximity to Homeobox gene clusters HoxA through HoxD, which produce DNA-binding proteins that may have co-evolved with IGFBPs. Some investigators speculate that Hox and IGFBP genes originated from common ancestral genes that repeatedly underwent coordinated duplications and translocations [20–22, Figure 2]. The genes for IGFBP-1 and IGFBP-3 are located on chromosome 7, next to the HoxA gene, while IGFBP-2 and IGFBP-5 are localized to the long arm of chromosome 2, by HoxD. In each of these IGFBP pairs, the former contains an RGD sequence and modulates carbohydrate metabolism (IGFBP-1 and -2), while the latter in each pair (IGFBP-3 and -5) primarily regulate growth [23]. Each

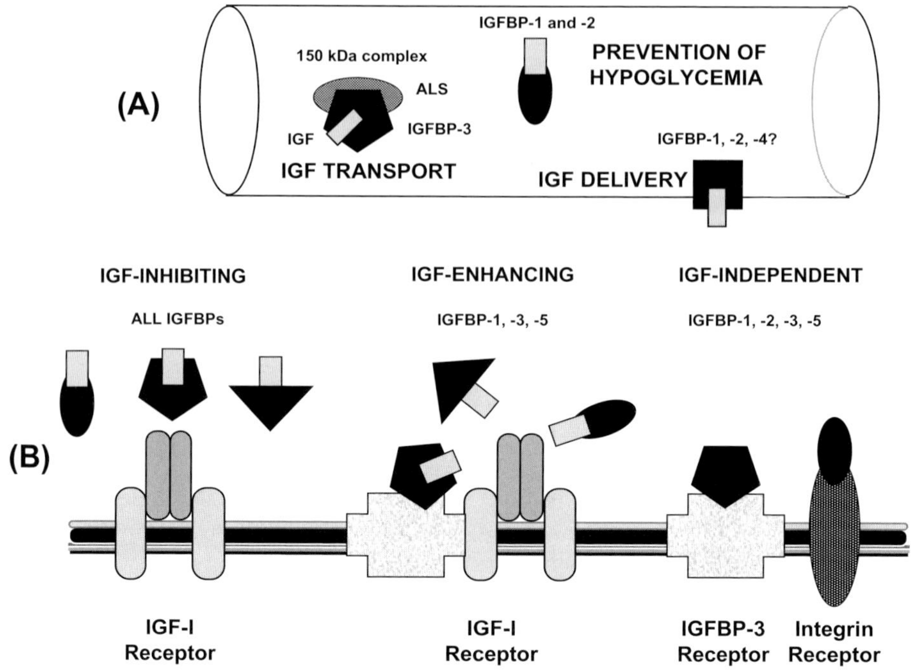

Figure 3. Roles of the IGFBPs in the intravascular space (A) and at the cell membrane (B). See text for details.

of these four binding proteins contains 18 cysteine residues. IGFBP-4, whose gene localizes to the long arm of chromosome 17, next to HoxB, contains 20 cysteines, and IGFBP-6, with 16 cysteines, is located on chromosome 12 next to HoxC [23].

The IGFBPs have several potential functions (Table I and Figure 3). Classically, IGFBPs exert their actions indirectly through modulation of IGFs. More recently, IGFBPs have been shown to act independently of IGFs, through specific cell-surface, cytosolic, and nuclear IGFBP association proteins, to impact crucial cell actions such as growth and apoptosis.

4. BIOCHEMISTRY, REGULATION, AND PHYSIOLOGY OF HIGH-AFFINITY IGFBPS AND THEIR PROTEASES

4.1. Insulin-like growth factor binding protein-1

IGFBP-1 is a 25 kDa protein with an RGD sequence in its structure [19,24]. Integrin receptors in the cell membrane recognize the RGD sequence, suggesting the possibility of an IGF-independent action via these receptors [24]. IGFBP-1 is produced in the liver, decidua, and kidneys and is the most abundant IGFBP in amniotic fluid. IGFBP-1 is found in various phosphorylated states that determine its affinity for IGFs [25]. Serum IGFBP-1 levels are predominantly regulated by insulin and corticosteroids. After meals, serum IGFBP-1 levels fall to less than 10 ng/mL, but during fasting, IGFBP-1 levels rise to over 100 ng/mL [12]. In children with ketotic hypoglycemia who underwent diagnostic fasting studies, IGFBP-1 levels were as high as 700 ng/mL at

Table I Functions of the insulin-like growth factor binding proteins.

Limit bioavailability of free IGFs to bind IGF receptors
Prevent IGF-induced hypoglycemia
Regulate transport of IGFs between intra- and extravascular spaces
Prolong half-life of IGFs in circulation
Enhance actions of IGFs by forming a slow-releasing pool of IGFs
Affect cellular proliferation/apoptosis via IGFBP receptors/protein partners
Nuclear actions

the time of hypoglycemia (serum glucose <60 mg/dL) [26].

Insulin and corticosteroids regulate serum IGFBP-1 levels through transcriptional control of hepatic IGFBP-1 synthesis [4]. Several studies have shown that insulin inhibits the synthesis of IGFBP-1, resulting in elevated levels during low insulin states such as intrauterine growth retardation, fasting, or poorly controlled type 1 diabetes mellitus [12,27–29]. Conversely, conditions associated with hyperinsulinemia decrease IGFBP-1 levels, including the post-prandial period, obesity, large for gestational age infants, insulinomas, and congenital hyperinsulinism with hypoglycemia [26,30]. Glucocorticoids and glucagon stimulate IGFBP-1 production in synergism with low levels of insulin [31–34]. Some cytokines (TNF-α, IL-1 ß, and IL-6) can also stimulate IGFBP-1 *in vitro* and *in vivo* [35].

In chronic renal failure, IGFBP-1 levels rise due to increased renal production [12]. In diabetes mellitus and renal failure, high IGFBP-1 levels are thought to have a pathophysiological role, presumably decreasing the levels of free IGFs to interact with the IGF-I receptor. Elevated IGFBP-1 may be responsible for the decreased linear growth seen in these conditions. Moreover, certain transgenic mice that overproduce IGFBP-1 have significantly lower birthweights, poorer postnatal weight gain, and disproportionately smaller brains compared to wild-type mice [36]. These mice also demonstrate fasting hyperglycemia, impaired glucose tolerance, and reduced fertility, suggesting that IGFBP-1 inhibits metabolic as well as growth-promoting effects of IGFs. However, other transgenic mouse models with equally high levels of serum IGFBP-1 did not share this phenotype [37].

4.2. Insulin-like growth factor binding protein-2

IGFBP-2 is a 31 kDa protein [19]; and, while like IGFBP-1, contains an RGD sequence, has not been demonstrated to bind integrin-type receptors. IGFBP-2 may associate with the cell surface via glycosaminoglycans, and presumably proteoglycans, if high concentrations of IGF-I or -II are present [38,39]. IGFBP-2 is neither phosphorylated nor glycosylated.

Concentrations of IGFBP-2 are age-dependent: high levels are seen in infancy and older age, and low levels are present in young adults [12]. The concentration of IGFBP-2 in seminal plasma, about 10,000 ng/mL, is the greatest level of any IGFBP in any biological fluid [40]. Produced in multiple neural tissues, IGFBP-2 is the major IGFBP in cerebrospinal fluid [41], and IGFBP-2 concentrations are elevated in the spinal fluid of patients with certain central nervous system tumors [42]. IGFBP-2 may mediate IGF-II activity during the acute phase following traumatic brain injury [43]. Targeted disruption of the IGFBP-2 gene, however, failed to demonstrate an altered phenotype [5].

The *in vitro* regulation of IGFBP-2 has been studied in multiple cell systems. In human neuroblastoma cells, IGFBP-2 expression is reduced by retinoic acid [41,44], while lymphocyte activation is associated with increased lymphoctye IGFBP-2 expression [45] and dexamethasone treatment increases IGFBP-2 concentrations in the pancreatic β-cell line βTC3 [46]. In contrast, the *in vivo* regulators of IGFBP-2 remain largely unknown. Growth hormone deficiency is associated with increased serum concentration of IGFBP-2, even though growth hormone itself seems to have no direct effect on IGFBP-2 gene expression in cultured cells [12,47]. Regulation of IGFBP-2 expression also depends on insulin and the metabolic state, albeit to a lesser degree than IGFBP-1 [48]. IGFBP-2 levels increase with prolonged fasting, malnutrition, and anorexia nervosa, and respond more sensitively to protein restriction than to caloric restriction [13]. Untreated (insulinopenic) insulin-dependent diabetics have elevated levels of IGFBP-2, which normalize with insulin therapy [28].

4.3. Insulin-like growth factor binding protein-3

The molecular weight of IGFBP-3 in its non-glycosylated form is 29 kDa [49]. IGFBP-3 has three glycosylation sites and exists in the circulation in various glycosylated forms, with a molecular weight between 40 and 44 kDa [12].

Serum concentrations of IGFBP-3 and IGFs vary similarly with age: starting low at birth, increasing during childhood, peaking during puberty, and decreasing thereafter [50]. IGFBP-3 levels are less susceptible to nutritional effects than IGF-I [12]. Depressed levels of IGF-I may be observed in chronic diseases and malnutrition, while IGFBP-3 remains relatively unchanged [11,13,27]. Growth hormone (GH) is the primary regulator *in vivo* of IGFBP-3 expression, although estrogens, parathyroid hormone (PTH), and glucocorticoids also regulate IGFs and IGFBPs in the circulation [51].

The mechanism by which GH stimulates IGFBP-3 production is still under investigation. Proposed mechanisms include: (a) a direct effect of GH on Kupffer cells, (b) an indirect effect mediated by IGF-I, and (c) stimulation of non-hepatic tissues. Support for the first mechanism came from *in vitro* studies of human hepatocarcinoma cells that demonstrated increased IGFBP-3 gene expression by GH independently of IGF-I [52]. Several *in vitro* studies support the second mechanism, but in a large population of growth hormone receptor deficient patients from Ecuador, administration of IGF-I stimulated growth but did not change IGFBP-3 levels by radioimmunoassay (RIA) [10,53]. However, in a similar population from Israel, IGF-I increased levels of IGFBP-3 by Western ligand blotting (unfortunately, this study did not utilize IGFBP-3 RIA) [54]. One possible explanation for these findings is that IGF-I does not stimulate production of IGFBP-3 but instead protects IGFBP-3 from proteolysis. Several animal models in which IGF-I induces serum IGFBP-3, as detected by ligand blotting, supports this hypothesis [55]. In support of the third mechanism, some *in vitro* studies have shown that: (a) GH can induce IGFBP-3 message expression in many tissues (rat liver, skin, and muscle); and (b) IGF-I can induce an increase in serum IGFBP-3 concentration in the absence of an increase in hepatic IGFBP-3 mRNA levels and despite a paucity of liver IGF-I type 1 receptor mRNA [56].

IGFBP-3 is also produced in a variety of non-hepatic tissues [57] and can be regulated by many compounds other than GH. Stimulation of bone cells with GH (but not IGF-I) increases IGFBP-3 expression [58]. Depending on the tissue studied, IGFBP-3 expression has been shown to be regulated *in vitro* by interleukin-1 [59], tumor necrosis factor-alpha (TNF-α) [59,60], transforming growth factor-beta (TGF-ß) [61], retinoic acid [62], PTH [12], osteogenic protein-1 [63], FSH [64], estradiol [65], prostaglandin E2 [66], glucocorticoids [51], Vitamin D [67], and p53 [68]. Though considered predominantly a serum and extracellular protein, IGFBP-3 has recently been shown to localize to the nuclei in multiple cell lines [69–73].

IGFBP-3 can promote or inhibit growth both *in vivo* and *in vitro*. Effects of IGFBP-3 can be either IGF-mediated or IGF-independent. Transgenic mice have been developed which express a human IGFBP-3 transgene in small bowel, colon, and kidney as detected by Northern analysis but do not have increased serum IGFBPs by Western ligand blotting [74]. Of note, the majority of serum hIGFBP-3 in these transgenic mice was not associated with ALS. Compared to genetically related wild-type mice and to non-transgenic littermates, the transgenic mice demonstrated selective organomegaly affecting heart, liver, and spleen [74]. Birth weight, body weight, brain and kidney weight, and litter size did not differ significantly between transgenic mice and nontransgenics. In a different model, rat ventral prostate rapidly increased gene expression of IGFBP-2 through -5 in association with apoptosis following castration [75]. In another model, in which the hIGFBP-3 gene was transfected into a cell line derived from an IGF receptor "knock-out" mouse that is non-IGF responsive, cell growth decreased significantly, proving definitively that IGFBP-3 inhibits cell growth independently of the IGF-IR [76]. Further evidence for the role of IGFBP-3 as an IGF-independent modulator of cell growth will be discussed later in this review.

4.4. Insulin-like growth factor binding protein-4

IGFBP-4 has been identified in all biological fluids [5] and exists at its predicted molecular weight of 24 kDa or in the glycosylated form weighing 28 kDa [12]. Different cell types produce IGFBP-4 locally, including fibroblasts, neuroblastoma cells, prostate cells, and bone cells [41,77]. IGFBP-4 can bind cell membranes (with as yet unknown function), but the majority of IGFBP-4 exists as extracellular molecule [3]. IGFBP-4 is the predominant IGFBP produced by vascular smooth muscle cells [78]. The effects of IGFBP-4 on the cellular level are modulated in part by specific IGFBP-4 proteases [79,80], providing an additional level of regulation by which local IGF actions may be modulated.

IGFBP-4 has been found to inhibit IGF actions in all cell systems studied to date, through high-affinity binding to both IGF-I and IGF-II, preventing interaction of either IGF molecule with its receptor [81]. Transgenic mice over-expressing IGFBP-4 selectively in smooth muscle have smooth muscle hypoplasia [82], as opposed to the smooth muscle hypertrophy associated with IGF-I overexpression, suggesting a functional inhibitory role for IGFBP-4 *in vivo* as well. Additional biological effects of IGFBP-4 include delaying of apoptosis in malignant prostate tumor cell line [83], inhibition of ovarian steroidogenesis [84], and blockage of skeletal myoblast differentiation [85,86]. Regulatory mechanisms affecting IGFBP-4 expression are poorly understood, although both Vitamin D and PTH regulate IGFBP-4 expression in bone [87,88], retinoic acid inhibits IGFBP-4 expression in neuroblastoma cells [41], and interleukin-6 stimulates hepatic IGFBP-4 expression [89].

4.5. Insulin-like growth factor binding protein-5

IGFBP-5 has a molecular weight of 29 kDa and can be found in several glycosylated forms, ranging between 29 and 32 kDa [12]. Like IGFBP-3, IGFBP-5 levels decrease with age, starting after puberty. Serum concentrations in older women are approximately 30% of those of teenagers [15,16]. Fetal tissues have high levels of IGFBP-5 during periods of rapid growth, [12] but tissue levels in adults vary. IGFBP-5 is the main IGFBP expressed in the kidney, and IGFBP-5 is found in substantial amounts in connective tissue and cerebrospinal fluid (CSF). IGFBP-5 can also form ternary complexes with IGFs and ALS [2].

Unlike other IGFBPs, IGFBP-5 strongly binds to bone cells due to its high affinity for hydroxyapatite [48]. Like IGFBP-3, IGFBP-5 binds to endothelial cell monolayers and is found in large concentrations in the extracellular matrix (ECM). Its binding to endothelia is competitively inhibited by heparin and heparan sulfate, and alterations near the C-terminus of IGFBP-5 inhibit its binding to cells but not to the ECM [90]. IGFBP-5 binds the ECM on an ionic basis. The affinity of IGFBP-5 for IGF-I, when IGFBP-5 is bound ionically to ECM, is seven- to twelve-fold reduced compared to intact IGFBP-5 in solution [91].

In vitro studies have demonstrated that IGFBP-5 possesses both stimulatory and inhibitory effects on IGF-I actions when compared to IGF-I alone [91–93]. Addition of molar excess of IGFBP-5 to cell lines that do not secrete IGFBP-5 proteases generally results in inhibition of IGF action [94]. IGF-enhancing actions of IGFBP-5 are observed particularly in osteoblast cells, but also in fibroblasts and smooth muscle [92,94]. These growth-potentiating effects may be dependent on IGFBP-5 binding to cell membrane or ECM, as IGFBP-5 has reduced affinity to IGF-I in these settings, facilitating IGF-I binding to its own receptors [92]. An additional role for IGFBP-5 in skeletal myocyte differentiation has also been found [93].

The regulatory mechanisms for IGFBP-5 are still under investigation. Serum concentrations of IGFBP-5 have recently been shown to be GH-dependent [15, 17]. Bone cells produce large amounts of IGFBP-5, whose levels decrease during maturation due to protease activity. Treatment of osteoblastic cells with fibroblast growth factor, TGF-β, and platelet-derived growth factor BB decrease IGFBP-5 expression [95], while in the same cell system, treatment with IGF-I, IGF-II, and retinoic acid increase levels of IGFBP-5 mRNA [95–97]. Glucocorticoids decrease IGFBP-5 levels in fibroblasts and osteoblast-like cells [51,98]. In vascular smooth muscle cells, IGFBP-5 mRNA expression is stimulated by IGF-I via a time-, dose-, and cell-type dependent mechanism, but is inversely related to culture density, which correlates closely with IGF-I concentration [99,100].

4.6. Insulin-like growth factor binding protein-6

IGFBP-6 is an O-glycosylated protein [101] with a predicted molecular weight of 34 kDa [5]. It is the only IGFBP that preferentially binds IGF-II over IGF-I by more than two orders of magnitude [101,102]. The major function of IGFBP-6 appears to be the regulation of IGF-II actions [103]. In osteoblastic cells, the addition of excessive concentration of IGFBP-6 inhibits IGF-II–stimulated DNA and glycogen synthesis, but had minimal effects on inhibiting IGF-I actions [14]. Similarly, IGFBP-6 reduces IGF-II binding and suppresses IGF-II-dependent myoblast differentiation and proliferation in culture, without affecting IGF-I-dependent functions [104, 105].

Human IGFBP-6 is found predominantly in CSF and serum [12]. IGFBP-6 is also expressed in ovarian cells, prostatic cells, fibroblasts and other cells [101]. Expression of IGFBP-6 is regulated by IGF-II and other hormones. Studies in breast carcinoma cell lines showed that only estrogen receptor-negative cells produced IGFBP-6, and IGFBP-6 expression was enhanced by retinoic acid but unchanged by IGF-I [106]. Retinoic acid also increases IGFBP-6 expression by more then 1000% in human osteoblast cells [107].

4.7. IGFBP proteases

IGFBP proteolysis was first described in pregnancy serum as proteolytic activity against IGFBP-3 [108], and prostate specific antigen (PSA) in seminal plasma was the first IGFBP protease to be identified biochemically [109]. Proteolysis of IGFBP-2 to -6 has been described in multiple clinical states and cellular systems [23]. Proteolysis of the IGFBPs in general results in fragments with decreased affinity for IGFs, allowing release of free IGFs in the circulation access to the extravascular space and interaction with cellular IGF receptors. Categories of IGFBP proteases include kallikreins [109,110], cathepsins [111–114], and matrix metalloproteinases (MMPs) [115–117].

Kallikrein-like serine proteases that cleave IGFBP-3 include PSA, gamma-nerve growth factor, and plasmin [118]. Plasmin has been found to degrade multiple IGFBPs [118], and thrombin, another serine protease, cleaves IGFBP-5 at physiologically relevant concentrations, i.e., within one order of magnitude of fibrinogen, its natural substrate [119]. An IGFBP-4-specific serine protease has recently been demonstrated in smooth muscle and neuronal cell lines [79].

Cathepsins are lysosomal proteinases that have been shown to be IGFBP proteases in multiple systems, including normal and malignant prostatic epithelial cells and seminal plasma [112, 113]. Cathepsins are active under acidic conditions and may be relevant to certain physiological and pathological processes, including neoplastic infiltration [111]. The increased release of hydrogen ions under these conditions *in vivo* may provide an acidic environment for extracellular cathepsin action which, through cleavage of IGFBPs and release of IGFs, may be related to increased cell growth rates [120].

MMPs, or matrixins, comprise a family of peptide hydrolases (2,800 to 92,000 kDa) that require a metal ion for their catalytic activity and are thus inactivated by metal chelators as well as by specific inhibitors. The MMP family, which includes a number of collagenases and gelatinases, are responsible for the degradation of extracellular matrix components and play important roles in tissue remodeling, inflammation, and cancer invasion and metastasis. The IGFBP protease activity in pregnancy serum is caused by members of the MMP family, and a zinc-dependent MMP produced by dermal fibroblasts and present in mouse pregnancy serum was demonstrated to have IGFBP protease activity, suggesting that MMPs may have a role in regulating cellular growth and proliferation via degradation of IGFBP-3 [116,121]. MMPs have been identified in prostatic fluid and cells [122]. The IGFBP protease induced in airway smooth muscle cell culture by inflammatory agents (leukotriene D4 and interleukin-1β) has been identified as MMP-1 by immunoblotting and immunoprecipitation techniques [123,124]. Different categories of proteases may act in a coordinated cascade. IGFBP-3 serine proteases in rat serum act sequentially to initiate activation of latent MMP precursors, which then directly degrade IGFBP-3 [125]. An MMP protease specific for IGFBP-4 has been identified as pregnancy-associated plasma protein-A, found in high concentrations in maternal circulation [80], and IGFBP-5 proteases of the MMP class are present in osteoblast-conditioned media following differentiation and maturation [126].

Proteolytic activity may play a role in normal and abnormal tissue proliferation by cleavage of IGFBPs into fragments with lower affinity for IGFs. Lower-affinity IGFBP fragments allow increased levels of free IGFs to activate IGF receptors (Figure 1). For example, PSA-generated IGFBP-3 fragments have decreased affinity for IGFs, and prostatic epithelial cells grown in the presence of PSA demonstrate reversal of the inhibitory effects of IGFBP-3 on IGF-stimulated cell growth [127]. Pregnancy serum protease may play a role in regulating IGF bioactivity by making IGFs more available to cells [128].

IGFBP proteases are also important autocrine/paracrine growth regulators. They have been implicated in physiologic processes such as ovarian follicular growth and atresia [129]. IGFBP proteases appear to play a role in benign proliferative diseases such as the airway smooth muscle hyperplasia of chronic asthma [117]. We have demonstrated *in vitro* that leukotriene D4 acts synergistically with IGFs to stimulate airway smooth muscle hyperplasia, and that this effect is mediated by the proteolytic activity of MMP-1 [123,124]. MMP-1 levels increase twelve-fold and are associated with increased MMP-1 proteolytic activity and IGFBP fragments in asthmatic airway extracts [130]. IGFBP proteases may be critical elements in autocrine IGF loops in malignant proliferative diseases such as prostate cancer [117,131].

Regulation of IGFBP proteolysis is an underexplored area of investigation with important physiological and therapeutic implications. The IGFBP-3 protease activity detectable in the serum of youngsters with newly diagnosed insulin-dependent diabetes mellitus appears to be reversible by insulin [132]. IGFs stimulate but IGFBPs inhibit an IGFBP-4 protease of MC3T3-E1 osteoblasts [133], and similar findings have been demonstrated in dermal fibroblasts [134]. The IGFBP-3 protease secreted by MCF-7 breast cancer cells is inhibited by IGFs, suggesting a unique loop by which IGFs can regulate their own activity [135]. Such findings implicate the relative proportion of IGFs to IGFBPs to be a critical regulator of IGFBP proteases.

5. IGF-DEPENDENT ACTIONS OF IGFBPS

IGFBPs clearly modulate the interactions between IGFs and type I IGF receptors (IGF-IR) on the cell surface (Figure 2). IGFs are potent mitogens, and many *in vitro* systems have demonstrated that all IGFBPs have growth-inhibitory effects by competitively binding IGFs and preventing their binding to the IGF-IR. The most powerful evidence supporting this sequestration mechanism has come from studies using the IGF-I analog des-(1-3)-IGF-I, which binds IGF-IR and stimulates DNA synthesis but does not bind IGFBP-3. In the human promyeloid cell line HL-60, adding IGFBP-3 to serum-free media inhibited cell proliferation induced by IGF-I and IGF-II but not by des-IGF-I [136]. Transfected Ishikawa endometrial cancer cells which overexpressed both IGF-IR and cell membrane-bound IGFBP-3 demonstrated equal IGF-IR binding by IGF-I and des-(1-3)-IGF-I [137]. In this cancer cell line, membrane-bound IGFBP-3 inhibited growth induced by IGF-I and not by des-(1-3)-IGF-I [137]. Similar sequestration effects have been demonstrated in human osteoarthritic chondrocytes [138] and human granulosa cells [139]. A comparison of serum levels of IGF-I and IGF-II versus IGFBP-2 through -5 in bovine first-wave dominant follicles showed that the IGF-I/IGFBP ratio decreased with atresia of the follicles [140]. *In vivo* studies have also suggested that IGFBPs inhibit growth by reducing the free IGF level [141,142]. IGFBP-3 may also directly inhibit IGF binding to the IGF-IR, an alternative to inhibition through extracellular sequestration [143].

IGFs can regulate IGFBP expression and action through paracrine or autocrine feedback loops. IGF-I has been shown to stimulate release of IGFBP-3 but not IGFBP-4 in a lung cancer cell line [144]. Under strict *in vitro* conditions, IGFBP-1, -3, and -5 can enhance IGF actions by enhancing IGF-I binding to IGF-IR (Figure 2) [145]. IGFBP-1, -2, and possibly -4 can also transport IGFs between the intra- and extravascular space [146]. Li *et al.* demonstrated in proliferating opossum kidney cells that fluorescent IGF-I and IGFBP-3 – whether added to media in combination or alone – were absorbed from the media and colocalized to the nucleus [147]. These fascinating observations suggest that IGFBP-3 may ferry IGF-I to the cell nucleus, providing another regulatory mechanism for IGF action.

6. IGF-INDEPENDENT ACTIONS OF IGFBPS AND IGFBP RECEPTORS

The discovery of IGF-independent modulation of growth by IGFBPs provides evidence for the presence of specific cell-surface IGFBP receptors and adds a further layer of complexity to the IGF axis (Figure 1). Multiple *in vitro* studies have demonstrated important IGF-independent effects of IGFBP-3 on growth regulation in various cells and tissues, although most *in vivo* studies to date (including those in transgenic animals or with infusions into human subjects) have failed to show direct inhibitory IGF-independent effects of IGFBP-3, despite high serum levels of free IGFBP-3. It has recently been shown, however, that subcutaneous injections of IGFBP-3 into tumor-bearing mice reduce spleen tumor size and occurrence of liver metastases [148].

Further evidence for IGF-independent actions of IGFBP-3 comes from *in vitro* studies targeting IGF-I/IGF-IR interactions. Transfection of the IGFBP-3 gene into murine fibroblasts inhibited cell growth by a mechanism that was not reversible by the addition of excess insulin, even though insulin has mitogenic activity in these cells, does not bind IGFBP-3, and would presumably saturate the IGF-IR [149]. A 16-kDa fragment of rhIGFBP-3, with negligible binding affinity for IGF-I and presumably none for insulin, inhibited insulin- and IGF-I–stimulated DNA synthesis in chick embryo fibroblasts [150]. This same fragment also inhibited mitogenesis in murine fibroblasts with a defective IGF-IR that could respond to bFGF but not IGF, epidermal growth factor, or platelet-derived growth factor PDGF [151]. Fibroblasts derived from IGF-IR knockout mice (that do not express the IGF-IR) demonstrated marked growth inhibition when transfected with a vector containing the IGFBP-3 gene [152]. This growth inhibition correlated with the magnitude of IGFBP-3 expression in these clones [152]. Because these cells did not express IGF receptors, the growth-inhibiting effects of IGFBP-3 clearly were mediated not through an IGF-IR pathway but presumably through a novel, specific, IGFBP-3-specific pathway.

The existence of cell-surface IGFBP-3 association proteins/receptors was first suggested by Oh et al., who demonstrated specific, dose-dependent binding of IGFBP-3 to breast cancer cell surface proteins of 20, 26, and 50 kDa [76,153], and this binding was subsequently shown to occur in the mid-region of the IGFBP-3 molecule [154]. In these estrogen-receptor-negative breast cancer cells, the inhibitory effects of IGFBP-3 on growth were shown to be dose-dependent and diminished by co-incubation with IGFs, but not by IGF analogs with reduced affinity for IGFBP-3. Recently, a protein biochemically identified as the type V TGF-ß receptor (by affinity cross-linking and immunoprecipitation techniques) was shown to bind IGFBP-3, and suggested to be the putative IGFBP-3 receptor [155]. However, this protein has not been characterized structurally, and its size is several-fold larger than other putative IGFBP-3 receptors. We detected specific IGFBP-3 association proteins in cell lysates (18- to 150-kDa) and membrane fractions (18-, 67-, and 150-kDa) of the prostate cancer cell line PC-3 [156]. The addition of exogenous IGFBP-3 to PC-3 cells resulted in a dose-dependent increase in the apoptotic index, which was only partially attenuated by the addition of IGF-I and unchanged by the addition of IGF analogs with reduced affinity for IGFBP-3 [156]. Confirmation of the direct action of IGFBP-3 on apoptosis was achieved by the induction of apoptosis by IGFBP-3 in IGF-IR–negative (knockout) murine fibroblasts. Using specific anti-sense oligonucleotides and neutralizing anti-IGFBP-3 antibodies to block IGFBP-3 expression and action, IGFBP-3 has been identified as the mediator of apoptosis induced by retinoic acid and TGF-ß in multiple cell types

[62,65,156–158]. More recently, IGFBP-3 has been implicated in the induction of apoptosis by topoisomerase inhibitors in retinoblastoma cells [159] and by protein kinase C-alpha in glioblastoma multiforme cells [160].

7. GROWTH-STIMULATORY ROLES OF IGFBPS

7.1. Insulin-like growth factor binding protein-1

Even though IGFBP-1 inhibits mitosis by removing free IGF-I from the extracellular space, early *in vitro* systems demonstrated that IGFBP-1 may stimulate growth. IGFBP-1, in the presence of low concentration of platelet-poor plasma and IGF-I, stimulated DNA synthesis in porcine aortic smooth muscle cells, chick embryo fibroblasts, and mouse embryo fibroblasts [161,162]. Koistinen et al. concluded that IGFBP-1 caused slow and steady release of IGF-I when they found that concentrations of IGFBP-1 that can inhibit binding of IGF-I to its receptor sometimes enhanced IGF-stimulated thymidine incorporation [163]. This inhibition did not occur when IGFBP-1 was added without IGF-I, suggesting an effect due to the slow release of IGF-I and not due to a direct effect of IGFBP-1 itself [163,164).

However, direct IGF-stimulatory actions of IGFBP-1 have also been demonstrated. IGFBP-1 binds specifically to the $\alpha5\beta1$-integrin receptor and stimulates cell migration of Chinese hamster ovarian cells in a monolayer wounding assay [165]. IGFBP-1 stimulates *in vivo* healing of rabbit ears when added with IGF-I, and this effect is dependent on binding of IGFBP-1 to the integrin receptor [166] (Figure 2). IGFBP-1 has also been shown to stimulate wound healing in diabetic rabbits [167].

Other investigators showed that IGFBP-1 phosphorylation alters its effects. Phosphorylated IGFBP-1 has high affinity for IGF-I and probably inhibits IGF-I action, while dephosphorylated forms, as found in pregnancy serum, have low affinity and stimulate IGF-I action [168,169]. Finally, IGFBP-1 may be involved in IGF-I transport through the capillary barrier, a process that in certain tissues appears to involve an insulin-dependent mechanism [170,171] (Figure 2).

7.2. Insulin-like growth factor binding protein-3

Pre-incubation of fibroblasts with IGFBP-3 followed by its removal potentiated IGF-I effects on DNA synthesis in a dose-dependent fashion [57], even though addition of IGFBP-3 to conditioned media in the same cell system caused inhibition of cell growth [172]. The conclusion was that the presence of large amounts of IGFBP-3 caused reduction in the free IGF levels while small amounts of IGFBP-3 protected IGFs, intensifying their effects. When cells were pre-incubated with IGFBP-3 and then washed, low molecular weight forms of IGFBP-3, probably representing proteolytic fragments, were bound to the cell membrane [57].

The proposed mechanism for the difference between the effects of pre-incubated and soluble IGFBP-3 is that IGFBP-3 fragments bound to the cell membrane have an order of magnitude lower affinity for IGF-I than does intact soluble IGFBP-3. Intact soluble IGFBP-3 has higher affinity for IGFs than to IGF-IR. Indeed, the affinity of IGF-I for the membrane-bound IGFBP-3 fragment is lower than that for the type I IGF receptor. Thus, IGFBP-3 might function as a reservoir of IGF-I, presenting and slowly releasing IGF-I to interact with the IGF-IR, while protecting the receptor from down-regulation. In support of this latter concept, Conover showed that the IGF-I receptor down-regulation induced by IGF-I can be prevented by IGFBP-3 via modulation of IGF-I availability for binding to IGF-IR [173].

Stimulatory effects of IGFBP-3 have also been demonstrated *in vivo*. Topical use of IGFBP-3 in association with IGF-I causes better wound healing then the use of IGF-I alone [174], and the administration of IGFBP-3 and IGF-I to growth hormone deficient rats caused better growth than the administration of IGF-I alone [175]. In the same study, additionally, IGFBP-3 protected the rats from the hypoglycemic effects of IGF-I. Incubation of bladder smooth muscle cells with either IGFBP-3 or IGFBP-5 increased rates of proliferation, and addition of IGFs potentiated this stimulatory effect [176,177]. Thus, it appears that IGFBP-3 targets IGF-I towards growth provocation and away from the insulin-sensitive glucose-consuming tissues.

7.3. Insulin-like growth factor binding protein-5

Several *in vitro* studies have shown that IGFBP-5 stimulates IGF-I actions when compared to IGF-I alone [57,92,145,176,177]. The IGF-enhancing actions of IGFBP-5 are particularly evident in bone cells [92]. Studies suggest a need for IGFBP-5 to be bound to the cell membrane or to extracellular matrix to cause this potentiating effect [57,92]. Compared to free IGFBP-5, IGFBP-5 bound to the extracellular matrix has lower affinity for IGF-I but a prolonged half-life [57, 91]. Thus, the binding of IGF-I to the matrix-bound IGFBP-5 may facilitate the subsequent binding of IGF-I to the IGF-IR [57,91,145]. Recently, IGF-independent stimulatory actions of IGFBP-5 have been demonstrated in an IGF-I knockout osteoblast cell culture system. In this model, in which IGF-I was absent and IGF-II was blocked by exogenous IGFBP-4 (a potent inhibitor of IGF-II), the addition of IGFBP-5 increased both cell proliferation and alkaline phosphatase activity in a dose-dependent manner and with a magnitude comparable to IGF-I [178]. Local injections of IGFBP-5 into the outer periosteum of IGF-I knockout mice increased alkaline phosphatase and osteocalcin levels [178]. The molecular pathways for these IGF-independent effects of IGFBP-5 are yet undetermined.

8. CELLULAR ROLES OF IGFBPS

Once considered predominantly extracellular carrier proteins, IGFBPs have emerged as molecules with important intracellular roles. IGFBP-3 and IGFBP-5 have recently been shown to be translocated into the nucleus compatible with the possession of a nuclear localization sequence (NLS) in their mid-region [69–71,179,180]. IGFBP-3 has also been shown to bind importin, a molecule that facilitates nuclear transport [69]. Nuclear transport of IGFBP-3 was observed only in actively dividing cells [147,179], raising the possibility that IGFBP-3 may directly control gene expression.

Newer potential functions of IGFBP-3 have emerged with the discovery of novel IGFBP-3 protein partners. Heparin and the latent TGF-β-binding protein (LTBP-1) have been observed to bind and interact with IGFBP-3 [181–183]. Employing a yeast two-hybrid system and coimmunoprecipiation methodologies, Gui and Murphy demonstrated binding of IGFBP-3 to fibronectin and the presence of IGF-I/IGFBP-3/fibronectin ternary complexes in human plasma [184].

Using similar methods, we have identified multiple IGFBP-3 protein partners. We reported three novel high-molecular-weight IGFBP-3 association proteins in human serum [185], and have characterized two of these proteins as transferrin [177] and collagen [186]. Utilizing biosensor interaction analysis, we have further demonstrated that IGFBP-3 mutated at the NLS has altered binding affinity for transferrin and collagen, suggesting that IGFBP-3 – transferrin and

IGFBP-3 – collagen binding occurs near or within this region. We showed that transferrin inhibited IGFBP-3-mediated growth of bladder smooth muscle cells and also inhibited IGFBP-3-mediated apoptosis in the prostate cancer cell line PC-3. Similarly, IGFBP-3 inhibited collagen-mediated smooth muscle cell migration.

We also demonstrated that IGFBP-3 and the retinoid X receptor-α (RXR-α) co-localize to the nucleus, where IGFBP-3 binds the RXR-RXR response element transcription factor-DNA complex and modulates RXR-α mediated signaling. Studies in RXR knockout cell lines confirm that RXR-α is necessary for IGFBP-3-induced apoptosis [187]. Most recently, IGFBP-3 has been demonstrated in preliminary studies to play a role in modulating insulin action. Acute IGFBP-3 infusion inhibited insulin-mediated suppression of hepatic glucose production in rats, and treatment with IGFBP-3 inhibited insulin-mediated glucose uptake in both isolated 3T3-L1 adipocytes and rats, by a mechanism that is yet unclear but may also involve IGFBP-3 – RXR interactions [188].

The discoveries that IGFBP-3 and IGFBP-5 interact with important viral oncoproteins implies additional roles for IGFBPs in the pathways of cell proliferation, apoptosis, and malignant transformation. Using several methods, we have shown that IGFBP-3 binds and interacts with the human papilloma virus oncoprotein E7 and may function as an E7 protease [189]. Simian virus 40-transformed fibroblasts have dramatically lower levels of IGFBP-5 expression compared to their normal counterparts [190,191]. Although this down-regulation may eventually prove to be an epiphenomenon, it still suggests an important role for IGFBP-5 in the maintenance of the normal phenotype. The structural determinants of IGFBP binding to their protein partners, including intracellular and extracellular ligands as well as cell surface and matrix molecules, will soon be characterized and surely illuminate the molecular mechanisms of IGFBP actions [192–194].

9. IGFBPS AND CANCER

The IGFs, with potent mitogenic and anti-apoptotic effects, have been widely studied for their roles in cancer. Emerging evidence strongly suggests that the IGFBPs play key roles in the regulation of human cancer. As noted above, IGFBP-3 has been shown to inhibit cell growth and induce apoptosis in many experimental and cellular systems. The antiproliferative effects of retinoic acid and p53 have been shown to be mediated at least in part through increased expression and action of IGFBP-3 [157,195]. IGFBP proteases, such as cathepsin D, PSA, and plasmin, can all be detected in human breast and prostate cancer cells. PSA, for example, has been shown to reverse the inhibitory effects of IGFBP-3 on IGF-stimulated prostate cell growth [122]. Rising levels of PSA during the natural course of prostate cancer likely facilitates disease progression by acting as a comitogen with IGFs in the presence of IGFBP-3, probably by degrading IGFBP-3 and releasing IGF for access to the IGF-IR.

Elevation in serum PSA has also been correlated with increased serum IGFBP-2 and decreased intact IGFBP-3. IGFBP-2 levels have been observed to be almost threefold higher in prostate cancer patients than in control subjects [196], suggesting that serum IGFBP-2 concentrations may be a useful prognostic indicator. IGFBP-2 has also been shown to potentiate growth of DU-145 prostate cancer cells in both the presence and absence of IGF-I, an effect blocked by the addition of anti-IGFBP-2 antibody [197]. Similarly, increased serum concentrations of IGFBP-2 and decreased levels of intact IGFBP-3 have been demonstrated in children with acute lymphoblastic leukemia [198,199], which normalized after chemotherapy [198]. Further evidence for the

Table II Taxonomy of the IGFBP-Related Proteins.

IGFBP-rP1	Mac 25, PSF, TAF
IGFBP-rP2	CTGF
IGFBP-rP3	NovH
IGFBP-rP4	Cyr61
IGFBP-rP5	L56
IGFBP-rP6	ESM-1
IGFBP-rP7	RCOP-1, WISP-2, CTGF-L
IGFBP-rP8	ELM1, WISP-1
IGFBP-rP9	WISP-3

anti-cancer effects of IGFBP-3 was demonstrated when subcutaneous injections of IGFBP-3 in mice with colorectal carcinoma reduced tumor size and metastases [148].

Recent attention has focused on the association between elevated serum IGF-I concentrations and cancer risk. Two large prospective studies found an approximately 8% increase in the serum levels of IGF-I in patients with prostate cancer versus matched control subjects [200,201]. Men with serum IGF-I concentrations in the highest quartile had a relative risk of prostate cancer of 4.3 compared to men in the lowest quartile for serum IGF-I [201]. Similarly, a strong positive correlation was found between circulating IGF-I concentrations and risk of breast cancer risk in premenopausal women [202]. Significantly, when controlling for IGF-I, serum IGFBP-3 was found to be inversely associated with cancer risk in prostate and breast cancer. Additionally, serum IGFBP-3 proteolysis was found to be higher in women with higher grade breast cancer as compared with those with lower grade cancer or benign breast lesions [203]. Similar detrimental associations of IGF-I and beneficial associations of IGFBP-3 were seen in lung cancer [204] and colorectal cancer [205]. Elevated serum free IGF-I levels [206], or IGF-I/IGFBP-3 ratio [207], rather than total IGF-I, may be a more useful marker for cancer risk. It remains to be seen whether IGF-I and IGFBP-3 are causal factors or surrogate markers of some other process, whether IGF-I or IGFPB-3 may be candidate markers for early detection of cancer, or whether IGF-I or IGFBP-3 may be the targets of novel therapeutic interventions.

10. IGFBP-RELATED PROTEINS

The IGFBP superfamily also encompasses an expanding group of IGFBP-related proteins (IGFBP-rPs) that are structurally similar to the IGFBPs but bind IGFs with low affinity [208–210]. IGFBP-rPs have been proposed to share an ancestral relationship to IGFBPs. Structurally, the IGFBPs and IGFBP-rPs share the highly-conserved cysteine-rich N-terminus, which appears to be crucial for several biological functions, including binding to IGFs [211,212]. A detailed description of the ever-expanding group of IGFBP-rPs is beyond the scope of this review, in which only an introduction will be made (Table II). The reader is invited to explore several recent comprehensive reviews [209,210], which will serve as the major references for the following section.

IGFBP-rP1 (mac 25) is a 31-kDa protein originally cloned from leptomeninges and breast cancer cells, although it has also been isolated from a variety of cell types. IGFBP-rP1 expression is increased by retinoids, TGF-β, PTH, and cortisol. Secreted IGFBP-rP1 has been identified in human urine, CSF, and amniotic fluid. Accumulating evidence suggests a growth-inhibitory role for IGFBP-rP1. Transduction of IGFBP-rP1 into breast cancer cells has been shown to inhibit growth and induce senescent morphology, and similar findings have also been seen in prostate cell culture. Loss of heterozygosity for IGFBP-rP1 expression is associated with initiation and progression of ductal breast cancer. In myoblasts, IGFBP-rP1 action may be growth-inhibitory or stimulatory. Given its low affinity for IGF-I, IGFBP-rP1 probably functions in an IGF-independent manner, a hypothesis supported by the recent identification of novel IGFBP-rP1 association proteins. IGFBP-rP2 (connective tissue growth factor) has been identified in conditioned media of breast cancer cells and in human serum, CSF, amniotic, follicular, and peritoneal fluids. IGFBP-rP2 may be a downstream effector of TGF-β, particularly in fibroblasts, where it may be a critical regulator of cellular regeneration and wound repair. Pathological conditions in which IGFBP-rP2 may play a role include leukemia, scleroderma, atherosclerosis, and diabetes complications. IGFBP-rP3 (NovH) has a widespread distribution, being present in serum CSF, follicular, and peritoneal fluid. It has been detected in several malignant cell lines, including breast and prostate. IGFBP-rP3 is associated with the developing kidney and may play a role in nephrogenesis, podocyte differentiation and function, and the development of Wilms' tumors. IGFBP-rP4 (Cyr61), a cysteine-rich secreted protein, associates with cell surfaces and extraceullar matrix, most likely through its heparin binding regions, where it may promote angiogenesis, fibroblast and epithelial cell adhesion, chemotaxis, and chondrogenesis. IGFBP-rP5 through –rP9 are only cDNAs cloned from libraries, whose natural tissues and functions are as yet undetermined. Based on sequence data, IGFBP-rP5 (L56) may be a serine protease. IGFBP-rP6 (ESM-1) was cloned from a human umbilical vein epithelial cell library. It has been expressed only in lung tissue. IGFBP-rP7 – rP9 may be linked to tumorigenesis.

In conclusion, the last two decades has witnessed an explosion of data in the scientific literature regarding the various components of the IGF axis. IGFBPs and related molecules are now believed to be critical elements in numerous cellular processes and key factors in the regulation of growth, proliferation, and apoptosis in both health and many disease states. Continuing research will undoubtedly reveal even more information on the molecular biology of these key cellular regulators, lead to better understanding of growth and cellular regulation, and ultimately to the development of novel therapeutic interventions in such diverse areas as growth disorders, cancer, inflammatory and vascular diseases, and diabetes.

REFERENCES

1. Baxter RC. Insulin-like growth factor binding proteins in the human circulation, a review. Horm Res 1994;42:140–144.
2. Twigg SM, Baxter RC. Insulin-like growth factor (IGF) binding protein-5 forms an alternative ternary complex with IGFs and the acid-labile subunit. J Biol Chem 1998;273:6074–6079.
3. Khosravi MJ, Diamandi A, Mistry J, Krishna RG, Khare A. Acid-labile subunit of human insulin-like growth factor-binding protein complex: measurement, molecular, and clinical evaluation. J Clin Endocrinol Metab 1997;82:3944–3951.
4. Hasegawa T, Cohen P, Rosenfeld RG. Characterization of the insulin-like growth factor axis in TM 3 Leydig cells. Growth Reg 1995;5:151–159.
5. Cohen P, Rosenfeld RG. The IGF axis. In: Rosenbloom AL, editor. Human Growth

Hormone, Basic and Scientific Aspects. Boca Raton: CRC Press, 1995;43–58.
6. Arany E, Afford S, Strain AJ, et al. Differential cellular synthesis of insulin-like growth factor binding protein 1 and IGFBP-3 within human liver. J Clin Endocrinol Metab 1994;79:1871–1876.
7. Chin E, Zhou J, Dai J, Baxter RC, Bondy CA. Cellular localization of gene expression for components of the IGF ternary complex. Endocrinol 1994;134:2498–2504.
8. Villafuerte BC, Koop BL, Pao CI, Gu L, Birdsong GG, Phillips LS. Coculture of primary rat hepatocytes and nonparenchimal cells permits expression of insulin-like growth factor binding protein-3 *in vitro*. Endocrinology 1994;134:2044–2050.
9. Holly JMP, Martin JL. IGFBPs: a review of methodological aspects of their purification, analysis and regulation. Growth Regul 1994;4:20–30.
10. Rosenfeld RG, Pham H, Cohen P, et al. Insulin-like growth factor binding proteins and their regulation. Acta Paediatr 1994;339:154–158.
11. Underwood LE, Thissen JP, Lemozy S, Ketelslegers JM, Clemmons DR. Hormonal and nutritional regulation of IGF-I and its binding proteins. Horm Res 1994;42:145–151.
12. Katz LEL, Rosenfeld RG, Cohen P. Clinical significance of insulin-like growth factor binding proteins (IGFBPs). Endocrinologist 1995;5:36–43.
13. Smith WJ, Underwood LE, Clemmons DR. Effects of caloric or protein restriction on insulin-like growth factor-I (IGF-I) and IGF-binding proteins in children and adults. J Clin Endocrinol Metab 1995;80:443–449.
14. Baxter RC, Saunders H. Radioimmunoassay of insulin-like growth factor-binding protein-6 in human serum and other body fluids. J Endocrinol 1992;134:133–139.
15. Mohan S, Farley JR, Baylink DJ. Age-related changes in IGFBP-4 and IGFBP-5 levels in human serum and bone: implications for bone loss with aging. Prog Growth Factor Res 1995;6:465–473.
16. Mohan S, Libanati C, Dony C, Lang K, Srinivasan N, Baylink DJ. Development, validation, and application of a radioimmunoassay for insulin-like growth factor binding protein-5 in human serum and other biological fluids. J Clin Endocrinol Metab 1995;80:2638–45.
17. Thoren M, Hilding A, Brismar T, et al. Serum levels of insulin-like growth factor binding proteins (IGFBP)-4 and -5 correlate with bone mineral density in growth hormone (GH)-deficient adults and increase with GH replacement therapy. J Bone Miner Res 1998;13:891–899.
18. Lamson G, Guidice LC, Rosenfeld RG. Insulin-like growth factor binding proteins: structural and molecular relationships. Growth Factors 1991;5:19–28.
19. Shimasaki S, Ling N. Identification and molecular characterization of insulin-like growth factor binding proteins (IGFBP-1, -2, -3, -4, -5, and -6). Prog Growth Factor Res 1991;3:243–266.
20. Hong YS, Kim SY, Bhattacharya A, Pratt DR, Hong WK, Tainsky MA. Structure and function of the HOX A1 human homeobox gene cDNA. Gene 1995;159:209–214.
21. Matsui T, Hirai M, Hirano M, Kurosawa Y. The HOX complex neighbored by the EVX gene, as well as two other homeobox-containing genes, the GBX-class and the EN-class, are located on the same chromosomes 2 and 7 in humans. FEBS Lett 1993;336:107–110.
22. Allander SV, Bajalica S, Larsson C, et al. Structure and chromosomal localization of human insulin-like growth factor-binding protein genes. Growth Regul 1993;3:3–5.
23. Collett-Solberg PF, Cohen P. The role of the insulin-like growth factor binding proteins and the IGFBP proteases in modulating IGF action. Endocrinol Metab Clin N Amer 1996;25:591–614.

24. Ruoslathi E, Pierschlaber MD. New perspectives in cell adhesions: RGD and integrins. Science 1987;238:491–497.
25. Coverley JA, Baxter RC. Phosphorylation of insulin-like growth factor binding proteins. Mol Cell Endocrinol 1997;128:1–5.
26. Levitt Katz LE, Satin-Smith MS, Collett-Solberg P, Thornton PS, Baker L, Stanley CA, Cohen P. Insulin-like growth factor binding protein-1 levels in the diagnosis of hypoglycemia caused by hyperinsulinism. J Pediatr 1997;131:193–9.
27. Counts DR, Gwirtsman H, Carlsson LM, Lesem M, Cutler GB, Jr. The effect of anorexia nervosa and refeeding on growth hormone binding protein, the insulin-like growth factors (IGFs) and the IGF- binding proteins. J Clin Endocrinol Metab 1992;75:762–767.
28. Strasser-Vogel B, Blum WF, Past R, et al. Insulin-like growth factor (IGF)-I and -II and IGF-binding proteins-1, -2, and -3 in children and adolescents with diabetes mellitus: correlation with metabolic control and height attainment. J Clin Endocrinol Metab 1995;80:1207–1213.
29. Unterman TG, Oehler DT, Murphy LJ, et al. Multihormonal regulation of insulin-like growth factor binding protein-1 in rat H4IIE hepatoma cells: the dominant role of insulin. Endocrinology 1991;128:2693–701.
30. Argente J, Caballo N, Barrios V, Munoz MT, Pozo J, Chowen JA, Hernandez M. Disturbances in the growth hormone-insulin-like growth factor axis in children and adolescents with different eating disorders. Horm Res 1997;48(Suppl 4):16-18.
31. Conover CA, Divertie GD, Lee PD. Cortisol increases plasma insulin-like growth factor binding protein-1 in humans. Acta Endocrinol 1993;128:140–143.
32. Lee PDK, Conover CA, Powell DR. Regulation and function of insulin-like growth factor binding protein-1. Proc Soc Exp Biol Med 1993;204:4–29.
33. Price WA, Stiles AD, Moats-Staats BM, D'Ercole AJ. Gene expression of insulin-like growth factors (IGFs), the type I IGF receptor, and the IGF-binding proteins in dexamethasone-induced fetal growth retardation. Endocrinology 1990;130:1424–1432.
34. Conover CA, Lee PD, Riggs BL, Powell DR. Insulin-like growth factor binding protein-1 expression in cultured human bone cells: regulation by insulin and glucocorticoids. Endocrinology 1996;137:3295–3301.
35. Samstein B, Hoimes ML, Fan J, Frost RA, Gelato MC, Lang CH. IL-6 stimulation of insulin-like growth factors binding proteins (IGFBPs) production. Bioch Biophy Res Commun 1996;228:611–5.
36. Murphy LJ, Rajkumar K, Molnar P. Phenotypic manifestations of insulin-like growth factor binding protein-1 (IGFBP-1) and IGFBP-3 overexpression in transgenic mice. Prog Growth Factor Res 1995;6:425–432.
37. Gutierrez-Ospina G, Calikoglu AS, Ye P, D'Ercole AJ. In vivo effects of insulin-like growth factor-I on the development of sensory pathways: analysis of the primary somatic sensory cortex (S1) of transgenic mice. Endocrinology 1996;137:5484–5492.
38. Russo VC, Bach LA, Fosang AJ, Baker NL, Werther GA. Insulin-like growth factor binding protein-2 binds to cell surface proteoglycans in the rat brain olfactory bulb. Endocrinology 1997;138:4858–4867.
39. Arai T, Busby W Jr, Clemmons DR. Binding of insulin-like growth factor (IGF) I or II to IGF-binding protein-2 enables it to bind to heparin and extracellular matrix. Endocrinology 1996;137:4571–4575.
40. Schwander J, Mary JL. The RIA for IGFBP-2 in man: a meagre catch? Growth Regul 1993;3:105–109.

41. Bernardini S, Cianfarani S, Spagnoli A, et al. Expression and down-regulation by retinoic acid of IGF binding protein-2 and -4 in medium from human neuroblastoma cells. J Neuroendocrinol 1994;6:409–413.
42. Müller HL, Oh Y, Lehrnbecher T, Blum WF, Rosenfeld RG. Insulin-like growth factor binding protein-2 concentrations in cerebrospinal fluid and serum of children with malignant solid tumors or acute leukemia. J Clin Endocrinol Metab 1994;79:428–434.
43. Walter H, Berry M, Hill D, Cwyfan-Hughes S, Holly J. Distinct sites of insulin-like growth factor (IGF)-II expression and localization in lesioned rat brain: possible roles of IGF binding proteins (IGFBPs) in the mediation of IGF-II activity. Endocrinol 1999;140:520–532.
44. Chambery D, de Galle B, Babajko S. Retinoic acid stimulates IGF binding protein (IGFBP)-6 and depresses IGFBP-2 and IGFBP-4 in SK-N-SH human neuroblastoma cells. J Endocrinol 1998;94:173–180.
45. Foll J, Dannecker L, Zehrer C, et al. Activation-dependent expression of the insulin-like growth factor binding protein-2 in human lymphocytes. Immunol 1998;94:173–180.
46. Katz L, Bhala A, Camron E, Nunn SE, Hintz RL. IGF-II, IGF-binding proteins and IGF receptors in pancreatic β-cell lines. J Endocrinol 1997;152:455–464.
47. Hardouin S, Gurmelem M, Noguiez P, et al. Molecular forms of serum insulin-like growth factor (IGF)-binding proteins in man: relationship with growth hormone and IGFs and physiologic significance. J Clin Endocrinol Metab 1989;69:1291–1301.
48. Kelley KM, Oh Y, Gargosky SE, et al. Insulin-like growth factor-binding proteins (IGFBPs) and their regulatory dynamics. Intern J Bioch Cell Biol 1996;28:619–637.
49. Sommer A, Spratt SK, Tatsuno GP, et al. Properties of glycosylated and nonglycosylated human recombinant IGF binding protein-3 (IGFBP-3). Growth Regul 1992;3:46–49.
50. Paolisso G, Ammendola S, Del Buono A, et al. Serum levels of insulin-like growth factor-I (IGF-I) and IGF-binding protein-3 in healthy centenarians: relationship with plasma leptin and lipid concentrations, insulin action, and cognitive function. J Clin Endocrinol Metab 1997;82:2204–2209.
51. Conover CA, Clarkson JT, Bale LK. Effect of glucocorticoid on insulin-like growth factor (IGF) regulation of IGF-binding protein expression in fibroblasts. Endocrinol 1995;136:1403–1410.
52. Gucev ZS, Oh Y, Kelley KM, et al. Evidence for insulin-like growth factor (IGF)-independent transcriptional regulation of IGF binding protein-3 by growth hormone in SKHEP-1 human hepatocarcinoma cells. Endocrinology 1997;138:1464–1470.
53. Vaccarello MA, Diamond FB, Guevara-Aguirre J, et al. Hormonal and metabolic effects and pharmacokinetics of recombinant insulin-like growth factor-I in growth hormone receptor deficiency/Laron syndrome. J Clin Endocrinol Metab 1993;77:273–280.
54. Kanety H, Karasik A, Klinger B, Silbergeld A, Laron Z. Long-term treatment of Laron type dwarfs with insulin-like growth factors increases serum IGFBP-3 in the absence of growth hormone activity. Acta Endocrinol 1993;128:144–149.
55. Zapf J, Hauri C, Waldvogel M, et al. Recombinant human IGF-I induces its own specific carrier proteins in hypophysectomized and diabetic rats. Proc Natl Acad Sci USA 1989;86:3813–3817.
56. Lemmey AB, Glassford J, Flick-Smith HC, Holly JM, Pell JM. Differential regulation of tissue insulin-like growth factor-binding protein (IGFBP)-3, IGF-I and IGF type 1 receptor mRNA levels, and serum IGF-I and IGFBP concentrations by growth hormone and IGF-I. J Endocrinol 1997;154:319–328.

57. Conover CA, Clarkson JT, Durham SK, Bale LK. Cellular actions of insulin-like growth factor binding protein-3. In: LeRoith D, Raizada MK, editors. Current Directions in Insulin-like Growth Factor Research. New York: Plenum Press, 1994;255–265.
58. Kanzaki S, Baxter RC, Knutsen R, Baylink DJ, Mohan S. Evidence that human bone cells in culture secrete insulin-like growth factor (IGF)-II and IGF binding protein-3 but not acid-labile subunit both under basal and regulated conditions. J Bone Mineral Res 1995;10:854–858.
59. Olney RC, Wilson DM, Mohtai M, Fielder PJ, Smith RL. Interleukin-1 and tumor necrosis factor-alpha increase insulin-like growth factor-binding protein-3 (IGFBP-3) production and IGFBP-3 protease activity in human articular chondrocytes. J Endocrinol 1995;146:279–286.
60. Besset V, Le Magueresse-Battistoni B, Collette J, Benahmed M. Tumor necrosis factor alpha stimulates insulin-like growth factor binding protein 3 expression in cultured porcine Sertoli cells. Endocrinology 1996;137:296–303.
61. Oh Y, M ller HL, Ng L, Rosenfeld RG. Transforming growth factor-ß-induced cell growth inhibition in human breast cancer cells is mediated through insulin-like growth factor-binding protein-3 action. J Biol Chem 1995;270:13589–13592.
62. Han GR, Dohi DF, Lee HY, et al. All-trans-retinoic acid increases transforming growth factor-beta2 and insulin-like growth factor binding protein-3 expression through a retinoic acid receptor-alpha-dependent signaling pathway. J Biol Chem 1997;272:13711–13716.
63. Hayden JM, Strong DD, Baylink DJ, Powell DR, Sampath TK, Mohan S. Osteogenic protein-1 stimulates production of insulin-like growth factor binding protein-3 nuclear transcripts in human osteosarcoma cells. Endocrinology 1997;138:4240–4247.
64. Smith EP, Dickson BA, Chernausek SD. Insulin-like growth factor binding protein-3 secretion from cultured rat Sertoli cells: dual regulation by follicle stimulating hormone and IGF-I. Endocrinol 1990;127:2744–2751.
65. Huynh H, Yang YF, Pollak M. Estradiol and antiestrogens regulate a growth inhibitory insulin-like growth factor binding protein 3 autocrine loop in human breast cancer cells. J Biol Chem 1996;271:1016–1021.
66. DiBattista JA, Dore S, Morin N, Abribat T. Prostaglandin E2 upregulates insulin-like growth factor binding protein-3 expression and synthesis in human articular chondrocytes by a c-AMP-independent pathway: role of calcium and protein kinase A and C. J Cell Biochem 1996;63:320–333.
67. Agarwal C, Lambert A, Chandraratna RAS, Rorke EA, Eckert RL. Vitamin D regulates human ectocervical epithelial cell proliferation and insulin-like growth factor-binding protein-3 level. Biol Reprod 1999;60:567–572.
68. Buckbinder L, Talbott R, Velasco-Miguel S, et al. Induction of the growth inhibitor IGF-binding protein 3 by p53. Nature 1995;377:646–649.
69. Jaques G, Noll K, Wegmann B, et al. Nuclear localization of insulin-like growth factor binding protein 3 in a lung cancer cell line. Endocrinol 1997;138:1767–1770.
70. Noll K, Wegmann B, Havemann K, Jacques G. Insulin-like growth factors stimulate the release of insulin-like growth factor binding protein-3 (IGFBP-3) and degradation of IGFBP-4 in nonsmall cell lung cancer cell lines. J Clin Endocrinol Metab 1996;81:2653–2662.
71. Li W, Fawcett J, Widmer HR, Fielder PJ, Rabkin R, Keller GA. Nuclear transport of insulin-like growth factor-I and insulin-like growth factor binding protein-3 in opossum kidney cells. Endocrinol 1997;138:1763–1766.

72. Schedlich LJ, Young TF, Firth SM, Baxter RC. Insulin-like growth factor binding protein (IGFBP)-3 and IGFBP-5 share a common nuclear transport pathway in T47D human breast carcinoma cells. J Biol Chem 1998;273:18347–18352.
73. Wraight CJ, Liepe IJ, Werther GA. Intranuclear localization of insulin-like growth factor binding protein-3 during cell division in human keratinocytes. J Invest Dermatol 1998;111:239–242.
74. Murphy LJ, Molnar P, Lu X, Huang H. Expression of human insulin-like growth factor-binding protein-3 in transgenic mice. J Mol Endocrinol 1995;15:293–303.
75. Nickerson T, Pollak M, Huynh H. Castration-induced apoptosis in the rat ventral prostate is associated with increased expression of genes encoding insulin-like growth factor binding proteins 2, 3, 4, and 5. Endocrinol 1998;139:807–810.
76. Oh Y, Müller HL, Lamson G, Rosenfeld RG. Insulin-like growth factor (IGF)-independent action of IGF-binding protein-3 in Hs578T human breast cancer cells. Cell surface binding and growth inhibition. J Biol Chem 1993;268:14964–71.
77. Mohan S, Nakao Y, Honda Y, et al. Studies on the mechanisms by which insulin-like growth factor (IGF) binding protein-4 (IGFBP-4) and IGFBP-5 modulate IGF actions in bone cells. J Biol Chem 1995;270:20424–20431.
78. Bayes-Genis A, Schwartz RS, Lewis DA, et al. Insulin-like growth factor binding protein-4 protease produced by smooth muscle cells increases in the coronary artery after angioplasty. Arterioscler Thromb Vasc Biol 2001;21:335–341.
79. Rees C, Clemmons DR, Horvitz GD, Clarkes JB, Busby WH. A protease-resistant form of insulin-like growth factor (IGF) binding protein 4 inhibits IGF-1 actions. Endocrinol 1998;139:4182–4188.
80. Lawrence JB, Oxvig C, Overgaard MT, et al. The insulin-like growth factor binding protein-4 protease secreted by human fibroblasts is pregnancy-associated plasma protein-A. Proc Natl Sci USA 1999;96:3149–3153.
81. Kelly KM, Oh Y, Gargosky SE, et al. Insulin-like growth factor-binding proteins (IGFBPs) and their regulatory dynamics. Int J Biochem Cell Biol 1996;28:619–637.
82. Wang J, Niu W, Witte DP, et al. Overexpression of insulin-like growth factor-binding protein-4 (IGFBP-4) in smooth muscle cells of transgenic mice through a smooth muscle alpha-actin-IGFBP-4 fusion gene induces smooth muscle hypoplasia. Endocrinol 1998;139:2605–2614.
83. Damon SE, Maddison L, Ware JL, Plymate SR. Overexpression of an inhibitory insulin-like growth factor-binding protein (IGFBP), IGFBP-4, delays onset of prostate tumor formation. Endocrinol 1998;139:3456–3464.
84. Mason HD, Cwyfan-Hughes S, Molly JMP, Franks S. Potent inhibition of human ovarian steroidogenesis by insulin-like growth factor binding protein-4 (IGFBP-4). J Clin Endocrinol Metab 1998;83:284–287.
85. Damon SE, Hauck KL, Birnbaum RS, Quinn LS. Retrovirally mediated overexpression of IGFBP-4: evidence that insulin-like growth factor is required for skeletal muscle differentiation. J Cell Phys 1998;175:109–120.
86. Ewton DZ, Coolican SA, Mohan S, Chernausek SD, Fiorini JR. Modulation of insulin-like growth factor binding protein-4 (IGFBP-4) and IGFBP-5: a dual role for IGFBP-5. J Cell Phys 1998;177:47–57.
87. Mohan S, Nakao Y, Honda Y, et al. Studies on the mechanisms by which insulin-like growth factor (IGF) binding protein-4 (IGFBP-4) and IGFBP-5 modulate IGF actions in bone cells. J Biol Chem 1995;270:20424–20431.

88. Scharla AH, Strong DD, Mohan S, Baylink DJ, Linkhart TA. 1,25-Dihydroxivitamin D3 differentially regulates the production of insulin-like growth factor I (IGF-I) and IGF-binding protein-4 in mouse osteoblasts. Endocrinol 1991;129:3139–3146.
89. Fernandez-Celemin L, Thissen JP. Interleukin-6 stimulates hepatic insulin-like growth factor binding protein-4 messenger ribonucleic acid and protein. Endocrinol 2001;142:241–248.
90. Booth BA, Boes M, Andress DL, et al. IGFBP-3 and IGFBP-5 association with endothelial cells: role of C-terminal heparin binding domain. Growth Regul 1995;5:1–17.
91. Jones JI, Gockerman A, Busby WH Jr, Camacho-Hubner C, Clemmons DR. Extracellular matrix contains insulin-like growth factor binding protein-5: potentiation of the effects of IGF-I. J Cell Biol 1993;121:679–687.
92. Andress DL, Birnbaum RS. Human osteoblast-derived insulin-like growth factor (IGF) binding protein-5 stimulates osteoblast mitogenesis and potentiates IGF action. J Biol Chem 1992;267:22467–22472.
93. Ewton DZ, Coolican SA, Mohan S, Chernausek SD, Florini JR. Modulation of insulin-like growth factor bindig protein (IGFBP)-4 and IGFBP-5: a dual role for IGFBP-5. J Cell Physiol 1998;177:47–57.
94. Clemmons DR. Insulin-like growth factor binding proteins and their role in controlling IGF actions. Ctyol Growth Factor Rev 1997; 22467–22472.
95. Canalis E, Gabbitas B. Skeletal growth factors regulate the synthesis of insulin-like growth factor binding protein-5 in bone cell cultures. J Biol Chem 1995;270:10771–10776.
96. Conover CA, Bale LK, Clarkson JT, Torring O. Regulation of insulin-like growth factor binding protein-5 messenger ribonucleic acid expression and protein availability in rat osteoblast-like cells. Endocrinol 1993;132:2525–2530.
97. Gabbitas B, Canalis E. Insulin-like growth factors sustain inulin-like growth factor-binding protein-5 expression in osteoblasts. Am J Physiol 1998; 275:E222–E228.
98. Okazaki R, Riggs BL, Conover CA. Glucocorticoid regulation of insulin-like growth factor-binding protein expression in normal human osteoblast-like cells. Endocrinol 1994;134:126–32.
99. Duan C, Hawes SB, Prevette T, Clemmons DR. Insulin-like growth factor-I (IGF-I) regulates IGF-binding protein-5 synthesis through transcriptional activation of the gene in aortic smooth muscle cells. J Biol Chem 1996;271:4280–4288.
100. Duan C, Clemmons DR. Differential expression and biological effects of insulin-like growth factor-binding protein-4 and -5 in vascular smooth muscle cells. J Biol Chem 1998;273:16836–16842.
101. Bach LA, Thotakura NR, Rechler MM. Human insulin-like growth factor binding protein-6 is O-glycosylated. Growth Regul 1993;3:59–62.
102. Claussen M, Buergisser D, Schuller AG, Matzner U, Braulke T. Regulation of insulin like growth factor (IGF)-binding protein-6 and mannose 6-phosphate/IGF-II receptor expression in IGF-II overexpressing NIH 3T3 cells. Mol Endocrinol 1995;9:902–912.
103. Bach L. Insulin-like growth factor binding protein-6: the forgotten binding protein? Horm Metab Res 1999;31:226–234.
104. Bach L, Hsieh S, Brown AL, et al. Recombinant human insulin-like growth factor (IGF)-binding protein-6 inhibits IGF-II induced differentiation of L6A1 myoblasts. Endocrinol 1994;135:2168–2176.
105. Bach LA, Salemi R, Leeding KS. Roles of insulin-like growth factor (IGF) receptors and IGF-binding proteins in IGF-II induced proliferation and differentiation of L6A1 rat myoblasts. Endocrinol 1995;136:5061–5069.

106. Sheikh MS, Shao ZM, Hussain A, et al. Regulation of insulin-like growth factor-binding protein -1, 2, 3, 4, 5, 6: synthesis, secretion, and gene expression in estrogen receptor-negative human breast carcinoma cells. J Cell Physiol 1993;155:556–67.
107. Zhou Y, Mohan S, Linkhart TA, Baylink DJ, Strong DD. Retinoic acid regulates insulin-like growth factor-binding protein expression in human osteoblast cells. Endocrinol 1996;137:975–983.
108. Hossenlopp P, Segovia B, Lassare C, Roghani M, Bredon M, Binoux M. Enzymatic evidence of degradation of insulin-like growth factor binding protein in 150 k complex during pregnancy. J Clin Endocrinol Metab 1990;71:797–805.
109. Cohen P, Graves HC, Peehl DM, Kamarei M, Giudice LC, Rosenfeld RG. Prostate-specific antigen (PSA) is an insulin-like growth factor binding protein-3 protease found in seminal plasma. J Clin Endocrinol Metab 1992;75:1046–1053.
110. Rajah R, Bhala A, Nunn SE, Peehl DM, Cohen P. 7S nerve growth factor is an insulin-like growth factor binding protein protease. Endocrinol 1996;137:2676–2682.
111. Conover C, DeLeon DD. Acid activated insulin-like growth factor binding protein-3 proteolysis in normal and transformed cells. Role of cathepsin D. J Biol Chem 1994;2691:7076–7080.
112. Conover C, Perry J, Tindall D. Endogenous cathepsin-D-mediated hydrolysis of insulin-like growth factor-binding proteins in cultured human prostatic carcinoma cells. J Clin Endocrinol Metab 1995;80:987–993.
113. Nunn S, Peehl DM, Cohen P. Acid-activated insulin-like growth factor binding protein protease activity of cathepsin D in normal and malignant prostatic epithelial cells and seminal plasma. J Cell Physiol 1997;171:196–204.
114. Claussen M, Kubler B, Wendland M, et al. Proteolysis of insulin-like growth factors (IGF) and IGF binding proteins by cathepsin D. Endocrinol 1997;138:3797–3803.
115. Fowlkes JL, Suzuki K, Nagase H, Thrailkill KM. Proteolysis of insulin-like growth factor binding protein-3 during rat pregnancy: a role for matrix metalloproteinases. Endocrinol 1994;135:2810–2813.
116. Fowlkes JL, Enghild JJ, Suzuki K, Nagase H. Matrix metalloproteinases degrade insulin-like growth factor binding protein-3 in dermal fibroblast cultures. J Biol Chem 1994;269:742–746.
117. Rajah R, Katz L, Nunn S, Solberg P, Beers T, Cohen P. Insulin-like growth factor binding protein proteases: functional regulators of cell growth. Prog Growth Factor Res 1996;6:273–284.
118. Angelloz-Nicoud P, Harel L, Binoux M. Recombinant human insulin-like growth factor (IGF) binding protein-3 stimulates prostate carcinoma cell proliferation via an IGF-dependent mechanism. Role of serine proteases. Growth Reg 1996;6:130–136.
119. Zheng B, Clarke JB, Busby WH, Duan C, Clemmons DR. Insulin-like growth factor-binding protein-5 is cleaved by physiological concentrations of thrombin. Endocrinol 1998;139:1708–1714.
120. Bereket A, Wilson TA, Kolasa AJ, Fan J, Lang CH. Regulation of the insulin-like growth factor system by acute acidosis. Endocrinol 1996;137:2238–2245.
121. Fowlkes JL, Suzuki K, Nagase H, et al. Proteolysis of insulin-like growth factor binding protein-3 during rat pregnancy: a role for matrix metalloproteinases. Endocrinol 1994;135:2810–2813.
122. Nunn S, Beers Gibson T, Rajah R, Cohen P. Regulation of prostate cell growth by the insulin-like growth factor binding proteins and their proteases. Endocrine

1997;7:115–118.

123. Cohen P, Noveral JP, Bhala A, et al. Leukotriene D_4 facilitates airway smooth muscle cell proliferation via modulation of the IGF axis. Am J Physiol 1995;269:L151–L157.
124. Rajah R, Nunn S, Herrick D, Grunstein M, Cohen P. LTD-4 induces matrix metalloproteinase-1 which functions as an IGFBP protease in airway smooth muscle cells. Am J Physiol 1996;271:L1014–1022.
125. Wu H, Lee C, Rechler M. Proteolysis of insulin-like growth factor binding protein-3 in serum from pregnant, non-pregnant and fetal rats by matrix metalloproteinases and serine proteinases. Horm Metab Res 1999;31:186–191.
126. Thrailkill KM, Quarles LD, Nagase H, et al. Characterization of insulin-like growth factor-binding protein 5-degrading proteases produced throughout murine osteoblast differentiation. Endocrinol 1995;136:3527–3533.
127. Cohen P, Peehl DM, Graves HC, Rosenfeld RG. Biological effects of prostate specific antigen as an insulin-like growth factor binding protein-3 protease. J Endocrinol 1994;142:407–415.
128. Blat C, Villaudy J, Binoux M. In vivo proteolysis of serum IGFBP-3 results in increased availability of IGF to target cells. J Clin Invest 1994;93:2286–2290.
129. Besnard N, Pisselet C, Zapf J, Hornebeck W, Monniaux D, Monget P. Proteolytic activity is involved in changes in intrafollicular insulin-like growth factor binding protein levels during growth and atresia of ovine ovarian follicles. Endocrinol 1996;137:1599–1607.
130. Rajah R, Nachajon R, Collins M, Hakonarson H, Grunstein M, Cohen P. Elevated levels of IGF-binding protein protease in asthmatic airway smooth muscle. Am J Respir Cell Mol Biol 1999;20:199–208.
131. Manes S, Llorente M, Lacalle R, et al. The matrix metalloproteinase-9 regulates the insulin-like growth factor-triggered autocrine response in DU-145 carcinoma cells. J Biol Chem 1999;274:6935–6945.
132. Bereket A, Lang CH, Blethen SL, Fan J, Frost RA, Wilson TA. Insulin-like growth factor binding protein-3 proteolysis in children with insulin-dependent diabetes mellitus: a possible role for insulin in the regulation of IGFBP-3 protease activity. J Clin Endocrinol Metab 1995;80:2282–2288.
133. Fowlkes JL, Thrailkill KM, George-Nascimento C, Rosenberg CK, Serra DM. Heparin-binding, highly basic regions within the thyroglobulin type-1 repeat of insulin-like growth factor (IGF)-binding proteins (IGFBPs) -3, -5, and -6 inhibit IGFBP-4 degradation. Endocrinol 1997;138:2280–2285.
134. Donnelly MJ, Holly JM. The role of IGFBP-3 in the regulation of IGFBP-4 proteolysis. J Endocrinol 1996;149:R1–R7.
135. Salahifar H, Baxter RC, Martin JL. Insulin-like growth factor binding protein (IGFBP)-3 protease activity secreted by MCF-7 breast cancer cells: inhibition by IGFs does not require IGF-IGFBP interaction. Endocrinol 1997;138:1683–1690.
136. Li YM, Schacher DH, Liu Q, et al. Regulation of myeloid growth and differentiation by the insulin-like growth factor I receptor. Endocrinol 1997;138:362–368.
137. Karas M, Danilenko M, Fishman D, LeRoith D, Levy J, Sharoni Y. Membrane-associated insulin-like growth factor-binding protein-3 inhibits insulin-like growth factor-I-induced insulin-like growth factor-I receptor signaling in ishikawa endometrial cancer cells. J Biol Chem 1997;272:16514–16520.
138. Olney RC, Tsuchiya K, Wilson DM, et al. Chondrocytes from osteoarthritic cartilage have increased expression of insulin-like growth factor I (IGF-I) and IGF-binding

protein-3 (IGFBP-3) and –5, but not IGF-II or IGFBP-4. J Clin Endocrinol Metab 1996;81:1096–1103.

139. Barreca A, Artini PG, Cesarone A, et al. Interrelationships between follicle stimulating hormone and the growth hormone – insulin-like growth factor – IGF-binding proteins axes in human granulosa cells in culture. J Endocrinol Invest 1996;19:35–42.
140. de la Sota RL, Simmen FA, Diaz T, Thatcher WW. Insulin-like growth factor system in bovine first-wave dominant and subordinate follicles. Biol Reprod 1996;55:803–812.
141. Hasegawa Y, Hasegawa T, Fujii K, et al. High ratios of free to total insulin-like growth factor-I in early infancy. J Clin Endocrinol Metab 1997;82:156–158.
142. Powell DR. Effects of renal failure on the growth hormone-insulin-like growth factor axis. J Pediatr 1997;131:S13–S16.
143. Mohseni-Zadeh S, Binoux M. Insulin-like growth factor (IGF) binding protein-3 interacts with the type 1 IGF receptor, reducing the affinity of the receptor for its ligand: an alternative mechanism in the regulation of IGF action. Endocrinol 1997;138:5645–5648.
144. Noll K, Wegmann B, Havemann K, Jaques G. Insulin-like growth factors stimulate the release of insulin-like growth factor-binding protein-3 (IGFBP-3) and degradation of IGFBP-4 in nonsmall cell lung cancer cell lines. J Clin Endocrinol Metab 1996;81:2653–2662.
145. Jones JI, Clemmons DR. Insulin-like growth factors and their binding proteins: biological actions. Endocrine Rev 1995;16:3–34.
146. Ranke MB, Elmlinger M. Functional role of insulin-like growth factor binding proteins. Horm Res 1997;48:9–15.
147. Li W, Fawcett J, Widmer HR, Fielder PJ, Rabkin R, Keller GA. Nuclear transport of insulin-like growth factor-I and insulin-like growth factor binding protein-3 in opossum kidney cells. Endocrinol 1997;138:1763–1766.
148. Portera CA, Shinohara H, Mima T, et al. Targeting insulin-like growth factor-I-receptor (IGF-I-R) signaling in the therapy of colorectal carcinoma (CRC) liver metastases. Proc Am Assoc Cancer Res 1999;40:2252.
149. Cohen P, Lamson G, Okajima T, Rosenfeld RG. Transfection of the human insulin-like growth factor binding protein-3 gene into Balb/c fibroblasts inhibits cellular growth. Mol Endocrinol 1993;7:380–386.
150. Lalou C, Lassarre C, Binoux M. A proteolytic fragment of insulin-like growth factor (IGF) binding protein-3 that fails to bind IGFs inhibits the mitogenic effects of IGF-I and insulin. Endocrinol 1996;137:3206–3212.
151. Zadeh SM, Binoux M. The 16-kDa proteolytic fragment of insulin-like growth factor (IGF) binding protein-3 inhibits the mitogenic action of fibroblast growth factor on mouse fibroblasts with a targeted disruption of the type 1 IGF receptor gene. Endocrinol 138: 3069–3072.
152. Valentinis B, Bhala A, DeAngelis T, Baserga R, Cohen P. The human insulin-like growth factor (IGF) binding protein-3 inhibits the growth of fibroblasts with a targeted disruption of the IGF-I receptor gene. Mol Endocrinol 1995;9:361–367.
153. Oh Y, Müller HL, Pham H, Rosenfeld RG. Demonstration of receptors for insulin-like growth factor binding protein-3 on Hs578T human breast cancer cells. J Biol Chem 1993;268:26045–26048.
154. Yamanaka Y, Fowlkes J, Wilson E, Rosenfeld RG, Oh Y. Characterization of insulin-like growth factor binding protein-3 (IGFBP-3) binding to breast cancer cells: kinetics of IGFBP-3 binding and identification of receptor binding domain on the IGFBP-3 molecule.

Endocrinol 1999;140:1319–1328.

155. Leal SM, Liu Q, Huang SS, Huang JS. The type V transforming growth factor beta receptor is the putative insulin-like growth factor-binding protein-3 receptor. J Biol Chem 1997;272:20572–20576.

156. Rajah R, Valentis B, Cohen P. Insulin-like growth factor (IGF)-binding protein-3 induces apoptosis and mediates the effects of transforming growth factor-ß1 on programmed cell death via a p53- and IGF-independent mechanism. J Biol Chem 1997;272:12181–12188.

157. Gucev ZS, Oh Y, Kelley KM, Rosenfeld RG. Insulin-like growth factor binding protein 3 mediaties retinoic acid- and transforming growth factor ß2-induced growth inhibition in human breast cancer cells. Cancer Res 1996;56:1545–1550.

158. Erondu NE, Dake BL, Moser DR, Lin M, Boes M, Bar RS. Regulation of endothelial IGFBP-3 synthesis and secretion by IGF-I and TGF-beta. Growth Reg 1996;6:1–9.

159. Giuliano M, Lauricella M, Vassallo E, Carabillo, Vento R, Tesoriere G. Induction of apoptosis in retinoblastoma cells by topoisomerase inhibitors. Invest Ophthalmol Vis Sci 1998;39:1300–1311.

160. Shen L, Dean N, Glazer R. Induction of p53-dependent, insulin-like growth factor-binding protein-3-mediated apoptosis in glioblastoma multiforme cells by a protein kinase C-alpha antisense oligonucleotide. Mol Pharmacol 1999;55:396–402.

161. Clemmons DR, Gardner LI. A plasma factor is required for IGFBP-1 to potentiate DNA synthesis. J Cell Physiol 1990;145:129–135.

162. Elgin RG, Busby WH, Clemmons DR. An insulin-like growth factor binding protein enhances the biologic response to IGF-I. Proc Natl Acad Sci USA 1987;84:3254–3258.

163. Koistinen R, Itkinen P, Selenius P, Seppala M. Insulin-like growth factor binding protein-1 inhibits binding of IGF-I on fetal skin fibroblasts but stimulates their DNA synthesis. Biochem Biophys Res Commun 1990;173:408–415.

164. Clemmons DR, Cascieri MA, Camacho-Hubner C, McCusker RH, Bayne ML. Discrete alterations of the IGF-I molecule which alter its affinity for IGF binding proteins result in changes in bioactivity. J Biol Chem 1990;265:12210–12216.

165. Jones JI, Gockerman A, Busby WH Jr, Wright G, Clemmons DR. Insulin-like growth factor binding protein 1 stimulates cell migration and binds to the alpha 5 beta 1 integrin by means of its Arg-Gly-Asp sequence. Proc Natl Acad Sci USA 1993;90:10553–10557.

166. Galiano RD, Zhao LL, Clemmons DR, Roth SI, Lin X, Mustoe TA. Interaction between the IGF family and the integrin receptor family in tissue repair processes. Evidence in a rabbit ear dermal ulcer model. J Clin Invest 1996;98:2462–2468.

167. Tsuboi R, Shi CM, Sato C, Cox GN, Ogawa H. Co-administration of insulin-like growth factor (IGF)-I and IGF-binding protein-1 stimulates wound healing in animal models. J Invest Derm 1995;104:199–203.

168. Jyung RW, Mustoe TA, Busby WH, Clemmons DR. Increased wound breaking strength induced by insulin-like growth factor-I in combination with IGF binding protein-1. Surgery 1994;115:133–239.

169. Clemmons DR. Role of post translational modifications in modifying the biologic activity of insulin like growth factor binding proteins. In: LeRoith D, Raizada MK, editors. Current Directions in Insulin-like Growth Factor Research. New York: Plenum Press, 1994;245–253.

170. Lewitt MS, Saunders H, Cooney GJ, Baxter RC. Effect of human insulin-like growth factor binding protein-1 on the half life and action of administered insulin-like growth factor-I in rats. J Endocrinol 1993;136:253–260.

171. Bar RS, Boes M, Clemmons DR, et al. Insulin differentially alters transcapillary movement of intravascular IGFBP-1, IGFBP-2 and endothelial cell IGF binding proteins in rat heart. Endocrinol 1990;127:497–499.
172. DeMellow JSM, Baxter RC. Growth hormone-dependent insulin-like growth factor (IGF) binding protein both inhibits and potentiates IGF-I stimulated DNA synthesis in human skin fibroblasts. Biochem Biophys Res Commun 1988;156:199–204.
173. Conover CA, Powell DR: Insulin-like growth factor (IGF)-binding protein-3 blocks IGF-I-induced receptor down-regulation and cell desensitization in cultured bovine fibroblasts. Endocrinol 1991;129:710–716.
174. Hamon GA, Hunt TK, Spencer EM. *In vivo* effects of systemic insulin-like growth-I alone and complexed with insulin-like growth factor binding protein-3 on corticosteroid suppressed wounds. Growth Regul 1993;3:55–56.
175. Clark RG, Mortensen D, Reifsynder D, Mohler M, Etcheverry T, Mukku V. Recombinant human insulin-like growth factor binding protein-3 (rhIGFBP-3): effects on the glycemic and growth promoting activities of rhIGF-I in the rat. Growth Regul 1993;3:50–52.
176. Weinzimer SA, Macarak E, Zhao H, Cohen P. The role of the IGF-IGFBP axis in the regulation of bladder smooth muscle growth. In: Program & Abstracts of the 80th Annual Meeting of The Endocrine Society, New Orleans. Bethesda MD: The Endocrine Society, 1998;316.
177. Weinzimer SA, Beers Gibson, T, Collett-Solberg PF, Zhao H, Khare A, Cohen P. Transferrin is an IGFBP-3 binding protein. J Clin Endocrinol Metab 2001;86:1806–1813.
178. Miyakoshi N, Richman C, Kasukawa Y, Linkhart TA, Baylink DJ, Mohan S. Evidence that IGF-binding protein-5 functions as a growth factor. J Clin Invest 2001;107:73–81.
179. Schedlich LJ, Young TF, Firth SM, Baxter RC. Insulin-like growth factor-binding protein (IGFBP)-3 and IGFBP-5 share a common nuclear transport pathway in T47D human breast carcinoma cells. J Biol Chem 1998;273:18347–18352.
180. Wraight CJ, Liepe IJ, Werther GA. Intracellular localization of insulin-like growth factor binding protein-3 during cell division in human keratinocytes. J Invest Dermatol 1998;111:239–242.
181. Hodgkinson S, Fowke P, Al Somai N, McQuoid M. Proteins in tissue extracts which bind insulin-like growth factor binding protein-3 (IGFBP-3). J Endocrinol 1995;145:R1–R6.
182. Yang YW-H, Yanagishita M, Rechler MM. Heparin inhibition of insulin-like growth factor binding protein-3 (IGFBP-3) binding to human fibroblasts and rat glioma cells: role of heparan sulfate proteoglycans. Endocrinol 1996;137:4363–4371.
183. Xu W, Murphy LJ: Interaction of IGFBP-3 with latent transforming growth factor-ß binding protein-1 identified using the yeast two-hybrid system. In: Program & Abstracts of the 80th Annual Meeting of The Endocrine Society, New Orleans. Bethesda MD: The Endocrine Society, 1998;313.
184. Gui Y, Murphy LJ. Insulin-like growth factor (IGF)-binding protein-3 (IGFBP-3) binds to fibronectin (FN): demonstration of IGF-I/IGFBP-3/fn complexes in human plasma. J Clin Endocrinol Metab 2001;86:2104–2110.
185. Collett-Solberg PF, Nunn SE, Beers Gibson T, Cohen P. Identification of novel high molecular weight insulin-like growth factor binding protein-3 association proteins in human serum. J Clin Endocrinol Metab 1998;83:2843–2848.
186. Liu B, Weinzimer SA, Collett-Solberg PF, Gibson TB, Mascarhenas D, Cohen P. Type Iα collagen is an IGFBP-3 binding protein. J Clin Endocrinol Metab 2001; in press.
187. Liu BR, Lee HY, Weinzimer SA, et al. Direct functional interactions between

IGFBP-3 and RXR-a regulate transcriptional signalling and apoptosis. J Biol Chem 2000;275:33607–13.

188. Vuguin PM, Shim ML, Cohen P, Barzilai N. IGFBP-3 induces hepatic and peripheral insulin resistance *in vivo*. In: Program & Abstracts of the 61st Annual Meeting of The American Diabetes Association, Philadelphia. Alexandria VA: The American Diabetes Association, 2001:A268.
189. Mannhardt B, Weinzimer SA, Wagner M, et al. Human papillomavirus type 16 E7 oncoprotein binds and inactivates growth-inhibitory insulin-like growth factor binding protein-3. Mol Cell Biol 2000;20:6483–95.
190. Reeve JG, Guadano A, Xiong J, Morgan J, Bleehen NM. Diminished expression of insulin-like growth factor (IGF) binding protein-5 and activation of IGF-I-mediated autocrine growth in simian virus 40-transformed human fibroblasts. J Biol Chem 1995;270:135–142.
191. Schenker T, Trueb B. Down-regulated proteins of mesenchymal tumor cells. Exp Cell Res 1998;239:161–168.
192. Spencer EM, Chan K. A 3-dimensional model for the insulin-like growth factor binding proteins (IGFBPs); supporting evidence using the structural determinants of the IGF binding site on IGFBP-3. Prog Growth Factor Res 1995;6:209–214.
193. Hashimoto R, Ono M, Fujiwara H, et al. Binding sites and binding properties of binary and ternary complexes of insulin-like growth factor-II (IGF-II), IGF-binding protein-3, and acid-labile subunit. J Biol Chem 1997;272:27936–27942.
194. Firth SM, Ganeshprasad U, Baxter RC. Structural determinants of ligand and cell-surface binding of insulin-like-growth factor-binding protein-3. J Biol Chem 1998;273:2631–2638.
195. Werner H, Karnieli E, Rauscher FJ, LeRoith D. Wild-type and mutant p53 differentially regulate transcription of the insulin-like growth factor I receptor gene. Proc Natl Acad Sci USA 1996;93:8318–8323.
196. Cohen P, Peehl DM, Stamey TA, Wilson KF, Clemmons DR, Rosenfeld RG. Elevated levels of insulin-like growth factor binding protein-2 in the serum of prostate cancer patients. J Clin Endocrinol Metab 1993;76:1031–1035.
197. Chatterjee S, Dai B, Owlia A, Singh P. Novel strategy for growth inhibition of human prostate cancer cells: Endogenous insulin-like growth factor binding protein-2 (IGFBP-2) suppression. In: Program and Abstracts of the 81st Annual Meeting of the Endocrine Society, San Diego, CA. Bethesda: The Endocrine Society, 1999;169.
198. Wex H, Vorwerk P, Mohnike K, et al. Elevated serum levels of IGFBP-2 found in children suffering from acute leukemia is accompanied by the occurrence of IGFBP-2 mRNA in the tumour clone. Br J Cancer 1998;78:515–520.
199. Petridou E, Dessypris N, Spanos E, et al. Insulin-like growth factor-I and binding protein-3 in relation to childhood leukemia. Int J Cancer 1999;80:494–496.
200. Wolk A, Mantzoros CS, Anderson S-O, et al. Insulin-like growth factor-I and prostate cancer risk: a population-based, case-control study. J Natl Cancer Inst 1998;90:911–915.
201. Chan JM, Stampfer MJ, Giovannucci E, et al. Plasma insulin-like growth factor-I and prostate cancer risk: a prospective study. Science 1998;279:563–566.
202. Hankinson SE, Willett WC, Colditz GA, et al. Circulating concentrations of insulin-like growth factor-I and risk of breast cancer. Lancet 1998;351:1393–1396.
203. Helle SI, Geisler S, Aas T, Paulsen T, Holly JM, Lonning PE. Plasma insulin-like growth factor binding protein-3 proteolysis is increased in primary breast cancer. Br J Cancer

2001;85:74–77.
204. Yu H, Spitz MR, Mistry J, Gu J, Hong WK, Xu X. Plasma levels of insulin-like growth factor-I and lung cancer risk: a case-control analysis. J Natl Cancer Inst 1999;91:151–156.
205. Ma J, Pollak MN, Giovannucci E, et al. Prospective study of colorectal cancer risk in men and plasma levels of insulin-like growth factor (IGF)-I and IGF-binding protein-3. J Natl Cancer Inst 1999;91:620–625.
206. Li BD, Khosravi MJ, Berkel HJ, et al. Free insulin-like growth factor-I and breast cancer risk. Int J Cancer 2001;91:736–739.
207. Cohen P, Clemmons DR, Rosenfeld RG. Does the GH-IGF axis play a role in cancer pathogenesis? GH & IGF Res 2000;10:297–305.
208. Baxter RC, Binoux MA, Clemmons DR, et al. Recommendations for nomenclature of the insulin-like growth factor binding protein superfamily. Endocrinol 1998;139:4036.
209. Hwa V, Oh Y, Rosenfeld RG. Insulin-like growth factor binding proteins: a proposed superfamily. Acta Pediatr Supp 1999;428:37–45.
210. Hwa V, Oh Y, Rosenfeld RG. The insulin-like growth factor-binding protein (IGFBP) superfamily. Endocr Rev 1999;20:761–787.
211. Yamanaka Y, Wilson EM, Rosenfeld RG, Oh Y. Inhibition of insulin receptor activation by insulin-like growth factor binding proteins. J Biol Chem 1997;272:30729–30734.
212. Vorwerk P, Yamanaka Y, Spagnoli A, Oh Y, Rosenfeld RG. Insulin and IGF binding by IGFBP-3 fragments derived from proteolysis, baculovirus expression and normal human urine. J Clin Endocrinol Metab 1998;83:1392–1395.

Edited by L. Matera and R. Rapaport

The Expression and Function of GH/IGF-I Receptors in the Immune System

ALFRED TENORE and GIULIANA VALERIO

Department of Pediatrics (DPMSC), University of Udine School of Medicine, Italy

ABSTRACT

In the last 70 years, an extensive body of literature has been published concerning the existence of a tight communicating network linking the neuroendocrine and immune systems through the presence of shared ligands and receptors. Much experimental evidence indicates that a similar communication link exists in the endocrine system between the GH/GH receptor/GH binding protein/IGF-I axis and the immune system, although the regulatory mechanisms differ somewhat.

Through endocrine, autocrine and paracrine mechanisms, GH and IGF-I contribute to regulate, in concert with cytokines, the development and function of the immune system. The age-related modifications occurring within the primary lymphoid tissues is likely to be influenced by the effect of GH and IGF-I on immune cells. Virtually all of the cells of the immune system are targets of GH and IGF-I actions, as has been demonstrated both by the expression of their specific receptors by radioligand assays and, more recently, by flow cytometry.

GH and IGF-I have stimulatory effects on the proliferation of immunocompetent cells and regulate several cellular and humoral functions. These growth factors have also been shown to have both thymopoietic and myelopoietic effects in immuno-deficient animals. Moreover, *in vitro* experiments indicate that these hormones are able to increase immunoglobulin secretion by normal and transformed B lymphocytes, to enhance the cytotoxic activity of natural killer cells, and to activate contact sensitivity reactions, graft rejections and graft versus host reactions.

Despite the pleiotropic effects of GH and IGF-I on the immune system, only minor alterations have been described in GH deficient subjects, probably due to a more pronounced redundancy in the hormonal control of the human immune system with respect to that in experimental animals.

Although much has been learned, many critical questions remain unanswered on the clinical importance of the interaction between GH, IGF-I and the immune system such as the potential role of these hormones as: (1) immunotherapeutic agents in clinical states of immunodeficiency, (2) immunomodulants in critically ill patients, and (3) enhancers of malignant hematopoiesis.

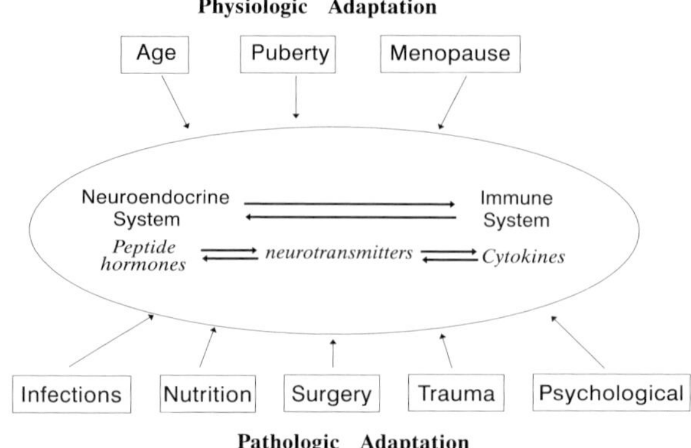

Figure 1. Interplay between neuroendocrine and immune systems and their adaptation to physiologic and pathologic conditions.

1. INTRODUCTION

A tight network of communication exists between the neuroendocrine and immune systems, involving shared ligands (peptide hormones, neurotransmitters and cytokines) and receptors in nervous, endocrine and immune tissues [1]. In order to maintain homeostasis, this complex system requires a very high level of integration. The interplay between the neuroendocrine and immune systems is associated with continuous adaptation to conditions which may be considered both physiological, such as those which depend on age and endocrine changes (e.g. puberty, menopause) and pathological (such as infectious, nutritional, surgical, traumatic, psychological factors) (Figure 1). Both physiological and pathological conditions may cause pronounced stress and require metabolic adaptation from the body. As a consequence, cytokines as well as hormones and neurotransmitters are important in eliciting the acute phase response and in regulating endocrine function and substrate metabolism [2]. Starting from the first observation made by Smith in 1930 [3] that hypophysectomized rats presented hypocellularity of primary lymphoid organs, a huge quantity of experimental work has been performed in subsequent years, which continued to show that endocrine-derived signals could regulate immune system development and function. It is now established that some hormones (e.g. ACTH, glucocorticoids, estrogens, progesterone and androgens) reduce immune responses, while others (GH, PRL, IGF-I and thyroid hormones) stimulate immune function.

The finding of hormone receptors on immune cells corroborates the existence of a functional role of their respective ligands. This review will focus on the expression and function of both the GH receptor (GH-R) and IGF-I receptor (IGF-I-R) in cells of the immune system under normal and pathological conditions.

2. THE GH/GH-R/GHBP/IGF-I AXIS AND THE IMMUNE SYSTEM.

Growth hormone, a protein of 191 amino acids, is mainly produced in the pituitary gland and is carried in the blood to target tissues, where it has multiple actions. Best known are the anabolic and metabolic effects exerted by GH, resulting in increased longitudinal bone growth and

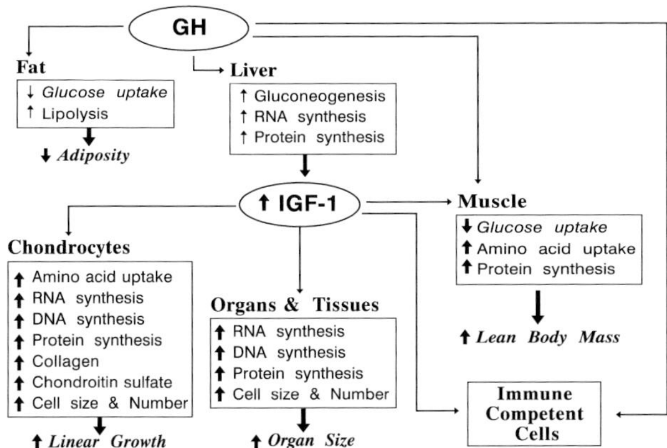

Figure 2. Anabolic and metabolic effects of GH and IGF-I in target cells.

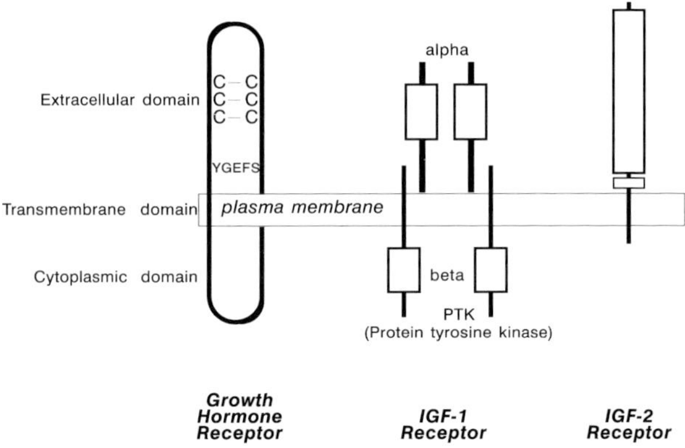

Figure 3. Schematic representation of the receptors for Growth hormone, IGF-I and IGF-II.

regulation of protein, lipid, carbohydrate, water and mineral metabolism [4] (Figure 2). Many other effects, such as those exerted on the regulation of immune functions in humans, are still poorly understood. Part of these effects are mediated by IGF-I, a small peptide of 70 amino acids presenting an extremely well conserved structure across various species, whose production mainly occurs in liver. It also seems possible that GH may directly act on target cells or induce local IGF-I production, with autocrine and paracrine effects.

Our understanding of the mechanism of action of GH has been greatly improved by the characterization of its receptor in 1987 from rabbit and human liver [5]. In previous studies, the identification of growth hormone receptor was based on measurements of specific, high affinity GH binding activity. The receptor was subsequently shown to be a member of the cytokine receptor superfamily which have in common the utilization of the JAK kinases, a group of non-receptor tyrosine kinases. The GH-R consists of a single chain of 638 amino acids comprising an 18 amino acid signal peptide, a 246 residue extracellular hormone-binding domain, a small 24 amino acid transmembrane domain and a 350 residue intracellular cytoplasmic domain [6] (Figure 3). The extracellular domain has a sequence motif characteristic of the hematopoietin receptor gene superfamily, which includes receptors for prolactin, IL-2, IL-3, IL-4, IL-5, IL-6,

IL-7, IL-9, IL-11, IL-13, CSF-I, G-CSF and erythropoietin. Members of this superfamily have four conserved cysteine residues and a tryptophan-serine-*any amino acid*-tryptophan-serine (WSXWS) motif in the extra-cellular region. The GH-R differs in that the WSXWS sequence is replaced by YGEFS (tyrosine-glycine-glutamate-phenylalanine-serine) [7–9]. Many of the full length receptors belonging to this family, including GH-R, exist in both a membrane-bound form and a soluble secreted form [10]. GH binding protein (GHBP) is the soluble form of the GH-R and corresponds to the extra-cellular portion of the membrane receptor [11] (Figure 3). In man, rabbit and most non-rodent species, GHBP is believed to be generated primarily by proteolysis of the membrane-bound GH-R [12], whereas in most rodent-species GHBP derives from an alternative splicing of a single primary transcript resulting in two distinct mRNA's [13,14].

The analysis of the three-dimensional crystal structure of the human GHBP-GH complex has shown that the hormone binds two receptor molecules, forming a dimer complex [15,16]. The binding of GH to its cell surface receptor rapidly promotes the binding of the JAK2 tyrosine kinase to the cytoplasmic portion of the GH receptor thus activating JAK2. Following activation of JAK2, several cellular proteins, including the cytoplasmic domain of the GH-R, undergo tyrosine phosphorylation. Activation of JAK2 appears to be a key mediator of signal transduction. Virtually all members of the cytokine receptor superfamily activate a family of transcription factors known as STATs (signal transducers and activators of transcription) with the transmission of a phosphorylated signal to the relevant nuclear transcription factor [17]. Alternatively, GH may directly act within the nucleus, generating a local transcriptional activating signal [18]. Following ligand-receptor binding, many cell surface receptors, including GH-R, undergo internalization and translocation to the nucleus where it may exert additional, although currently unknown, effects.

The IGF-I receptor is a transmembrane-spanning ligand-stimulated heterotetrameric tyrosine kinase, structurally and functionally homologous to the insulin receptor. The IGF-I receptor and the insulin receptor are both able to bind insulin and IGF-I, albeit at lower affinity for the heterologous ligand [19]. Both belong to the so-called "insulin receptor family" and are characterized by having clustered cysteines in the ligand-binding region and an intact kinase domain. These receptors are cleaved in the extra-cellular domain to give rise to a membrane-bound intracellular subunit (β) attached by disulfide bonds to an exclusively extra-cellular subunit (α) (Figure 3). The actual binding region of the receptor is centered around a single cysteine cluster that appears to recognize a common core structure in both IGF-I and insulin. Both hormones have similar metabolic effects, but IGF-I has additional effects promoting cell proliferation and differentiation, acting like other tyrosine kinase receptors, which are known to regulate growth and differentiation of hematopoietic cells. The IGF-I-R is different from the IGF-II receptor, in that the latter receptor is a single chain polypeptide with no tyrosine kinase activity, binds IGF-II with high affinity and IGF-I with very low affinity [20] and its role is not yet well defined post-natally (Figure 3).

Immune cells are able to produce GH [21,22], IGF-I [23–25] and IGF-binding proteins [25]. The existence of receptors for GHRH, somatostatin, GH and IGF-I on immune cells strongly indicates that endocrine, autocrine and paracrine mechanisms link the endocrine and the immune systems. However, there is evidence to indicate that the mechanisms that regulate the GH/IGF-I axis in immune cells differ from those in the endocrine system.

Numerous clinical and experimental studies indicate that GH and/or IGF-I participate, in concert with cytokines, in the development of lymphocytes in primary (thymus and bone marrow) and secondary (spleen, lymph nodes) lymphoid organs. These hormones also participate in the regulation of innate immunity (neutrophils, macrophages, natural killer) (Figure 4). In this chap-

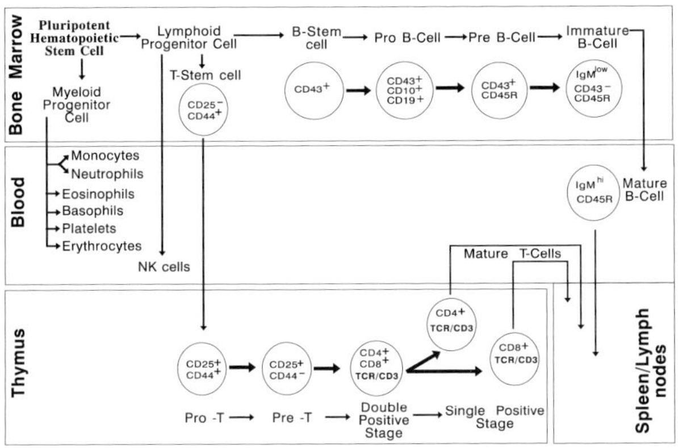

Figure 4. B and T cell development in bone marrow and thymus.

Legend: CD3: TCR-associated (transduces signals from TCR)

CD4: Helper T-cells (receptor for HLA class II antigens)

CD8: Cytotoxic T-cells (receptor for HLA class I antigens)

CD10: Common acute lymphoblastic leukemia antigen (cLLA)

CD19: Regulates B-cell activation

CD24: Heat-stable antigen (HAS) expression

CD25: IL-2 receptor α chain expression

CD43: Sialophorin; Leukosialin expression

CD44: Hyaluronate receptor expression

CD45R: Leukocyte common antigen -LCA (isoform of CD45; tyrosine phospahatase that regulates lymphocyte activation)

IgM and IgD: Surface immunoglobulin expression

TCR: T-cell receptor

Double positive and single positive stages are characterized by expression of TCR/CD3

ter we will treat the expression and function of GH-R and IGF-I-R in immune cells separately.

3. GROWTH HORMONE RECEPTORS AND FUNCTION ON IMMUNE CELLS

3.1. Circulating immune cells

It is now well established that GH-R is ubiquitously expressed on peripheral blood mononuclear cells (PBMC). The first demonstration of high affinity GH binding sites on immune cells was made by Lesniack et al. [26] on a human B cell lymphoma (IM-9) lymphocyte cell line, which has become the prototype for investigation of the GH/GH-R/IGF-I axis [27]. Subsequently, GH-R was also demonstrated on normal human PBMC [28], but results obtained with radioligand assays were very difficult to reproduce. The affinity constant of GH-R, calculated by Scatchard analysis, was similar in both studies ($1.3–1.5 \times 10^{-9}$ l/mol) and the number of receptors was 4000–7000 per cell. However, the real presence of GH-R on PBMC was obscured by the demonstration that hGH could bind to and induce signal transduction through the structurally related human PRL receptor [29,30]. With the use of a specific monoclonal antibody (mAb 263)

Table I Effects of GH and IGF-I on the immune system.

	GH	IGF-I
T- and B-lymphocytes	T-lymphocyte engraftment in immunodeficient mice Stimulates proliferation of T and B cell lines (IGF-I mediated) Stimulates immunoglobulin synthesis in EBV transformed B-cell lines Stimulates cytokine production	Stimulates proliferation of T and B cell lines Stimulates primary B-lymphopoiesis and B cell differentiation (from pro-B to pre-B) Mediates T cell chemotaxis Protects from apoptosis
Thymocytes	Weight gain of spleen and thymus in congenitally GH deficient mice Increases thymic cellularity in aged mice Stimulates IGF-I production	Weight gain of spleen and thymus in congenitally GH deficient mice
Natural Killers	Enhances cytotoxic activity in GH deficient children and normal adults	Enhances cytotoxic activity
Monocytes	Increases migration of circulating monocytes Priming for enhanced in vitro production of H_2O_2 in response to phorbol esters (non IGF-I mediated)	No effect on the priming of monocytes to release increased amounts of H_2O_2
Neutrophils	Enhances the proliferation of human myeloid progenitor cells and their maturation towards mature granulocytes Stimulates superoxide anion secretion and phagocytosis (non IGF-I mediated) Reverses the impaired neutrophil function in the elderly	Enhances the proliferation of human myeloid progenitor cells and their maturation towards mature granulocytes Stimulates superoxide anion secretion and phagocytosis

raised against the GH-R and shown to be unreactive against the PRL receptor, the presence of GH-R has been confirmed by cytofluorometry on both IM-9 cells [31] and human PBMC [32]. In these latter studies the use of fluorescein isothiocyanate (FITC)-conjugated anti GH-receptor antibody (mAb263) gave binding patterns similar to those obtained with the fluoresceinated ligand. The use of flow cytometry, with respect to radioligand assay, has the advantage of providing further information about the expression of the GH-R in different cell types within the heterogeneous leukocyte population. In fact, using two-color flow cytometry it was possible to demonstrate a differential expression of GH-R on immune cells. A higher GH-R expression was found on $CD20^+$ cells (B cells), corresponding approximately to that found on IM-9 cells, whereas $CD2^+$ cells (T cells, NK) expressed much lower amounts [32]. Similar data were

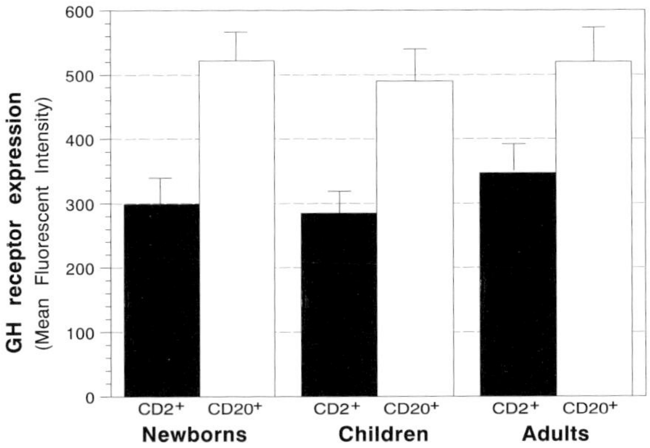

Figure 5. GH receptor expression (mean ± sd) on CD2+ (T and NK) and CD20+ (B) lymphocytes in normal newborns (n = 7; cord blood), prepubertal children (n=18; age range 3–11. years) and adults (n=13; age range 18–50. years) (modified from Ref. [42]).

reported by Rapaport et al. using FITC-GH [33]. These studies showed that GH-R was expressed in more than 90% of B-lymphocytes and monocytes and in only 2–20% of T cells. The differential GH-R expression on B-cells compared to T-cells could signify a preferential role of GH in the control of the differentiation and /or function of these cells. Studies of cDNA sequence have established that the GH-R from lymphocytes is identical to that reported in human liver and placenta [34].

Although GH-R is differently expressed on circulating immune cells, GH has been shown to influence cellular as well as humoral immunity [35]. With regard to T cell function, GH can stimulate DNA synthesis in human T-lymphocytes as well as the *in vitro* and *in vivo* proliferation of both phytohemagglutinin-activated normal and leukemic human T-lymphocytes. This latter effect, however, seems to be mediated by IGF-I [36]. Moreover, GH may have stimulatory effects on cytokine (IL-2, IL-6) production, enhances the cytotoxic activity of human natural killer (NK) cells [37,38] and activates contact sensitivity reactions, graft rejection and graft versus host reactions [39]. With regard to B cell function, GH may stimulate immunoglobulin synthesis and B-cell proliferation [40] as well as IGF-I synthesis by B-cell lines [41] (Table I).

The inaccessibility of target tissues of GH action has forced most clinical studies to rely on measurements of GHBP as an indirect index of GH-R expression. The use of lymphocytes, as a model of a GH target tissue, might represent the best non-invasive model to study GH-R status both in physiological and pathological conditions in man. In order to explore whether GH-R expression on PBMC followed an age-related pattern, cytofluorometric analysis was employed to compare the GH-R expression on PBMC from newborns, children and adults [42]. This study demonstrated that there was a higher expression of the GH-R on $CD20^+$ than on $CD2^+$ cells from PBMC of the three age groups studied. An age-related regulation of the GH-R was observed on $CD2^+$ cells, which reached the highest levels of expression in the adult (Figure 5). This finding is in agreement with other reports that indicate an age-dependent increase of serum GHBP, reaching highest levels in adult life [43,44]. A positive correlation between GHBP levels and GH-R expressed on B cells was found by Rapaport et al. [33]. However, the analysis in this study was restricted to a small and heterogeneous group of subjects that included 8 patients presenting with various growth disorders (precocious puberty, GH deficiency, Turner syndrome) and only

2 normal subjects.

However, according to other investigations, the percentage of PBMC bearing GH-R was significantly lower in the 20–40 year age group than that found in subjects between 12–20 and > 60 years, while mean fluorescent intensity and consequently the number of GH-R expressed per cell, did not change [45]. In this study a negative correlation was found between GH-R expression and plasma levels of GHBP suggesting that the regulation of GH-R expression differs between liver and blood cells. Since there is evidence for up to 7 alternative exon 1 sequences for the hGH-R [14], each with different regulatory elements, different regulatory responses of GH-R may be present in different tissues, according to the specific physiological role of the cell.

Mechanisms responsible for GH-R regulation on immune cells are still debated. The number of GH-R receptors was shown to be down-regulated by hGH in a dose dependent manner on IM-9 cells [46,47], on the contrary maximal binding capacity was obtained by pre-incubation of mononuclear cells in hGH free-medium [28]. It seems reasonable to assume that GH-R expression may depend on locally produced cytokines and GH in an autocrine/paracrine model, since it has been demonstrated in mice that GH-R expression was dependent on the proliferative phase of the cell cycle [48].

Only a few investigations directly explored GH-R expression in pathological conditions of the somatotropic axis. Stewart et al. [49] in 1982 performed GH binding studies on lymphocytes from 18 GH deficient children, demonstrating an increased binding after GH injection *in vivo*, in contrast to what was reported by Lesniack et al. [47] on IM-9 cells *in vitro*. A decreased expression of GH-R on B cells has been reported after GH replacement treatment [42,50,51]. No substantial difference was found in human GH-R mRNA detected in lymphocytes of patients with GH-deficiency or acromegaly by the reverse transcription–polymerase chain reaction (RT-PCR) method. The results of these studies may be questionable because they are semi-quantitative and performed on a limited number of subjects [52], and no correlation between GH-R gene expression and circulating GH levels was found [53].

GH-R expression on T cells and NK cells ($CD2^+$) was found to be higher in short children than in normal stature children or in girls with Turner syndrome [42]. However, no substantial difference was found with regard to the mean fluorescent intensity level of GH-R expression on B cells ($CD20^+$) [42] or the percentage of B-cells expressing GH-R [33] in clinical conditions affecting growth, such as idiopathic short stature, GH deficiency and Turner syndrome.

Experimental conditions of GH deficiency have been associated with disorders of the immune and hematopoietic systems (reviewed in 54), while only minor alterations have been described in GH deficient subjects. In untreated GH deficient subjects T-lymphocytes subsets were normal, while the number of B-cells was either increased, normal or reduced; interleukin-2 production and NK activity were variously reported as normal or reduced [55–57]. GH treatment in clinical conditions of GH deficiency has been associated with no alterations at all or only with a transient decrease in circulating T and B cells and variable effects in the lymphoproliferative responses to mitogens. GH restored the decreased NK activity in GH-deficient children and adults [37,38]. The finding of these slight and not consistent effects of GH deficiency in the human immune system may be explained by the hypothesis that, in contrast with animal models, there is a more pronounced redundancy in hormonal control in humans, probably associated with an intact paracrine/autocrine axis at the lymphoid level. Therefore, IGF-I expression may be less dependent on GH levels in humans or, alternatively, a vicarious role of the IGF-II factor may occur.

3.2. Primary lymphoid tissues: thymus and bone marrow

GH exerts an important role on the development of immune cells in thymus, bone marrow, spleen and lymph nodes. The pattern of GH exposure seems to be responsible for different effects observed at the tissue level: experimental animals lacking GH show leukopenia and reduced size of all lymphoid organs [58], while transgenic mice, who present with high GH levels and a continuous pattern of exposure to this hormone have enlarged lymphoid organs, particularly the spleen [35]. The thymus plays a central role in the development of cellular mediated immune functions. T cell precursors migrate from bone marrow to the thymus, where they undergo a complex process of maturation (Figure 4). Within the thymus, T cell clones specific for exogenous antigens are selectively stimulated, whereas auto-reactive T cell clones are suppressed. Thymic microenvironment, represented by thymic epithelial cells and bone marrow-derived macrophages and dendritic cells, influences the early events of T-cell differentiation by producing thymic hormones and cytokines (IL-7). Physiologically, the thymus reaches its maximal size at puberty and involutes by 45–50 years. Interestingly, this growth pattern follows the secretion pattern of the GH/IGF-I axis, with IGF-I values peaking in puberty and declining with advancing age [59,60]. Extensive data in the literature confirm the important role that GH treatment exerts on the proliferation of thymocytes. Thymic involution normally found in aged rats as well as in hypophysectomized rats and Snell Bagg dwarf mice (that are congenitally deficient in GH, PRL and TSH) was reversed by the implantation of GH3 pituitary adenoma cells [61,62]. GH treatment had thymopoietic effects in vivo in dwarf mice, causing the reappearance of the $CD4^+/CD8^+$ double positive-cells within the gland. The presence of GH-R on thymocytes unequivocally confirms a causal relationship. Initially, GH binding was demonstrated in murine thymic epithelial cells [63], which secrete thymic hormones and cytokines and influence the differentiation of thymocytes. It has been demonstrated that GH as well as PRL are able to up-regulate thymulin secretion *in vivo*, whereas low thymulin levels characterize clinical conditions associated with GH deficiency [64]. Subsequently, the presence of GH-R mRNA was also demonstrated in rat and human thymus (RT-PCR and Southern analysis, in situ hybridization), both in thymic epithelial cells and thymocytes [65–67]. Finally, Gagnerault et al. [68] demonstrated the expression of GH-R on murine thymocytes by cytofluorometry, while De Mello-Coelho et al. [67], using the biotinylated anti-GH receptor mAb 263, found that GH-R was predominantly expressed by the most immature thymocytes ($CD3^-CD4^-CD8^-CD19^-CD34^+CD2^+$ cells). In addition to these data, the finding that thymocytes produce GH reinforces the assumption for the existence of a paracrine GH-mediated lympho-epithelial interplay in the differentiation of T cells, which could also be mediated by IGF-I [69]. Thymocytes produce IGF-I following GH stimulation [64]. Human GH has significant thymopoietic and myelopoietic effects in immunodeficient animals, such as the severe combined immunodeficiency mouse and the azidothymidine treated young mouse [70,71].

In humans the other primary lymphoid organ is bone marrow, which, apart from the formation of the precursors of the T-cells, is specifically responsible for the production of erythroid, myeloid and B cells, all of which express GH-R (Figure 4). Murphy et al. [72] reported a reduced frequency of cells of the B lineage ($CD45R^+$) in the bone marrow of dwarf mice. GH and IGF-I enhanced the in vitro proliferation of human myeloid progenitor cells and their maturation towards mature granulocytes [73]. This effect required the presence of marrow adherent cells and appeared to be mediated by IGF-I receptors, since it was inhibited by the addition of anti-IGF-I receptor monoclonal antibody.

With regard to the function of other bone marrow derived immune cells, GH may affect the activity of polymorphonuclear phagocytes, since the "oxidative burst" of these cells was strongly affected by the state of GH secretion, being reduced in GH deficient children and increased in acromegalic patients [74]. Furthermore the impaired neutrophil function in the elderly was reversed by GH stimulation *in vitro* [74]. GH has been shown to induce priming of superoxide anion production of macrophages and neutrophils, which is necessary for phagocytosis [30,75]. In hypophysectomized rats infected with lethal *Salmonella typhimurium*, GH treatment enabled macrophages to generate reactive oxygen intermediates enhancing the survival of these rats [76]. In man, the administration of rh-GH significantly increased migration of circulating monocytes *in vivo* [74], while it primed monocytes to release increased amounts of H_2O_2 *in vitro* [77]. The latter effect was not mediated through the production of endotoxin, IFN-γ or IGF-I. Monocytes specifically bound radiolabeled GH with low affinity and contained mRNA for GH-R [77,78]. Since human GH may also bind to the human PRL receptor [30], it is still unclear whether its role as a macrophage-activating factor is exerted directly through its own receptor or through the PRL receptor.

4. IGF-I RECEPTOR EXPRESSION AND FUNCTION ON IMMUNE CELLS

4.1. Circulating immune cells

The discovery of the binding of insulin on resting and activated lymphocytes [79,80] lead immunologists to consider the insulin receptor as a universal marker for lymphocyte activation. Subsequently it was demonstrated that insulin could bind to IGF-I receptors as well, although with less affinity. The specific binding of ^{125}I-somatomedin, distinct from insulin, to human mononuclear cells was first reported in 1977 [81]. The finding that depletion of monocytes (> 90%) resulted in only a 38% decrease of the IGF-I binding, while their enrichment by 50% increased it of only 18% indicated that monocytes were not the unique population being able to bind IGF-I. Subsequently, it was shown that IGF-I bound specifically to both resting and PHA-activated T-lymphocytes with high affinity (K_d 1.2×10^{-10} M) and that the receptor sites/cell were higher in activated than resting cells (330 versus 45, respectively) [82]. Peak receptor number occurred at the same time as maximal thymidine incorporation [83]. Stuart et al. [84], using two-color flow cytometry, found specific binding of monoclonal antibodies directed against the type I IGF-R (αIR3) or the insulin receptor (αIR1) on nearly all monocytes and B-lymphocytes, but on only 2% of T-lymphocytes. On the contrary, Koojiman et al. [85] reported a relatively high number of receptors on monocytes, natural killer cells and T-helper cells (CD4$^+$) an intermediate number of receptors on T-suppressor cells (CD8$^+$) cells and a relatively low number on B-cells. The results were confirmed by binding studies with ^{125}I-IGF-I to purified subpopulations of PBMC. Although the discrepancy between these results is unclear, a possible explanation may reside with the fact that Koojiman et al. claimed that they employed a different methodology in the staining technique, in that they used whole blood, whereas Stuart et al. used purified PBMC. The increased expression of IGF-I receptor mRNA on resting and activated human lymphocytes has also been confirmed by different studies [25,86], while the expression of mRNA for IGF occurred only during cellular activation, indicating that at variance with IGF-I production, the IGF-I receptors are constitutively expressed, independently of the functional status of the cell. In order to explain whether IGF-I-R expression was dependent on different stages of activation and maturation of T lymphocytes, the binding of mAbαIR3 on naive, memory and antigen-activated T cells was investigated by flow cytometry [87]. It appeared that 87% of the naive subpopulations CD4$^+$ CD45RA$^+$ cells and 6% of the CD8$^+$ CD45RA$^+$ were αIR3$^+$ whereas only 37% of the memory CD4$^+$ CD45RO$^+$ and 38% of the CD8$^+$ CD45RO$^+$ bound αIR3. The demonstration that a down-regulation of the IGF-I receptor occurs early during the activation process and is increased by IGF-I [88] contributed to clarify the discrepancies found among studies. The different regulation of IGF-I-R during T cell proliferation and the different expression found in T, B and monocytes [82,87] might be explained by the different experimental conditions used. Other studies, which explored the IGF-I-R characteristics on cells of different lineages (erythrocytes vs. PBMC), demonstrated a positive correlation between IGF-I-R and free IGF-I levels only for erythrocytes and no correlation with total IGF-I, insulin and IGFBP-1 levels for both lineages [89].

As it has been already reported above, many of the immunological effects exerted by GH on leukocytes are shared by IGF-I, even though part of these functions are not necessarily mediated by autocrine/paracrine IGF-I production. IGF-I induces chemotaxis and thymidine incorporation in both resting and activated T-cells, augments basal colony formation of normal, mitogen stimulated human T-lymphocytes, virally transformed T-lymphoblasts and EBV-transformed B-lymphoblast cell lines [90,91] (Table I). IGF-I participates in the maturation of B-cells by stimulating the transition of CD45R$^-$ precursors to CD45R$^+$/cμ$^+$ cells [92]; moreover, in association with IL-7 it acts as a B-cell proliferation factor [93].

IGF-I has a marked anti-apoptotic effect in several tissues and cell types, including the hematopoietc system [94]. Enhanced cell survival in many cell types has been associated with an increased expression of IGF-I receptors [95]. Stimulation through the CD28 receptor, whose main role is supposed to be the enhancement of T cell survival following activation, transiently increased the expression of IGF-I receptors on T cells, providing them with essential survival signals. Antibodies that block signaling through the IGF-I receptor decreased the survival of T cells activated through the T cell receptor and CD28 and enhanced susceptibility to Fas-induced apoptosis [96].

4.2. Primary lymphoid tissues: thymus and bone marrow

Analogously to GH, IGF-I treatment increases thymus and spleen size [97] in GH deficient animals. Moreover, the high IGF-I levels obtained in transgenic mice produce selective lymphoid organomegaly [98]. IGF-I may influence T cell development in human thymus. In fact, IGF-I-Rs have been detected on human thymocytes by radioligand binding (Kd 0.12±0.01, 257±28 receptors/cell) and flow cytometric analysis [99]. The number of IGF-I-R per cell was constant in three different samples derived from children aged 10 days, 2 months and 5 years. The immature CD4$^-$CD8$^-$ cells expressed 3-4 times more receptors per cell compared with CD4$^+$CD8$^+$, CD4$^+$CD8$^-$, and CD4$^-$CD8$^+$ cells.

The expression of IGF-I-R has also been reported in neutrophils [100] and basophils [101]. Analogously to GH, IGF-I stimulates neutrophils to secrete superoxide anions, but unlike GH it is not able to prime monocytes for enhanced production of hydrogen peroxide in response to phorbol 12-myristate 13-acetate [77,102].

Studies on the expression of IGF-I receptor have also been performed in clinical conditions of GH deficiency by analyzing the IGF-I specific binding both on leukocytes after 4 days of GH treatment [103] and on erythrocytes after 6 months of hormonal therapy [104]: no change in affinity was shown in either case. After 6 months of rh-GH treatment, IGF-I receptor expression on erythrocytes was inversely correlated with increased serum levels of IGF-I (r – 0.88, $p < 0.001$) [104]. In clinical conditions of GH resistance (i.e. Laron Syndrome) increased IGF-I receptor expression was demonstrated on lymphocytes [105], erythrocytes [106] and neutrophils [107], as a compensatory response to reduced circulating IGF-I levels.

Given the stimulatory effect of both GH and IGF-I on immune cells, the potential role of these hormones as immunotherapeutic agents in clinical states of immunodeficiency, such as HIV-1 infection [108], or as immunomodulators in critically ill patients [109] is under investigation. Apart from this effect, their role is nevertheless fundamental as anabolic and stress-modulating agents in most cells, including those of the immune system.

4.3. GH and IGF-I in the regulation of malignant hematopoiesis

GH and IGF-I may also have a possible effect in enhancing malignant hematopoiesis. GH has been found to promote proliferation of leukemic clones and IGF-I was found to be mitogenic for several malignant cell lines [110,111]. The expression of IGF-I-R has been reported in nearly all neoplastic immune cells and in particular, myeloid leukemic cells [112], erythroleukemic cell line K562 [113], human leukemic T-lymphoblasts [114] and multiple myeloma and B lymphoblastoid cell lines [115]. A decrease in the number and affinity of IGF-I-Rs usually occurs during the differentiation of human promyelocytic leukemia cell lines (HL-60) to macrophage-like cells [116]. Children with acute leukemia presented a significantly lower number of high affinity IGF-I binding sites on circulating mononuclear cells in comparison to normal children [117]. However, the peripheral blast cells of these patients expressed higher numbers of IGF-I binding sites than peripheral lymphocytes and monocytes (536 ± 98.6 vs. 254 ± 43.6, $p = 0.02$). Future research is still required to understand the role of GH and IGF-I in the occurrence or progression of neoplastic hematopoietic disorders. According to two recent surveys on the safety and efficacy of GH treatment in GH-deficient patients, the incidence of leukemia in GH-treated patients without risk factors is not greater than that in the general population aged 0–15 yr [118,119]. The conclusions reached in these surveys indicate that a possible increased occurrence of leukemia with GH treatment appears to be limited to patients with known risk factors.

5. SUMMARY AND CONCLUSIONS

Growth hormone, released from the pituitary gland stimulates IGF-I production in the liver. Secretory balance between these hormones is preserved through a feed-back loop, since IGF-I levels modulate the amount of GH secreted. This was the extent of conventional endocrinology not too long ago. Today we know that GH and IGF-I are virtually made in every tissue in the body and what has become more fascinating is the finding of receptors for these hormones on immunologic competent cells, opening a communication link between the immune and endocrine systems.

It is now known that the immune, endocrine and nervous systems communicate with each other in a bi-directional fashion through a network of molecules which collectively produce a coordinated response to immune challenges. This bi-directional communication occurs as a result of the immune and neuroendocrine systems sharing a common set of hormones as well as receptors. In fact, the molecular characterization of shared ligands, receptors and second messengers, provides evidence on the structural and functional basis of this immune-neuroendocrine interaction.

GH and IGF-I receptors are widely distributed in human and animal tissues. In the immune system, GH (and prolactin) is required for bone marrow function, for growth of the thymus gland

and spleen and for lymphocyte proliferation and differentiation. In the presence of these hormones, bone marrow can be stimulated to produce lymphocytes some of which migrate and differentiate in the thymus. Although humans with GH deficiency have been found to have no clinically significant evidence of immune defects, rodents with GH deficiency demonstrate abnormal cellularity of the thymus and bone marrow as well as reduced T-cell function and natural killer cell activity. These immune defects in the GH deficient rodent are abolished by the administration of exogenous GH. In GH deficient humans, many immune parameters are normal and unaffected by GH treatment; however, natural killer cell activity which is decreased in GH deficient individuals is restored to normal with GH treatment. Furthermore, the association between GH deficiency and immune malfunction, primarily due to B cell involvement, in children with X-linked hypogammaglobulinemia and the correction of some of their immune deficits with GH treatment, indirectly substantiates a causal relationship.

Studies *in vivo* and *in vitro* have shown that GH can influence various aspects of thymocyte migration and differentiation, probably via an intrathymic autocrine circuit involving IGF-I, as well as interacting (directly or via IGF-I) with the immune system through specific high-affinity receptors on monocytes and B- and T-lymphocytes. IGF-I enhances bone marrow B-cell proliferation and may be involved in the stimulation of both T and B lymphopoietic organs. Human thymic cells have also been shown to produce GH as well as to express its transcription factor.

Although much information has and continues to be accumulated on the interactions between GH, IGF-I and the immune system, many important questions remain, at least for the time being, unanswered: (a) How does the information obtained from *in vitro* and animal studies relate to the human? (b) What exactly is the role of GH and IGF-I in immune activation? (c) What is the clinical importance of the interaction between GH, IGF-I and the immune system?

REFERENCES

1. Blalock JE. The syntax of immune-neuroendocrine communication. Immunol Today 1994;504–510.
2. Mandrup-Poulsen T, Nerup J, Reimers JI, et al. Cytokines and the endocrine system. I. The immunoendocrine system network. Eur J Endocrinol 1995;660–671.
3. Smith PE. The effect of hypophysectomy upon the involution of the thymus in the rat. Anat Rec 1930;119–129.
4. Isaksson OGP, Edén S, Jansson J-O. Mode of action of pituitary growth hormone on target cells. Ann Rev Physiol 1985;483–499.
5. Leung DW, Spencer SA, Cachianes G, Waters MJ, Wood WI. Growth hormone receptor and serum binding protein: purification, cloning and expression. Nature 1987;537–543.
6. Postel-Vinay MC, Finidori J, Goujon L, et al. Molecular biology of the growth hormone receptor. Endocrin Metab 1994 (Suppl A);3–7.
7. Kelly PA, Djiane J, Postel-Vinay MC, Edery M. The prolactin/growth hormone receptor family. Endocr Rev 1991;235–251.
8. D' Andrea AD, Fasman GD, Lodish HF. A new hematopoietic growth factor receptor superfamily: structural features and implications for signal transduction. Curr Opin Cell Biol 1990;648–651.
9. Bazan JF. Structural design and molecular evolution of a cytokine receptor superfamily. Proc Natl Acad Sci USA 1990;6934–6938.
10. Baumann G, Shaw MA, Winter RJ. Absence of the plasma growth-hormone binding

protein in Laron-type dwarfism. J Clin Endocrinol Metab 1987;814–816.
11. Spencer SA, Hammonds RG, Henzel WJ, Rodriguez H, Waters MJ, Wood WI. Rabbit liver growth hormone receptor and serum binding protein; purification, characterization and sequence. J Biol Chem 1988;7862–7867.
12. Sotiropoulos A, Goujon L, Simonin G, Kelly PA, Postel-Vinay MC, Finidori J. Evidence for generation of the growth hormone-binding protein through proteolysis of the growth hormone membrane receptor. Endocrinology 1993;1863–1865.
13. Baumbach WR, Horner DL, Logan JS. The growth hormone-binding protein in rat serum is an alternatively spliced form of the rat growth hormone receptor. Gene Dev 1989;1199–1205.
14. Pekhletsky R, Chernov BK, Rubtsov PM. Variants of the 5'-untranslated sequence of human growth hormone receptor mRNA. Mol Cell Endocrinol 1992;103–109.
15. De Vos AM, Ultsch M, Kossiakoff AA. Human growth hormone and extracellular domain of its receptor: crystal structure of the complex. Science 1992;306–312.
16. Fuh G, Cunningham BC, Fukunaga R, Nagata S, Goeddel DV, Wells JA. Rational design of potent antagonists to the human growth hormone receptor. Science 1992;1677–1680.
17. Argetsinger LS, Campbell GS, Yang X, et al. Identification of JAK2 as a growth hormone receptor-associated tyrosine kinase. Cell 1993;237–244.
18. Waters MJ, Rowlinson SW, Clarkson RW, et al. Signal transduction by the growth hormone receptor. Proc Soc Exp Biol Med 1994;216–220.
19. Siddle K, Soos MA, Field CE, Nave BT. Hybrid and atypical insulin/insulin like growth factor-1 receptors. Horm Res 1994;56–65.
20. LeRoith D, Werner H, Beitner-Johnson D, Roberts CT. Molecular and cellular aspects of the insulin-like growth factor I receptor. Endocr Rev 1995;143–163.
21. Varma S, Sabharwal P, Sheridan JF, Malarkey WB. Growth hormone secretion by human peripheral blood mononuclear cells detected by an enzyme-linked immunoplaque assay. J Clin Endocrinol Metab 1993;49–53.
22. Wu H, Devi R, Malarkey WB. Localization of growth hormone messenger ribonucleic acid in the human immune system: a clinical research center study. J Clin Endocrinol Metab 1996;1278–1282.
23. Abbound SL, Bethel CR, Aron SL. Secretion of insulin-like growth factor I and insulin-like growth factor -binding proteins by murine bone marrow stromal cells. J Clin Invest 1991;470–475.
24. Yamada M, Hato F, Kinoshita Y, Tominaga K, Tsuji Y. The indirect participation of growth hormone in the thymocyte proliferation system. Cell Mol Biol 1994;111–121.
25. Nyman T, Pekonen F. The expression of insulin-like growth factors and their binding proteins in normal human lymphocytes. Acta Endocrinol 1993;168–172.
26. Lesniak MA, Gorden P, Roth J, Gavin JR. Binding of ^{125}I-human growth hormone to specific receptors in human cultured lymphocytes. J Biol Chem 1974;1661–1667.
27. Yang Y, Hoeflich A, Kessler U, et al. Human IM-9 lymphoblasts as a model of the growth hormone-insulin-like growth factor axis: gene expression, and interactions of ligands with receptors and binding proteins. Regul Pept. 1993;41–53.
28. Kiess W, Butenandt O. Specific growth hormone receptors on human peripheral mononuclear cells: reexpression, identification and characterization. J Clin Endocrinol Metab 1985;740–746.
29. Cunningham BC, Bass G, Fuh G, Wells JA. Zinc mediation of the binding of human growth hormone to the human prolactin receptor. Science 1990;1709–1712.

30. Fu YK, Arkins S, Fuh G, et al. Growth hormone augments superoxide anion secretion of human neutrophils by binding to the prolactin receptor. J Clin Invest 1992;451–457.
31. Barnard R, Budensen PG, Rylatt DB, Waters MJ. Evidence from the use of monoclonal antibody probes for structural heterogeneity of the growth hormone receptor. Biochem J 1985;459–468.
32. Badolato R, Bond HM, Valerio G, et al. Differential expression of surface membrane growth hormone receptor on human peripheral blood lymphocytes detected by dual fluorochrome flow cytometry. J Clin Endocrinol Metab 1994;984–990.
33. Rapaport R, Sills IN, Green L, et al. Detection of human growth hormone receptors on IM-9 cells and peripheral blood mononuclear cell subsets by flow cytometry: correlation with growth hormone-binding protein levels. J Clin Endocrinol Metab 1995;2612–2619.
34. Hansen B, Hjorty S, Welinder BS, Skriver L, De Meyts P. The growth hormone (GH)-binding protein cloned from human IM-9 lymphocytes modulates the down-regulation of GH receptors by 22- and 20 kilodalton human GH in IM-9 lymphocytes and the biological effects of the hormone in Nb2 lymphoma cells. Endocrinology 1993;2809–2817.
35. Clark R. The somatogenic hormones and insulin-like growth factor-1: stimulators of lymphopoiesis and immune function. Endocr Rev 1997;157–175.
36. Geffner ME, Bersch N, Lippe BM, Rosenfeld RG, Hintz RL, Golde DW. Growth hormone mediates the growth of T-lymphoblast cell lines via locally generated insulin-like growth factor I. J Clin Endocrinol Metab 1990;464–469.
37. Bozzola M, Valtorta A, Moretta A, Cisternino M, Biscaldi I, Schimpff RM. In vitro and in vivo effect of growth hormone on cytotoxic activity. J Pediatr 1990;596–599.
38. Bidlingmaier M, Auernhammer CJ, Feldmeier H, Strasburger CJ. Effects of growth hormone and insulin-like growth factor I binding to natural killer cells. Acta Paediatr Suppl 1997;80–81.
39. Berczi I. Pituitary hormones and immune function. Acta Paediatr Suppl 1997;70–75.
40. Yoshida A, Ishioka C, Kimata H, Mikawa H. Recombinant human growth hormone stimulates B cell immunoglobulin synthesis and proliferation in serum-free media. Acta Endocrinol 1992;524–529.
41. Merimee TJ, Grant MB, Broder CM, Cavalli-Sforza LL. Insulin-like growth factor secretion by human B-lymphocytes: a comparison of cells from normal, pygmy subjects. J Clin Endocrinol Metab 1989;978–984.
42. Valerio G, Bond HM, Badolato R, et al. Expression of growth hormone receptor by peripheral blood lymphocytes in children: evaluation in clinical conditions of impaired growth. Clin Endocrinol 1997;329–335.
43. Baumann G, Shaw MA, Amburn K. Regulation of plasma growth hormone-binding proteins in health and disease. Metabolism 1989;683–689.
44. Holl RW, Snehotta R, Siegler B, Scherbaum W, Heinze E. Binding protein for human growth hormone: effects of age and weight. Horm Res 1991:190–197.
45. Bresson JL, Jeay S, Gagnerault MC, et al. Growth (GH) hormone and prolactin receptors in human peripheral blood mononuclear cells: relation with age and GH-binding protein. Endocrinology 1999;3203–3209.
46. Smal J, Closset J, Hennen G, De Meyts P. Receptor-binding and down-regulatory properties of 22000-M human growth hormone and its natural 20000-M variant on IM-9 human lymphocytes. Biochem J 1985;283–289.
47. Lesniack MA, Roth J. Regulation of receptor concentration by homologous hormone. Effect of human growth hormone on its receptor in IM-9 lymphocytes. J Biol Chem

1976;3720–3729.
48. Dardenne M, de Mello-Coelho V, Gagnerault MC, Postel-Vinay MC. Growth hormone receptors and immunocompetent cells. Ann NY Acad Sci 1998;510–517.
49. Stewart C, Clejan S, Fugler L, Cheruvanky T, Collipp PJ. Growth hormone receptors in lymphocytes of growth hormone-deficient children. Arch Biochem Biophys 1983;309–313.
50. Rapaport M, Oleske J, Ahdieh H, Solomon S, Delfaus C, Denay T. Suppression of immune function in growth hormone deficient children during treatment with human growth hormone. J Pediatr 1986;434–439.
51. Petersen BH, Rapaport R, Henry DP, Huseman C, Moore WV. Effect of treatment with biosynthetic human growth hormone (GH) on peripheral blood lymphocyte populations and function in growth hormone deficient children. J Clin Endocrinol Metab 1990;1756–1760.
52. Kaji H, Ohashi S, Abe H, Chihara K. Regulation of the growth hormone (GH) receptor and GH-binding protein mRNA. Proc Soc Exp Biol Med 1994;257–262.
53. Fukuda I, Hizuka N, Takano K, et al. Growth-hormone (GH) receptor and IGF-I receptor messenger RNA expression in peripheral lymphocytes from patients with GH deficiency and acromegaly. Endocrin J 1998: S109–S112.
54. Dorshkind K, Horseman ND. The roles of prolactin, growth hormone, insulin-like growth factor-I, and thyroid hormones in lymphocyte development and function: insights from genetic models of hormone and hormone receptor deficiency. Endocr Rev 2000:292–312.
55. Kiess W, Malozowski S, Gelato M, et al. Lymphocyte subset distribution and natural killer activity in growth hormone deficiency before and during short term treatment with growth hormone releasing hormone. Clin Immunol Immunopathol 1988;85–94.
56. Kiess W, Doerr H, Butenand O, Belohradsky BH. Lymphocyte subsets and natural-killer activity in growth hormone deficiency. New Engl J Med 1986;321.
57. Casanova S, Repellin AM, Schimpff RM. Production of Interleukin-1a and Interleukin-2 by mononuclear cells from children with growth delay in relation to the degree of growth hormone deficiency: effects of substitutive treatment. Horm Res 1990;209–214.
58. Nagy E, Berczi I. Pituitary dependence of bone marrow function. Br J Hematol 1989;457–462.
59. Corpas E, Harman SM, Blackman MR. Human growth hormone and human aging. Endocr Rev 1993;20–39.
60. Kelley KW, Arkins S, Minshall C, Liu Q, Dantzer R. Growth hormone, growth factors and hematopoiesis. Horm Res 1996;38–45.
61. Kelley KW, Brief S, Westly HJ, et al. GH_3 pituitary adenoma cells can reverse thymic aging in rats. Proc Natl Acad Sci USA 1986;5663–5667.
62. Li YM, Brunke DL, Dantzer R, Kelley KW. Pituitary epithelial cell implants reverse the accumulation of $CD4^-CD8^-$ lymphocytes in thymus glands of aged rats. Endocrinology 1992;2703–2709.
63. Ban E, Gagnerault MC, Jammes H, Postel-Vinay MC, Haour F, Dardenne M. Specific binding sites for growth hormone in cultured mouse thymic epithelial cells. Life Sci 1991;2141–2148.
64. Dardenne M, Savino W. Control of thymus physiology by peptidic hormones and neuropeptides. Immunol Today 1994;518–523.
65. Mertani HC, Delehaye-Zervas MC, Martini JF, Postel-Vinay MC, Morel G. Localization of growth hormone receptor messenger RNA in human tissues. Endocrine 1995;135–142.

66. Maggiano N, Plantelli M, Ricci R, Larocca LM, Capelli A, Ranelletti FO. Detection of growth hormone-producing cells in human thymus by immunohistochemistry and non radioactive *in situ* hybridization. J Histochem Cytochem. 1994;1349–1354.
67. De Mello-Coelho V, Gagnerault MC, Souberbielle JC, et al. Growth hormone and its receptor are expressed in human thymic cells. Endocrinology 1998;3837–3842.
68. Gagnerault MC, Postel-Vinay MC, Dardenne M. Expression of growth hormone receptor in murine lymphoid cells analyzed by flow cytofluorometry. Endocrinology 1996;1719–1726.
69. Sabharwal P, Varma S. Growth hormone synthesized and secreted by human thymocytes acts via insulin-like growth factor I as an autocrine and paracrine growth factor. J Clin Endocrinol Metab 1996;2663–2669.
70. Murphy WJ, Durum SK, Longo DL. Human growth hormone promotes engrafment of murine or human T cells in severe combined immunodeficient mice. Proc Natl Acad Sci USA 1992;4481–4485.
71. Murphy WJ, Tsarfaty G, Longo DL. Growth hormone exerts hematopoietic growth-promoting effects in vivo and partially counteracts the myelosuppressive effects of azidothymidine. Blood 1992;1433–1447.
72. Murphy WJ, Durum SK, Anver MR, Longo DL. Immunologic and hematologic effects of neuroendocrine hormones. Studies on dw/j dwarf mice. J Immunol 1992;3799–3805.
73. Merchav S, Tatarsky I, Hochberg Z. Enhancement of human granulopoiesis in vitro by biosynthetic insulin-like growth factor I/somatomedin C and human growth hormone. J Clin Invest 1988;791–797.
74. Wiedermann CJ, Reinisch N, Kahler C, Braunsteiner H. Regulation of myeloid phagocyte development and function by growth hormone: a review. J Pediatr Endocrinol Metab 1993;85–91.
75. Edwards III CK, Ghiasuddin SM, Schepper JM, Yunger L, Kelley K. A newly defined property of somatotropin: priming of macrophages for production of superoxide anion. Science 1988;769–771.
76. Edwards III CK, Ghiasuddin SM, Yunger LM, et al. In vivo administration of recombinant growth hormone or gamma interferon activates macrophages: enhanced resistance to experimental *Salmonella typhimurium* infection is correlated with generation of reactive oxygen intermediates. Infect Immun 1992;2514–2521.
77. Warwick-Davies J, Lowrie DB, Cole PJ. Growth hormone is a human macrophage activating factor. Priming of human monocytes for enhanced release of H_2O_2. J Immunol 1995;1909–1918.
78. Jafari P, Khansari DN. Detection of somatotropin receptors on human monocytes. Immunol Lett 1990;199–202.
79. Gavin JR, Buell DN, Roth J. Water soluble insulin receptors from human lymphocytes. Science 1972;168–169.
80. Helderman JH, Strom TB. Specific insulin binding site on T and B lymphocytes as a marker for T cell activation. Nature 1978;62–63.
81. Thorsson AV, Hintz RL. Specific ^{125}I-somatomedin receptors on circulating human mononuclear cells. Biochem Biophys Res Commun 1977;1566–1573.
82. Tapson VF, Boni-Schnetzler M, Pilch PF, Center DM, Berman JS. Structural and functional characterization of the human T lymphocyte receptor for insulin-like growth factor I in vitro. J Clin Invest 1988;950–957.
83. Kozak RW, Haskell JF, Greenstein LA, Rechler MM, Waldmann TA, Nissley SP. Type

I and II insulin-like growth factor receptors on human phytohemagglutinin-activated T lymphocytes. Cell Immunol 1987;318–331.

84. Stuart CA, Meehan RT, Neale LS, Cintron NM, Furlanetto RW. Insulin-like growth factor-I binds selectively to human peripheral blood monocytes and B-lymphocytes. J Clin Endocrinol Metab 1991;1117–1122.
85. Kooijman R, Willems M, De Haas CJC, et al. Expression of type I insulin-like growth factor receptors on human peripheral blood mononuclear cells. Endocrinology 1992;2244–2250.
86. Hartmann KK, Baier TG, Papa V, et al. A monoclonal antibody to the T-cell receptor increases IGF-I receptor content in normal T-lymphocytes: comparison with phytohemagglutinin. J Cell Biochem 1992;81–85.
87. Kooijman R, Scholtens LE, Rijkers GT, Zegers BJM. Differential expression of type I insulin-like growth factor receptors in different stages of human T cells. Eur J Immunol 1995;931–935.
88. Schillaci R, Brocardo M, Galeano A, Roldán A. Down regulation of insulin-like growth factor-I receptor (IGF-IR) expression in human T lymphocyte activation. Cell Immunol 1998;157–161.
89. Janssen JA, Uitterlinden P, Hofland LJ, Lamberts SW. Insulin-like growth factor I receptors on blood cells: their relationship to circulating total and "free" IGF-I, IGFBP-1, IGFBP-3 and insulin levels in healthy subjects. Growth Horm IGF Res 1998;47–54.
90. Cortez AB, Van Dop C, Bailey RC, et al. IGF-I resistance in virus-transformed B-lymphocytes from African Efe Pygmies. Biochem Mol Med 1996;31–36.
91. Geffner ME, Bersch N, Golde DW. Insulin and IGF-I stimulate normal and virally-transformed T-lymphocyte cell growth *in vitro*. Brain Behav Immun 1992;377–386.
92. Landreth K, Narayanan R, Dorshkind K. Insulin-like growth factor I regulates pro-B cell differentiation. Blood 1992;1–6.
93. Gibson LF, Piktel D, Landreth KS Insulin-like growth factor I potentiates expansion of interleukin-7 dependent pro-B cells. Blood 1993;3005–3011.
94. Rodriguez-Tarduchy G, Collins MKL, Garcia I, Lopez-Rivas A. Insulin-like growth factor-I inhibits apoptosis in IL-3 dependent hemopoietic cells. J Immunol 1992;535–540.
95. O' Connor R, Kauffmann-Zeh A, Liu Y, et al. Identification of domains of the IGF-IR that are required for protection from apoptosis. Mol Cell Biol 1997;427–435.
96. Walsh PT, O' Connor R. The insulin-like growth factor-1 receptor is regulated by CD28 and protects activated T cells from apoptosis. Eur J Immunol 2000:1010–1018.
97. Guler H-P, Zapf J, Scheiwiller E, Froesch ER. Recombinant human insulin-like growth factor I stimulates growth and has distinct effects on organ size in hypophysectomized rats. Proc Natl Acad Sci USA 1988;4889–4893.
98. Wood TL. Gene-targeting and transgenic approaches to IGF and IGF binding protein function. Am J Physiol 1995: E613–622.
99. Kooijman R, Scholtens LE, Rijkers GT, Zegers BJM. Type I insulin-like growth factor receptor expression in different developmental stages of human thymocytes. J Endocrinol 1995;203–209.
100. Kooijman R, Hooge-Peters EL, Hooghe R. Prolactin, growth hormone and insulin-like growth factor-1 in the immune system. Adv Immunol 1996;377–454.
101. Hirai K, Miyamasu M, Yamaguchi M, et al. Modulation of human basophil histamine release by insulin-like growth factors. J Immunol 1993;1503–1508.
102. Fu Y-K, Arkins S, Shang Wang B, Kelley KW. A novel role of growth hormone and

insulin-like growth factor-I. Priming neutrophils for superoxide anion secretion. J Immunol 1991;1602–1608.
103. Rosenfeld RG, Kemp SF, Gaspich G, Hintz RL. *In vivo* modulation of somatomedin receptor sites: effects of growth hormone treatment of hypopituitary children. J Clin Endocrinol Metab 1981;759–764.
104. Mandel S, Moreland E, Nichols V, Hanna C, Lafranchi S. Changes in insulin-like growth factor-I (IGF-I), IGF-binding protein-3, growth hormone (GH)-binding protein, erythrocyte IGF-I receptors, and growth rate during GH treatment. J Clin Endocrinol Metab 1995;190–194.
105. Eshet R, Werner H, Klinger B, et al. Up-regulation of insulin-like growth factor-1 receptor gene expression in patients with reduced serum IGF-I levels. J Mol Endocrinol 1993;115–120.
106. Eshet R, Klinger B, Silbergel A, Laron Z. Modulation of insulin-like growth factor-1 binding sites on erythrocytes by IGF-I treatment in patients with Laron syndrome. Regul Pept 1993;233–239.
107. Bierknes R, Vesterhus P, Aarskog D. Increased neutrophil insulin-like growth factor-1 (IGF-I) receptor expression and IGF-I induced functional capacity in patients with untreated Laron syndrome. Eur J Endocrinol 1997;92–95.
108. Nguyen BY, Clerici M, Venzon DJ, et al. Pilot study of the immunologic effects of recombinant human growth hormone and recombinant insulin-like growth factor in HIV-infected patients. AIDS 1998;895–904.
109. Zarkesh-Esfahani SH, Kolstad O, Metcalfe RA, et al. High-dose growth hormone does not affect proinflammatory cytokine (Tumor Necrosis Factor-α, Interleukin-6, and Interferon-γ) release from activated peripheral blood mononuclear cells or after minimal to moderate surgical stress. J Clin Endocrinol Metab 2000;3383–3390.
110. Estrov Z, Meir R, Barak Y, Zaizov R, Zadik Z. Human growth hormone and insulin-like growth factor-I enhance the proliferation of human leukemic blasts. J Clin Oncol 1991;394–399.
111. Shimon H, Shpilberg O. The insulin-like growth factor system in regulation of normal and malignant hematopoiesis. Leukemia Res 1995;233–240.
112. Pepe MG, Ginzton NH, Lee DK, Hintz RL, Greenberg PL. Receptor binding and mitogenic effects of insulin and insulin-like growth factor I and II for human myeloid leukemic cells. J Cell Physiol 1987;219–225.
113. Hizuka N, Sukegawa I, Takano K. Characterization of insulin-like growth factor I receptors on human erythroleukemia cell line (K562 cells). Endocrinol Jpn 1987;81–90.
114. Lee PDK, Rosenfeld RG, Hintz RL, Smith SD. Characterization of insulin, insulin-like growth factors I and II, and growth hormone receptors on human leukemic lymphoblasts. J Clin Endocrinol Metab 1986;28–35.
115. Freund GG, Kulas DT, Way BA, Mooney RA. Functional insulin and insulin-like growth factor-1 receptors are preferentially expressed in multiple myeloma cell lines as compared to B-lymphoblastoid cell lines. Cancer Res 1994;3179–3185.
116. Sukegawa I, Hizuka N, Takano K, Asakawa K, Shizume K. Decrease in IGF-I binding sites on human promyelocytic leukemia cell line (HL-60) with differentiation. Endocrinol Jpn 1987;365–372.
117. Eshet R, Silbergerd A, Zaizov R, et al. Decreased insulin-like growth factor I receptor sites on circulating mononuclear cells from children with acute leukemia. Pediatr Hematol Oncol 2000: 253–260.

118. Nishi Y, Tanaka T, Takano K, et al. Recent status in the occurrence of leukemia in growth hormone-treated patients in Japan. GH Treatment Study Committee of the Foundation for Growth Science, Japan. J Clin Endocrinol Metab 1999:1961–1965.
119. Maneatis T, Baptista J, Connelly K, Blethen S. Growth hormone safety update from the National Cooperative Growth Study. J Pediatr Endocrinol Metab 2000;1035–1044.

Growth and Lactogenic Hormones
Edited by L. Matera and R. Rapaport

Growth Hormone and Insulin-like Growth Factor-1 Production by Cells of the Immune System

DOUGLAS A. WEIGENT

University of Alabama at Birmingham, Department of Physiology and Biophysics, 1918 University Blvd MCLM 894, Birmingham, AL 35294-0005, USA

ABSTRACT

In recent years, it has become apparent that there are numerous sites of growth hormone (GH) and insulin-like growth factor-1 (IGF-1) production in the body outside of the pituitary gland and the liver, respectively. Studies show that various cells of the immune system from humans, rats, mice, chickens, dogs, and the bovine all produce a GH molecule indistinguishable from pituitary GH. A similar scenario has been observed for immune cell-derived IGF-1. The evidence supporting the idea that cells of the immune system produce these hormones has been derived from analysis of specific messenger RNA by Northerns, RT-PCR, and in situ hybridization. The specific proteins have been identified by immunofluorescence, RIA, Western blot, and bioactivity. The mechanisms involved in the regulation of synthesis and release of these molecules from cells of the immune system are both similar and different than that observed in the pituitary and liver. Although much work is yet to be done defining the mechanisms of synthesis and secretion of these hormones, it is clear that small amounts of hormones are produced and secreted. It is suggested that the local production of small amounts of these hormones may prove to be regulated by immunologically relevant mediators and function by paracrine/autocrine or intracrine pathways to participate in a successful immune response.

1. INTRODUCTION

Growth hormone (GH) and insulin-like growth factor-1 (IGF-1) expression in mammalian tissues has been traditionally thought to be restricted to the pituitary and the liver, respectively. However, within the last 20 years, it has become increasingly clear that additional sites possess the ability to produce GH and IGF-1. In the case of GH, such sites include the brain [1,2], mammary gland [3], placenta [4], skin [5], ovary [6], and cells of the immune system [7] (Table I). In the case of IGF-1, such sites include the brain [8], muscle [9], adipose tissue [10], lung [11], pancreas [11], kidney [12], heart [11], testes [13], pituitary [14], and cells of the immune system [15] (Table II). This article focuses on cells of the immune system and summarizes developments in our understanding of GH and IGF-1 production and the regulation of their synthesis.

Table I Extrapituitary sites reported to synthesis GH.

Source of tissue	Nature of evidence	References
Brain	RT-PCR, RIA, Western blot, Northern blot	1,2
Mammary gland	Immunohistochemistry RT-PCR, RIA	3
Placenta	Immunochemistry, RT-PCR	4
Skin	RT-PCR	5
Ovary	Immunochemistry, RT-PCR	6
Leukocytes	Immunofluorescence, bioactivity, RT-PCR, Northern blot	7, 21–40

Table II Extrahepatic sites reported to synthesize IGF-1.

Source of tissue	Nature of evidence	References
Brain	In situ hybridization, RT-PCR	8
Muscle	RNase protection assay	9
Adipose	RIA	10
Lung	RIA	11
Pancreas	RIA	11
Kidney	HPLC, RIA, radioreceptor assays	12
Heart	RIA	11
Testes	RIA	13
Pituitary	Northern analysis, RIA immunoperoxidase staining	14
Leukocytes	HPLC, RIA, immunofluorescence bioactivity	15,55–65

Excellent earlier reviews are available on this subject along with the potential role of these hormones during inflammation and hematopoiesis [7,16–18]. Although the functional effects of exogenous GH and IGF-1 have been well studied, the exact function of endogenous or leukocyte-derived hormones is less clear. Since cells of the immune system produce significantly less GH and IGF-1 than the pituitary and liver, respectively, it is logical to suggest that their local production acts by a paracrine/autocrine or intracrine mechanism of action in support of the immune response.

2. GH EXPRESSION IN THE IMMUNE SYSTEM

Approximately 20 years ago, the pioneering studies of Blalock and coworkers discovered that lymphocytes could produce corticotropin and endorphin-like substances [19]. Shortly after this, thyroid stimulating hormone was also shown to be secreted by lymphocytes after stimulation with mitogen [20]. A prolactin/GH-related mRNA species was also detected in mitogen-stimulated lymphocytes [21]. Previously, it had been shown that GH was a single-chain 22 kD polypeptide primarily produced within the pituitary with a broad range of physiological actions, including effects on somatic growth, metabolism, and immunity [16,17]. In the light of these

Table III Detection of GH mRNA transcripts and proteins in cells of the immune system.

Origin	GH mRNA			GH protein				Refs
	RT-PCR	Northern	In situ	Immuno-fluorescence	RIA	Western	Bioactivity	
1. Cell lines								
H-9 T cell	✓				✓		✓	25
IM-9 B cell	✓			✓			✓	24,25
Sf Ramos	✓	✓			✓	✓	✓	30
Hut-78	✓							5
P388	✓							51
EL4	✓							51
U937	✓							5
HL-60	✓					✓		26
2. Primary tissues								
Human PBL		✓		✓			✓	22,27,34
Human thymocyte	✓			✓				27,38
Rat spleen, thymus				✓				32,37,41
Mouse bone marrow	✓		✓	✓				35
Mouse spleen, thymus	✓							48
Chicken spleen, thymus	✓				✓	✓		41
Bovine spleen, thymus	✓	✓						39
Canine lymph nodes	✓		✓	✓				40

Analysis of GH mRNA transcripts were primarily done by RT-PCR, Northern analysis, and in situ hybridization. Analysis of GH proteins were done by immunofluorescence, partial purification, Western blotting, and bioactivity.

observations, we examined in detail the potential for cells of the immune system to produce GH [22]. Our results provided evidence that mononuclear leukocytes can synthesize GH and that the molecule produced was identical to pituitary GH in terms of antigenicity and molecular weight. The de novo synthesis of GH was confirmed by our ability to radiolabel the hormone and block its synthesis with actinomycin-D and cycloheximide [22]. In this same report, we included data showing that conditioned medium from human PBL purified on immunoaffinity columns coupled with antibodies to hGH could stimulate the proliferation of Nb2 rat node lymphoma cells confirming the purified material was bioactive similar to pituitary-derived GH. Hattori and colleagues came to the same conclusion, confirming the production of GH in normal human PBL, mitogen-stimulated lymphocytes, and Epstein-Barr virus transformed B lymphocytes [23]. Also, the human B-cell lymphoma cell line IM-9, myeloid cell line HL-60, and the human T-cell line H9 were shown to synthesize and secrete hGH [24–26] (Table III).

We also examined the in vivo production of GH-related RNA and protein by rat leukocytes after intraperitoneal injection or treatment with different inducing agents known to activate the immune system, including bacterial lipopolysaccharide (LPS) [27]. The data in rats after exposure to LPS showed that leukocytes obtained from the spleen, thymus, and peritoneum all showed a dose-dependent increase in GH-related RNA content. We also evaluated the ability of LPS-sensitive (C3HeB/FeJ) and resistant (C3H/HeJ) inbred mice treated with LPS to produce GH-related RNA. The LPS-sensitive mice presented with a typical pathophysiologic response pattern and higher levels of GH-related RNA in the spleen and thymus than the LPS-resistant mice. An increase in the production of immunoreactive GH (irGH) was also observed in spleen cells by direct immunofluorescence with specific antibodies to rat GH. We validated that the GH-related RNA produced in vivo by leukocytes was similar in structure to pituitary GH RNA using reverse transcription and the polymerase chain reaction (PCR). In other studies with normal nontreated animals, the GH RNA levels in spleen were higher in the evening hours and early on during the first month of life than during the daytime or in older animals, respectively. Taken together, our data were the first demonstration that GH RNA and immunoreactive protein could be detected in leukocytes in vivo both in normal and stimulated animals and supported the idea that GH may be active in an immune response [27].

From the beginning, studies examining the production of GH by cells of the immune system have been hampered by the small amounts of hormone produced. The percentage of cells positive for GH production have ranged from 1–10% in primary cell cultures and cell lines and the levels detected depended on the type of cells examined and the conditions of study employed (10–100 pg/mL/10^6 cells/24h) [22–25]. Two groups, however, have studied GH production in cells of the immune system by plaque assay and obtained interesting results. In one study in IM-9 cells, the authors compared the percentage of cells positive for GH production by immunofluorescence to the number of cells secreting GH by the reverse hemolytic plaque assay [24]. The data showed that 10% of the cells were positive by immunofluorescence, whereas only about 30 cells out of 50,000 showed signs of secretion suggesting that the majority (99 out of 100) of cells did not secrete GH. In another study with human PBL using an enzyme-linked immunoplaque assay, the authors showed that 10% of the total population of nonstimulated cells actively secreted GH and that treatment with a T-cell mitogen increased the GH plaque area as well as the plaque number [28]. These data suggest that mitogen-stimulated cells may secrete factors that recruit other cells to secrete GH. The low levels of GH produced have also made it difficult to study the mRNA by Northern analysis; however, we and others have performed RT-PCR to validate the presence of GH mRNA. In our initial study, we amplified by RT-PCR a predicted 600 bp fragment encompassing a part of exons 1 to 5 [29]. Most importantly, the identity and analysis of this fragment obtained from the rat spleen by restriction enzyme analysis was confirmed to be similar to pituitary GH. The cloning and sequencing of this fragment revealed its complete identity to that reported for pituitary GH [30]. The secretion of GH by the Burkitt B lymphoma cell line of Ramos has also been reported and the mRNA studied by RT-PCR [31]. In this thorough analysis of hGH expression, the authors reported the partial sequence analysis of a 248 bp RT-PCR Ramos cell line generated fragment and found it to be identical to the human pituitary GH sequence. The expression of the hGH-N gene by RT-PCR was also reported in the cell lines of lymphoid (Hut-78) and of myelomonocytic type (U937), whereas expression of hGH-V or chorionic somatomammotropin genes was not observed [5]. In another report, hGH-V gene expression in human PBL was detected by RT-PCR both in men and women, as well as pregnant women [32]. In this later study, the GH protein was not studied and the primers selected to amplify the closely related gene transcripts were different from the work of other investigators. Taken together, these findings summarized in Table III strongly support the presence and identity of GH molecules in cells of the immune system.

Although there are numerous reports now that many different cell lines, including both T and

B cells as well as primary lymphoid cells can produce GH in vitro (Table III), the situation in vivo is less clear. Very early, a number of investigators showed in vitro that both rat, mouse, and human primary lymphoid cells and cell lines could produce GH and that mitogen treatment generally enhanced the levels of GH produced [5,22–25,27–32]. Our initial work to identify the subpopulations of cells producing GH mRNA and protein was conducted in rat tissues [33]. The data demonstrated that mononuclear leukocytes from various tissues, including spleen, thymus, bone marrow, Peyer's patches, and peripheral blood, all have the ability to produce GH mRNA and secrete GH. Data obtained with cells separated by adherence, nylon wool columns, and positive and negative sorting with monoclonal antibodies that define B, monocyte, T helper, and T cytotoxic cells showed that several different cell types have the ability to produce GH mRNA. The results suggested that B cells, macrophages, and T helper cells produce more GH mRNA and protein than that of T cytotoxic cells. Natural killer cells also produce detectable levels of GH mRNA and protein [33]. We have also investigated the subpopulation of lymphoid cells from normal and hypophysectomized rats producing GH and IGF-1 in vitro [34]. The data show that removal of the pituitary results in depression of GH production in spleen, thymus, and bone marrow and an increase in the peripheral blood leukocytes. The changes in the percentage of cells producing GH in hypophysectomized animals are not due to a single cell type but appear to influence the T-helper, T-cytotoxic, and B-cell subsets. Interestingly, no significant changes in the levels of GH RNA were detected between control and hypophysectomized animals after in vitro culture. We also found that the increase in GH production in spleen cell cultures after mitogen stimulation could be accounted for by an increase in the percentage of T cells producing GH. Lastly, we demonstrated by immunofluorescence that the cells positive for GH production were also positive for IGF-1 production. This later finding coupled with our previous results suggests that an autocrine regulatory circuit may be important for the production of leukocyte-derived GH and IGF-I within the immune system [34]. In studies by others, GH mRNA has been found by in situ hybridization in normal lymphoid tissues as well as endothelial cells and distributed diffusely throughout T and B cell lymphomas and a thymoma [35]. GH expression in murine bone marrow cells and in particular granulocytes was detectable in normal and dwarf mice by immunocytochemistry and in situ hybridization [36]. Using these same techniques, another group showed a positive GH mRNA signal in the septa, capsular and subcapsular cortex in the human thymus but not in thymocytes [37]. An earlier report by Binder found GH mRNA in the bone marrow and thymus of neonatal rats but not in the spleen [38]. More recently, the presence of hGH messenger RNA was shown by RT-PCR in both human thymocytes and in primary cultures of thymic epithelial cells [39]. The investigation of GH production has been studied in systems other than human and murine, and these include the bovine [40], the spleen, thymus, and bursa of fabricious of the chicken [41], and normal lymph nodes and lymphomas of the dog [42]. The results of these important studies (Table III) confirm the lymphoid presence of GH mRNA transcripts and protein and as discussed below, although the coding regions are identical to pituitary GH, the 5'-UTR regions may be different in these species suggesting a different manner of regulation.

In the beginning of the studies on GH production by cells of the immune system, a controversy regarding the factors involved in controlling its production emerged. Although there was a general consensus that mitogens enhanced the levels of leukocyte GH, the effect of different hormones that influence pituitary GH production on lymphocyte GH production were conflicting. In our first report, we showed that mononuclear leukocytes immediately begin to express GH mRNA and GH protein after removal of lymphoid tissues from animals [22] and peak about 8 hr after in vitro culture. The mechanism of this spontaneous induction and whether it is due to removal of negative inhibitors and the presence of positive regulation is unclear. All that is known is that protein synthesis is required for GH mRNA induction. In other studies, the in vitro stimulation of rat mononuclear leukocytes with GHRH caused a dose-dependent increase of cytoplasmic GH mRNA levels and thymidine incorporation in rat mononuclear leukocytes [43]. These effects could be blocked by GHRH antisense oligodeoxynucleotides, but not by antibodies to GHRH [44], thus suggesting an intracrine effect of GHRH on leukocyte-derived GH expression. Also, human IM-9 cells have been shown to increase hGH secretion during incubation with GHRH [24]. In contrast, Hattori et al. did not find any in vitro effect of GHRH and SRIH upon hGH secretion of human PBMC [23,45]. We also have not observed any effect of somatostatin on lymphoid cell GH expression. Exogenous IGF-I has been reported to decrease the levels of rat leukocyte GH-related RNA and the secretion of immunoreactive GH in vitro [15]. Our data [15] and those of Sabharwal [46] suggests that GH secreted by thymocytes acts as an autocrine/paracrine growth factor via IGF-I to promote proliferation. On the other hand, another group suggested that exogenous IGF-I did not affect hGH secretion of human lymphocytes while exogenous hGH was demonstrated to up-regulate hGH secretion in vitro [47]. These data suggest that the mechanisms of regulation of GH secretion in lymphocytes are both similar and different from those in the endocrine system. Some of the differences observed in leukocyte studies may reflect species differences as well as cultural conditions and cell types. Unfortunately, in some cases, only secretion was examined, whereas in other studies only synthesis was studied.

Another area of interest over the years regarding the synthesis of GH by cells of the immune system has been the role, if any, for the homeodomain transcription factor GHF-1 or Pit-1, which serves an important role in the trans-activation of the GH gene in the pituitary [48]. It has been convincingly shown that Pit-I is expressed in hemopoietic and lymphoid tissues [49] and even in the same leukocytes producing GH [16]. However, the idea that Pit-1 may not be involved in GH expression in the murine system was first suggested by our work showing near-normal levels of GH mRNA and protein in dwarf spleen cells compared with those in the pituitary of these animals [50]. This work was essentially confirmed and extended to bone marrow cells, where in situ hybridization, immunocytochemistry, and RT-PCR analysis showed that GH expression does not depend on Pit-1 [36]. In preliminary experiments using GH-promoter fragments and EMSA, we have not been able to block or shift any band formed by complexes with proteins from P-388 nuclear extracts and GHF oligonucleotides or GHF-specific antibodies. A similar situation has been described for human and monkey trophoblasts, in that although GHF-1 expression was detectable, supershift analysis could not detect GHF-1 binding to this region [51]. It may be that under certain conditions, selected cells of the immune system may use Pit-1 to regulate GH, whereas it does not appear to be required for basal expression. Further, it has been shown that SP1 may displace Pit-1 from a binding site and stimulate transcription from the GH gene promoter [52] supporting the idea that other factors may function at this site in cells of the immune system. The findings in dwarf mice also suggest that cells of the immune system do not compensate by upregulating GH synthesis for the lack of GH production in animals lacking the transcription factor GHF-1 in the pituitary.

The molecular mechanisms involved in transcriptional regulation of GH in cells of the immune system are in their infancy. Most studies to date that have studied the gene have sequenced the coding region and shown it to be the same as that reported for pituitary [30,31,51]. Identification of the extrapituitary GH transcription initiation start site and promoter in lymphoid cells has been reported for the rat [53], dog [42], and bovine [40]. Interestingly, in the bovine, the analysis of the 5'-untranslated region of the lymphocyte GH mRNA showed that transcription began 336 nucleotides upstream from the start site in the pituitary gland, suggesting differences in regulation in these tissues. In the dog, analysis of the transcriptional start sites of the GH gene using 5'-RACE (rapid amplification of cDNA ends) showed that the canine lymphoid transcripts contained a 33–85 bp enlarged 5'-untranslated region compared to the pituitary and mammary GH transcripts. Part of the lymphoid GH transcripts contained intron 1, which would result in early termination of the translation due to an in-frame stopcodon. Since mitogen stimulation activates splicing, the authors suggest that removal of the intron may indicate gene regulation at a late step in mRNA processing [42]. Our studies in a mouse monocyte cell line suggest that these cells utilize the same exon 1 as the pituitary and therefore probably the same promoter sequence for expression of monocyte GH mRNA as the pituitary [53]. Although more work needs to be done, the data indicate that almost the same GH promoter region is used in canine and murine lymphoid tissues whereas the bovine appears to initiate transcription much farther upstream. Our data in a mouse cell line also show that the region between –299 bp and –193 bp may play an important positive role, whereas the region between –193 bp and –107 bp may play a critical negative role in mediating the expression of monocyte GH [53]. The overexpression of Pit-1 showed an unexpected modest inhibition of the full GH promoter construct. Finally, we have extended these results by determining that two members of the SP family of transcription factors, SP1 and SP3, bind to the region at –138/–133 bp containing a GGGAGG motif [54]. Confirmation that this region of the monocyte GH promoter-bound SP1 and SP3 was accomplished using electrophoretic mobility shift assays with SP1 consensus and mutant probes as well as specific antibodies to SP1 and SP3. Selective mutation of the SP1/SP3 site increased basal transcription by 73%, indicating that this site is important in transcriptional inhibition. Overexpression of SP1 had no demonstrable effect on the GH promoter, whereas overexpression of SP3 caused inhibition of expression in P-388 monocyte cells. Cotransfection of P-388 cells with overexpression vectors for both SP1 and SP3 transcription factors also resulted in inhibition of basal expression. Taken together, the results demonstrate that basal expression of monocyte GH may be negatively regulated by SP3 [54].

3. IGF-1 EXPRESSION IN THE IMMUNE SYSTEM

IGF-1 is a 70-amino acid mitogenic polypeptide that circulates in the blood and plays a major role in the growth and development of a variety of tissues and cell types [55]. Although IGF-1 was originally identified as being produced in the liver as the major mediator of the anabolic functions of GH, it is now quite clear that IGF-1 can be produced by other cell types as well (see Table II). The first hint that cells of the immune system may produce an IGF-like growth factor was provided by Bitterman and colleagues in a 1982 report of a "progression" factor from activated alveolar macrophages, they termed alveolar macrophage-derived growth factor (AMDGF) [56]. Subsequent studies by Rom and associates involving partial purification of the material, receptor-displacement studies and solution hybridization provided convincing evidence that the material originally termed AMDGF was in fact IGF-1 [57]. Shortly later, IGF-1 molecules were reported in Epstein-Barr virus transformed B lymphocytes [58], transformed human T cell lines [59], and the human lymphoid cell line, IM-9 [60]. Most of the evidence included specific radioimmunoassays for IGF-1 and growth inhibition by antibodies specific for IGF-1 and/or its receptor. The results in T cell lines supported a role for locally generated IGF-1 in the mediation of GH action on T-lymphocytes and indicated the effect was mediated via the type 1 IGF receptor. Our own studies in primary rat spleen and thymus confirmed the earlier results [15]. We could detect IGF-1 by direct immunofluorescence with specific IGF-1 antibodies and using immunoaffinity purification, HPLC, and a fibroblast proliferation bioassay showed that IGF-1 was de novo synthesized and similar to serum IGF-1 in molecular weight, antigenicity, and bioactivity [15]. Furthermore, treatment of leukocytes with GH increased the levels of IGF-1 whereas treatment of leukocytes with IGF-1 reduced the levels of leukocyte-derived GH. Taken together, the results suggested a regulatory circuit for GH and IGF-1 within the immune system [15]. Most, but not all, of the studies reported support of the ability of GH to stimulate leukocyte IGF-1 production. In some instances, the levels of IGF-1 produced are low and difficult to measure, although the biological effects of GH can be blocked with specific IGF-1 antibodies. In the IM-9 cell line, both positive [15,61], and negative [60] findings have been reported. The discrepant results may stem from the type of cells being examined, the abundance, and function of the GH receptor, the stimulus and/or the sensitivity and methods employed to detect IGF-1. In this regard, by RT-PCR IGF transcripts could be amplified from splenocytes, thymocytes, and macrophages; however, only macrophages were positive on Northern blots or RNase protection assays [62]. Likewise, IGF-1 expression by RT-PCR was detectable in PHA-stimulated but not in freshly isolated human peripheral blood lymphocytes [63].

Although the IGF-1 peptide is a relatively simple molecule, it has a large and complex gene structure that gives rise to an array of messenger RNA (mRNA) transcripts varying at both the 5' and 3' ends [55]. Class 1 or exon-1 containing transcripts are expressed by many tissues and thought to encode the "paracrine" form of IGF-1 whereas class 2 or exon-2 containing transcripts are liver-enriched, sensitive to GH, and thought to be responsible for "endocrine" IGF-1 [55]. At the 3'-end, transcripts lacking exon 5 (Ea) and containing exon 5 (Eb) have been described [55]. The major work examining the expression of the IGF-1 mRNA in the mouse lymphohemopoietic system has been done by Kelley and colleagues in macrophages [62]. Their results establish that murine macrophages express abundant insulin-like growth factor-1 class 1 Ea and Eb transcripts. A 26 kilodalton prepro IGF-1 peptide was detected in macrophage cell lysates by Western blotting [62]. Further, their data suggest myeloid rather than lymphoid cells are the major source of IGF-1 that is associated with differentiation of bone marrow macrophages [62]. Thus, in macrophages, initiation of transcription is primarily within exon 1 that is typical of extrahepatic tissues with a higher percentage of Eb transcripts that is typical of hepatic tissues. The significance of different leader peptides and E terminal domains on IGF-1 is unknown but may influence targeting, processing or stability.

The regulation of expression of IGF-1, except for studies on GH dependence discussed above, has been investigated mostly in stimulated primary macrophages and macrophage cell lines. The data in the macrophage-like cell line U937 showed that, although the transcription rate increased 4-5-fold after phorbol myristate acetate (PMA) or Ca++ ionophore treatment, there was a significant time and dose-dependent reduction in the steady-state IGF-1 mRNA levels [64]. In this same report, surface stimulation appeared to cause IGF-1 release from a preformed cellular storage pool. Circulating monocytes do not normally express IGF-1 mRNA or peptide but can be stimulated to do so after exposure to asbestos, acetylated glycosylated endproducts and other inflammatory mediators, including IL-1 and TNF [57,65]. The colony-stimulating factors also have been shown to induce expression of IGF-1 mRNA during hematopoiesis [66]. Three different populations of murine macrophages have been used to show that IFN potently inhibits IGF-1 synthesis at the transcriptional level [67].

In plasma, IGFs circulate primarily in a ternary complex with IGF-binding proteins (IGF-BP) and an acid-labile subunit [68]. Although the importance of IGF-binding proteins is well established, their expression by cells of the immune system has received little attention. They were first detected in 1991 by Western blot analysis in six out of 12 lymphoblast cell lines. In this initial report, neither IGF-BP1 or 3 was identified in conditioned media; however, IGF-BP2 and 4 were detected in both T and B cells [69]. In another study, unstimulated human lymphocytes by RT-PCR expressed IGF-BP2 and 3 whereas PHA stimulated lymphocytes expressed IGF-BP-2,3,4 and 5. The addition of a number of hormones, including estrogen, progesterone, IGF-1 or GH did not affect secretion of IGF-BPs by lymphocytes as measured by Western ligand blotting of conditioned medium [63]. More recently, Kelley and coworkers have shown that mature adherent myeloid cells synthesize and secrete a substantial amount of IGF-BPs, whereas less differentiated or nonadherent myeloid cells produce fewer IGF-BPs. Premyeloid cells, mature T cells, and primary murine thymocytes did not synthesize detectable IGF-BPs [70]. Additional gel-shift, Northern blotting, and sequencing analysis showed that the IGF-BP secreted by mature adherent macrophages was IGF-BP4 [70]. Taken together, the presence of IGF-1, the IGF-1 receptor and IGF-BPs, particularly in myeloid cells, strongly supports the suggestion that the IGF system is important in hematopoiesis and inflammation.

4. SUMMARY AND SPECULATION

The purpose of this chapter was to summarize the evidence that GH and IGF-1 are, in fact, produced by cells of the immune system. After a review of the published data, it is now apparent that GH and IGF-1 production are not confined to the pituitary and liver, respectively, but extend to other tissues, including the immune system. It is also apparent that small amounts of each hormone are produced relative to the sites they were originally identified. Structurally, they are similar to their respective counterparts whereas the regulation of synthesis and secretion may be different. We have argued for a long time now, as others have, that the local production most likely acts by a paracrine/autocrine, and/or intracrine mechanism to facilitate the immune response. There are data that support a role for endogenous lymphocyte GH in promoting proliferation and possibly protecting cells from undergoing apoptosis. There are also some data to suggest that exogenous and endogenous GH act in a similar fashion [71]. There are other data that identify GH receptors on the nuclear membrane [72]. We speculate that a significant percentage of endogenous GH remains within the producer cell, whereas detectable amounts are secreted. Secreted GH may interact at GH receptors on the same or neighboring cells and induce a functional response via the signaling pathways previously described for GH (i.e., JAK/stats, PKC) [73]. The GH remaining within cells we speculate may serve a different role that bypasses the plasma membrane GH receptor and acts at the nuclear receptor to modulate nuclear second messengers and/or the activity of transcription factors. It is our opinion that this additional mechanism affords the GH-producing cell a measure of autonomy and most likely serves a protective and/or metabolic survival function, when a decision is required at a local site (i.e., independent of pituitary) such as during an immune response.

ACKNOWLEDGEMENTS

I thank Diane Weigent for excellent editorial assistance and for typing the manuscript.

REFERENCES

1. Martinoli MG, Ouellet J, Rheaume E, Pelletier G. Growth hormone and somatostatin gene expression in adult and aging rats as measured by quantitative in situ hybridization. Neuroendocrinology 1991;54:607–615.
2. Render CL, Hull KL, Harvey S. Neural expression of the pituitary GH gene. J Endocrinology 1995;147:413–422.
3. Mol JA, VanGardesen E, Seman PJ, Wolswinkel J, Rijinberk A, Rutteman GR. Growth hormone mRNA in mammary gland tumours of dogs and cats. J Clin Invest 1995;95:2028–2034.
4. Ogilvie S, Buhl WC, Olson JA, Shiverick KT. Identification of a novel family of growth hormone-related proteins secreted by rat placenta. Endocrinology 1990;126:3271–3273.
5. Palmetshofer A, Zechner D, Luger TA, Barta A. Splicing variants of the human growth hormone mRNA – Detection in pituitary, mononuclear cells and dermal fibroblasts. Mol Cell Endocrin 1995;113:225–234.
6. Izadyar F, Zhao J, Van Tol HTA, Colenbrander B, Bevers MM. Messenger RNA expression and protein localization of growth hormone in bovine ovarian tissue and in cumulus oocyte complexes (COCs) during in vitro maturation. Mol Reprod Dev 1999;53:398–406.
7. Weigent DA, Blalock JE. Production of peptide hormones and neurotransmitters by the immune system. In: Blalock JE, editor. Neuroimmunoendocrinology. Basel: Karger, 1997;1–30.
8. Helzenberger M, Lapointe F. Expression of insulin-like growth factor-1 (IGF-I) and IGF-II in the avian brain: relationship of in situ hybridization patterns with IGF type 1 receptor expression. J Dev Neurosci 2000;18:69–82.
9. Lewis AJ, Wester TJ, Burrin DG, Dauncey MJ. Exogenous growth hormone induces somatotrophic gene expression in neonatal liver and skeletal muscle. Am J Physiol 2001;278:R838–R844.
10. Vasilators-Younken R, Dunnington EA, Siegel PB, McMurtry JP. Tissue-specific alterations in insulin-like growth factor-1 concentrations in response to 3,3',5-triiodo-L-thyronine supplementation in the growth hormone receptor-deficient sex-linked dwarf chicken. Gen Comp Endocrin 1997;105:31–39.
11. D'Ercole AJ, Stiles AD, Underwood LE. Tissue concentrations of somatomedin-C: Further evidence for multiple sites of synthesis and paracrine or autocrine mechanisms of action. Proc Natl Acad Sci USA 1984;81:935–939.
12. Conti FG, Striker LJ, Elliot SJ, Andreani D, Striker GE. Synthesis and release of insulin-like growth factor 1 by mesangial cells in culture. Am J Physiol 1988;255:F1214–F1219.
13. Palmero S, Prati M, Barreca A, Minuto F, Giordano G, Fugassa E. Thyroid hormone stimulates the production of insulin-like growth factor 1 (IGF-1) by immature rat Sertoli cells. Mol Cell Endocrin 1990;68:61–65.
14. Fagin JA, Pixley S, Slanina S, Ong J, Melmed S. Insulin-like growth factor 1 gene expression in GH3 rat pituitary cells: Messenger ribonucleic acid content, immunocytochemistry, and secretion. Endocrin 1987;120:2037–2043.
15. Baxter JB, Blalock JE, Weigent DA. Characterization of immunoreactive insulin-like growth factor-I from leukocytes and its regulation by growth hormone. Endocrinology 1991; 129(4):1727–1734.
16. Hooghe-Peters EL, Hooghe R. Growth Hormone, Prolactin and IGF-1 as Lymphohemopoietic Cytokines. Austin: R.G. Landes Co., 1990.

17. Harvey S, Azumaya Y, Hull KL. Pituitary and extrapituitary growth hormone: Pit-1 dependence? Can J Phys Pharm 2000;78:1013–1028.
18. Venters HD, Dantzer R, Freund GG, Broussard SR, Kelley KW. Growth hormone and insulin-like growth factor as cytokines in the immune system. Psychoneuroimmunology. San Diego: Academic Press, 2001;339–359.
19. Smith EM, Blalock JE. Human lymphocyte production of corticotropin and endorphin-like substances: association with leukocyte interferon. Proc Natl Acad Sci USA 1981;78(12):7530–7534.
20. Smith EM, Phan M, Kruger TE, Coppenhaver DH, Blalock JE. Human lymphocyte production of immunoreactive thyrotropin. Proc Natl Acad Sci USA 1983;80(19):6010–6013.
21. Hiestand PC, Mekler P, Nordmann R, Grieder A, Permmongkol C. Prolactin as a modulator of lymphocyte responsiveness provides a possible mechanism of action for cyclosporine. Proc Natl Acad Sci USA 1986;83(8):2599–2603.
22. Weigent DA, Baxter JB, Wear WE, Smith LR, Bost KL, Blalock JE. Production of immunoreactive growth hormone by mononuclear leukocytes. FASEB J 1988;2(12):2812–2818.
23. Hattori N, Shimatsu A, Sugita M, Kumagai S, Imura H. Immunoreactive growth hormone (GH) secretion by human lymphocytes: augmented release by exogenous GH. Biochem Biophy Res Commun 1990;168(2):396–401.
24. Kao TL, Harbour DV, Meyer WJ3d. Immunoreactive growth hormone production by cultured lymphocytes. Ann NY Acad Sci 1992;650:179–181.
25. Kao TL, Supowit SC, Thompson EA, Meyer WJ3d. Immunoreactive growth hormone production by human lymphocyte cell lines. Cell Mol Neurobiol 1992;12(5):483–498.
26. Costoya JA, Vidal A, Garcia-Barros M, Arce V, Devesa J. Expression of the human growth hormone normal gene (hGH-N) in proliferating and differentiated HL-60 cells. Exp Cell Res 1996;228:164–167.
27. Baxter JB, Blalock JE, Weigent DA. Expression of immunoreactive growth hormone in leukocytes in vivo. J Neuroimmunol 1991;33(1):43–54.
28. Varma S, Sabharwal P, Sheridan JF, Malarkey WB. Growth hormone secretion by human peripheral blood mononuclear cells detected by an enzyme-linked immunoplaque assay. J Clin Endocrinol Metab 1993;76(1):49–53.
29. Weigent DA, Riley JE, Galin FS, LeBoeuf RD, Blalock JE. Detection of growth hormone and growth hormone-releasing hormone- related messenger RNA in rat leukocytes by the polymerase chain reaction. Proc Soc Exp Biol Med 1991;198(1):643–648.
30. Rohn WA, Weigent DA. Cloning and nucleotide sequencing of rat lymphocyte growth hormone cDNA. Neuroimmunomodulation 1995;2:108–114.
31. Lytras A, Quan N, Vrontakis ME, Shaw JE, Cattini PA, Friesen HG. Growth hormone expression in human Burkitt lymphoma serum-free Ramos cell line. Endocrinology 1993;132(2):620–628.
32. Melen L, Hennen G, Dullaart RPF, Heinen E, Igout A. Both pituitary and placental growth hormone transcripts are expressed in human peripheral blood mononuclear cells (PBMC). Clin Exp Immunol 1997;110:336–340.
33. Weigent DA, Blalock JE. The production of growth hormone by subpopulations of rat mononuclear leukocytes. Cell Immunol 1991;135(1):55–65.
34. Weigent DA, Baxter JB, Blalock JE. The production of growth hormone and insulin-like growth factor- I by the same subpopulation of rat mononuclear leukocytes. Brain Behav

Immun 1992;6(4):365–376.
35. Wu H, Devi R, Malarkey WB. Localization of growth hormone messenger ribonucleic acid in the human immune system: A clinical research center study. J Clin Endocrin Metab 1996;81:1278–1282.
36. Kooijman R, Malur A, Van Buul-Offers SC, Hooghe-Peters EL. Growth hormone expression in murine bone marrow cells is independent of the pituitary transcription factor Pit-1. Endocrinology 1997;138:3949–3955.
37. Maggiano N, Piantelli M, Ricci R, Larocca LM, Capelli A, Ranelletti O. Detection of growth hormone-producing cells in human thymus by immunohistochemistry and non-radioactive in situ hybridization. J Histochem Cytochem 1994;42:1349–1354.
38. Binder G, Revskoy S, Gupta D. In vivo growth hormone gene expression in neonatal rat thymus and bone marrow. J Endocrinol 1994;140:137–143.
39. Mello-Coelho V, Gagnerault M, Souberbielle J, Strasburger CJ, Savino W, Dardenne M et al. Growth hormone and its receptor are expressed in human thymic cells. Endocrinology 1998;139:3837–3842.
40. Chen HT, Schuler LA, Schultz RD. Growth hormone and Pit-1 expression in bovine fetal lymphoid cells. Dom Animal Endo 1997;14:399–407.
41. Render CL, Hull KL, Harvey S. Expression of the growth hormone gene in immune tissues. Endocrine 1995;3:729–735.
42. Van Leeuwen ISL, Teske E, Van Garderen E, Mol JA. Growth hormone gene expression in normal lymph nodes and lymphomas of the dog. Anticancer Res 2000;20:2371–2376.
43. Guarcello V, Weigent DA, Blalock JE. Growth hormone releasing hormone receptors on thymocytes and splenocytes from rats. Cell Immunol 1991;136(2):291–302.
44. Payne LC, Rohn W, Weigent DA. Lymphocyte-derived growth hormone releasing hormone is an autocrine modulator of lymphocyte-derived growth hormone. Endocrinology A1255;514:1994.
45. Hattori N, Ikekubo K, Ishihara T, Moridera K, Hino M, Kurahachi H. Spontaneous growth hormone (GH) secretion by unstimulated human lymphocytes and the effects of GH-releasing hormone and somatostatin. J Clin Endo Metab 1994;79:1678–1680.
46. Sabharwal P, Varma S. Growth hormone synthesized and secreted by human thymocytes acts via insulin-like growth factor I as an autocrine and paracrine growth factor. J Clin Endo Metab 1996;81:2663–2669.
47. Hattori N, Shimormura K, Ishihara T, Moridera K, Hino M, Ikekubo K, et al. Growth hormone (GH) secretion from human lymphocytes is up-regulated by GH, but not affected by insulin-like growth factor-1. J Clin Endocinol Metab 1993;76:937–939.
48. Bodner M, Karin M. A pituitary-specific trans-acting factor can stimulate transcription from the growth hormone promoter in extracts of nonexpressing cells. Cell 1987;50:267–275.
49. Delhase M, Vergani P, Malur A, Hooghe-Peters EL, Hooghe RJ. The transcription factor Pit-1/GHF-1 is expressed in hemopoietic and lymphoid tissues. Eur J Immunol 1993;23(4):951–955.
50. Weigent DA, Blalock JE. Effect of the administration of growth hormone-producing lymphocytes on weight gain and immune function in dwarf mice. Neuroimmunomodulation 1994;1:50–58.
51. Schanke JT, Durning M, Johnson KJ, Bennett LK, Golos TG. SP1/SP3-binding sites and adjacent elements contribute to basal and cyclic adenosine 3',5'-monophosphate-stimulated transcriptional activation of the rhesus growth hormone-variant gene in trophoblasts. Mol Endocrin 1998;12:405–417.

52. Lemaigre FP, Lafontaine DA, Courtois SJ, Durviaux SM, Rousseau GG. Sp1 can displace GHF-1 from its distal binding site and stimulate transcription from the growth hormone gene promoter. Mol Cell Biol 1990;10:1811–1814.
53. Weigent D, Vines CR, Long J, Blalock JE, Elton TS. Characterization of the promoter directing expression of rat growth hormone in a monocyte cell line. Neuroimmunomodulation 2000;7:126–134.
54. Vines CR, Weigent DA. Identification of SP3 as a negative regulatory transcription factor in the monocyte expression of growth hormone. Endocrinology 2000;141:938–9946.
55. Rotwein P. Structure, evolution, expression and regulation of insulin-like growth factors I and II. Growth Factors 1991;5(1):3–18.
56. Bitterman PB, Rennard SI, Adelberg S, Crystal RG. Role of fibronectin as a growth factor for fibroblasts. J Cell Biol 1983;97:1925–1932.
57. Rom WN, Basset P, Fells GA, Nukiwa T, Trapnell BC, Crystal RG. Alveolar macrophages release an insulin-like growth factor I-type molecule. J Clin Invest 1988;82:1685–1693.
58. Merimee TJ, Grant MB, Broder CM, Cavalli-Sforza LL. Insulin-like growth factor secretion by human B-lymphocytes: A comparison of cells from normal and pygmy subjects. J Clin Endocrin Metab 1989;69:978–984.
59. Geffner ME, Bersch N, Lippe BM, Rosenfeld RG, Hintz RL, Golde DW. Growth hormone mediates the growth of T-lymphoblast cell lines via locally generated insulin-like growth factor-I. J Clin Endocrin Metab 1990;71(2):464–469.
60. Clayton PE, Day RN, Silva CM, Hellmann P, Day KH, Thorner MO. Growth hormone induces tyrosine phosphorylation but does not alter insulin-like growth factor 1 gene expression in human IM-9 lymphocytes. J Mol Endocrin 1994;13:127–136.
61. Palmer JM, Wallis M. Human growth hormone stimulates somatomedin C/insulin-like growth factor 1 production by the human lymphoid cell line, IM-9. Mol Cell Endocrin 1989;63:167–173.
62. Arkins S, Rebeiz N, Biragyn A, Reese DL, Kelley KW. Murine macrophages express abundant IGF-I class I, Ea and Eb transcripts. Endocrinology 1993;133:2334–2343.
63. Nyman T, Pekonen F. The expression of insulin-like growth factors and their binding proteins in normal human lymphocytes. Acta Endocrinol 1993;128:168–172.
64. Nagaoka I, Trapnell BC, Crystal RG. Regulation of insulin-like growth factor 1 gene expression in the human macrophage-like cell line U937. J Clin Invest 1990;85:448–455.
65. Kirstein M, Aston C, Hintz R, Vlassara H. Receptor-specific induction of insulin-like growth factor 1 in human monocytes by advanced glycosylation end product-modified proteins. J Clin Invest 1992;90:439–446.
66. Arkins S, Rebeiz N, Brunke-Reese DL, Minshall C, Kelley KW. The colony-stimulating factors induce expression of insulin-like growth factor-I messenger ribonucleic acid during hematopoiesis. Endocrinology 1995;136:1153–1160.
67. Arkins S, Rebeiz N, Brunke-Reese DL, Biragyn A, Kelley KW. Interferon-gamma inhibits macrophage insulin-like growth factor-1 synthesis at hte transcriptional level. Mol Endocrin 1995;9:350–360.
68. Baxter RC, Dai J. Purification and characterization of the acid-labile subunit of rat serum insulin-like growth factor binding protein complex. Endocrinology 1994;134:848–852.
69. Neely EK, Smith SD, Rosenfeld RG. Human leukemic T and B lymphoblasts produce insulin-like growth factor binding proteins 2 and 4. Acta Endocrinol 1991;124:707–714.
70. Li YM, Arkins S, McCusker RH, Jr., Donovan SM, Liu Q, Jayaraman S et al. Macrophages synthesize and secrete a 25 KDa protein that binds insulin-like growth factor-I. J Immunol

1996;156:64–72.
71. Niandong L, Hichem C, Mertani CNG, Tornell J, Lobie PE. Mode of the autocrine/paracrine mechanism of growth hormone action. Exp Cell Res 1997;237:196–206.
72. Lobie PE, Mertani H, Morel G, Morales-Bustos O, Norstedt G, Waters MJ. Receptor-mediated nuclear translocation of growth hormone. J Biol Chem 1994;269:21330–21339.
73. Argetsinger LS, Carter-Su C. Mechanism of signaling by growth hormone receptor. Physiol Rev 1996;76:1089–1107.

Growth and Lactogenic Hormones
Edited by L. Matera and R. Rapaport

Potential Applications of Growth Hormone in Promoting Immune Reconstitution

LISBETH WELNIAK, RUI SUN and WILLIAM J. MURPHY

Laboratory of Molecular Immunoregulation, National Cancer Institute-Frederick Cancer Research and Development Center, Frederick, MD. Intramural Research Support Program, SAIC, NCI-Frederick, Frederick, MD, USA

ABSTRACT

With the increasing use of bone marrow transplantation (BMT) in cancer and in the advent of AIDS, it has been realized that successful reconstitution of the immune system of the adult is of paramount concern. Naive T cell production in the host requires T cell development in the thymus of the adult. Due to the impairment of thymus function with age, there has been renewed interest in utilizing neuroendocrine hormones (i.e. growth hormone or GH) to restore thymopoietic function. GH has been previously demonstrated to improve T cell function and affect thymopoiesis in mice. Recent studies indicate that GH is not an obligate growth factor for thymopoiesis but instead acts to counteract the effects of stress on the thymus. Thus, GH may be of potential use to enhance thymus function and T cell repopulation, particularly after myeloablative procedures such as BMT or where peripheral T cell pools are depleted as in AIDS.

1. THE CLINICAL SPECTRUM WHERE IMMUNE RECONSTITUTION NEEDS TO OCCUR

Bone marrow transplantation (BMT) is being increasingly used as a treatment for both neoplastic and non-neoplastic diseases. Unfortunately, serious obstacles currently limit the efficacy of BMT. During the BMT regimen, the patient undergoes myeloablative conditioning, usually in the form of chemotherapy or irradiation, leaving the patient severely immunosuppressed and at risk for opportunistic infections. This need to protect the patient from opportunistic infections until myeloid recovery (i.e. platelets and neutrophils) occurs, partially accounts for the high cost of BMT. Immune reconstitution occurs much later with T cell recovery being the most delayed. When BMT is used in cancer, relapse from the original tumor is also of concern in this highly immune-suppressed state. Thus, deciphering means to promote myeloid and lymphoid reconstitution would be significantly advantageous in improving the efficacy of BMT.

A similar situation occurs in HIV, in which the mature peripheral T cell pool is depleted. In fact, in late stage HIV infection, the thymus of the affected individual is atrophied to an extent more profound then a normal atrophic thymus [1–3]. Studies have shown that redistribution of T cells from tissues to the peripheral blood accounts for the much of the early rise in CD4+ cell counts during highly active anti-retroviral therapy (HAART) [4,5] and that the CD4+ T cell pool in HIV infected adults is maintained primarily through proliferation of mature CD4+ T cells and not thymopoiesis [6,7]. Even with the advent of HAART, where the virus is rendered relatively undetectable, restoration of the T cell arm must occur for ultimate success. Thus, in both BMT and HIV infection, there is ablation of mature T cells in the periphery, followed by delayed or incomplete immune reconstitution, leaving the individual highly susceptible for infection until restoration can occur.

2. T CELL DEVELOPMENT AND THE PIVOTAL NATURE OF THE THYMUS

It has long been thought that the thymus, the site of T cell generation and differentiation, is largely ineffective in the adult. The thymus significantly atrophies with age and long-lived T cells exist in the periphery, rendering the thymus relatively obsolete in the adult. However, the total ablation of these peripheral T cells in HIV or BMT has renewed debate as to whether the adult thymus could function and renew the supply of T cells. Indeed, it has been demonstrated that T cell reconstitution is significantly delayed in the adult BMT recipient when compared to pediatric patients [8]. The thymus can still function, as evidenced by the detection of naive T cells and T-cell receptor excision circles (TRECs) in adult BMT patients [8–10], but the ability of the thymus to repopulate the entire T cell pool appears severely limited. The advent of HIV and the intrusive measures of BMT have created a situation that nature had not originally considered when devising a means to establish and maintain the immune system; the need to completely replace the T cell compartment in an adult. Thus, it is imperative to devise means or characterize factors that can promote thymopoiesis in the adult. Simply promoting mature T cell function will not suffice as the entire T cell pool needs to be replenished.

In order to determine if agents, such as growth hormone, can be used to promote T cell recovery, it is important to understand the various mechanisms involved in T cell differentiation. T cell development begins in the bone marrow with multi-potential hematopoietic stem cells (HSC) which are capable of self-renewal and capable of giving rise to all lineages of hematopoietically-derived cells (i.e. lymphocytes). Upon receiving the appropriate stimuli, the HSC can differentiate into lymphoid stem cells. Growth hormone (GH) has been reported to play a role in myeloid development [11,12], thus potentially affecting T cell development indirectly. The precise stages at which GH can affect hematopoietic development is unclear (Figure 1). It is also possible that GH can affect hematopoiesis via the production of IGF-1 which has also been shown to affect myeloid growth both in vitro and in vivo [12,13]. GH has been shown to promote hematopoietic recovery after BMT in mice [14], making it potentially attractive to promote recovery after clinical BMT. The precise effects of GH on lymphoid stem cells is still undetermined. These cells can emigrate from the bone marrow to the thymus where they undergo differentiation into pre-T cells for eventual T cell receptor (TCR) rearrangement. It is also possible that GH and or IGF-1 can affect the homing of these pro-T cells to the thymus as GH has been shown to affect T cell trafficking in vivo [15] and IGF-1 potentiates thymic colonization [16]. Once in the thymus, the pre-T cells under receptor rearrangement and proceed in differentiation where they undergo "education" in the form of positive and negative selection based on their TCR specificity (in the context of self MHC). Those T cells surviving positive and negative selection then emigrate out of the thymus and proceed to secondary lymphoid organs (i.e. lymph nodes and

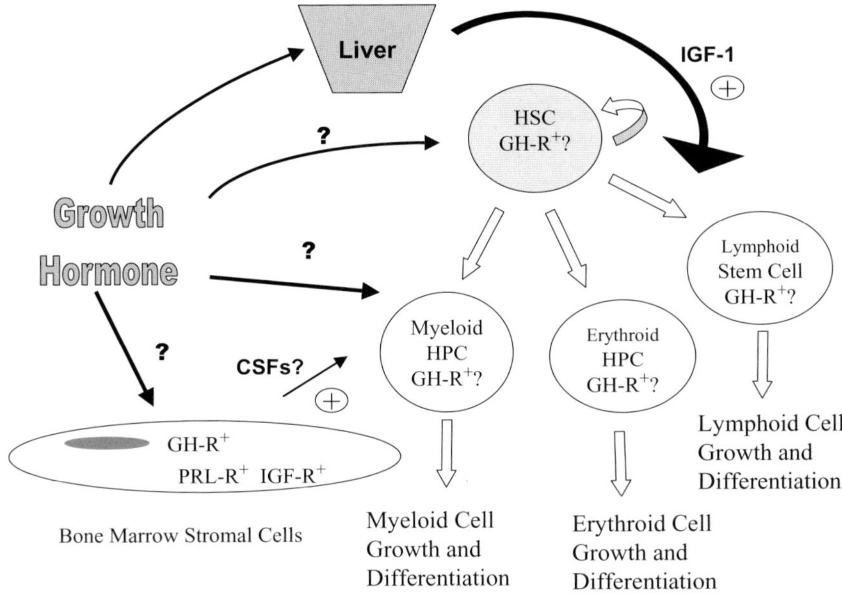

Figure 1. Potential role of GH in hematopoiesis. Abbreviations: GH-R+, cells expressing GH receptors; PRL-R+, cells expressing prolactin receptors; IGF-R+, cells expressing IGF receptors; HSC, hematopoietic stem cells; HPC, hematopoietic progenitor cells; CSF, colony-stimulating factors.

spleen) where these naive T cells wait for encounter with the appropriate antigens. It is therefore possible that GH may affect T cell recovery and thymopoiesis by affecting early hematopoietic development and lymphoid stem cell output and subsequent homing to the thymus.

3. FACTORS AFFECTING THYMOPOIESIS: CONFLICTING DATA ABOUT GROWTH HORMONE

There have been numerous growth factors that have been demonstrated to affect immune cell survival, differentiation, and/or function. However, there are very few that have been demonstrated to affect thymic function. Numerous cytokines (i.e. IL-2, interferons, etc.) have been shown to affect mature T cell function both in vitro and in vivo. However, agents that can promote thymic function, either in aged situations or in BMT, have been difficult to definitively demonstrate. One issue revolves around the unique nature of the thymus itself. The thymus is exquisitely sensitive to stress stimuli via corticosteroids [17]. As opposed to B cells, the production of corticosteroids in response to stress stimulus can result in the complete ablation of pre-T cells (i.e. CD4/CD8 double-positive cells) which comprise the majority of the cells in the thymus (reviewed in [18]). Therefore, there may be agents that can affect thymopoiesis directly but also agents that may affect overall thymic output by merely making the thymocyte resistant to the apoptotic effects of glucocorticoids. IL7 is a cytokine that has been under intense scrutiny as a potential thymopoietic agent as it has been demonstrated to be critical in TCR rearrangement and early T cell survival [19]. Another agent of interest is growth hormone (GH). GH has been long suggested to play a role in T cell development. Initial studies had centered on the char-

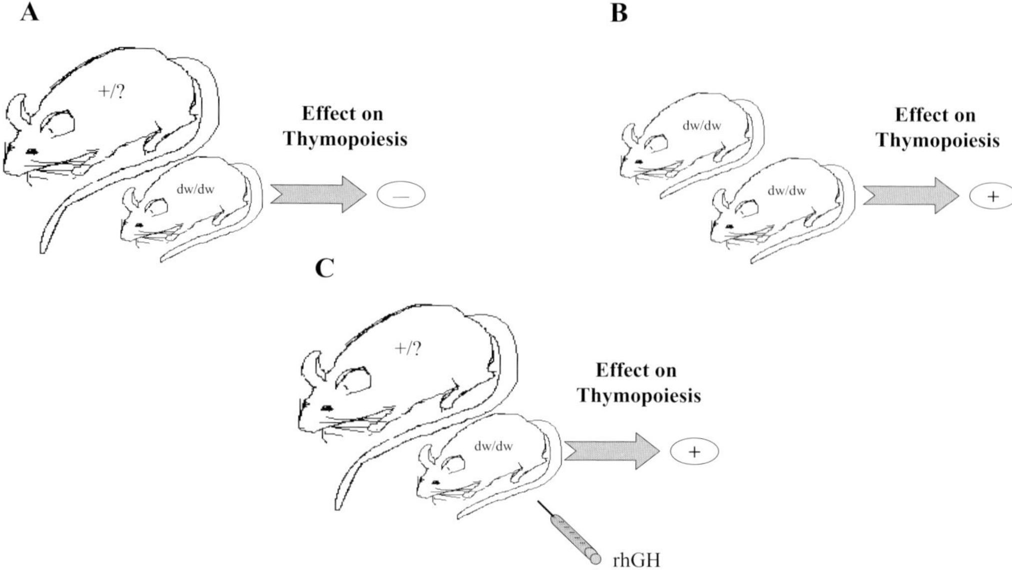

Figure 2. Effects of housing conditions on the thymus of dwarf mice. (A) If dwarf mice are housed with their normal littermates a reduction of thymus size is observed. (B) However, if dwarf mice are housed separate from their normal-sized littermates, a normal thymus (relative to body weight) is observed. (C) Treatment of the dwarf mice housed with littermates with GH can restore thymus cellularity.

acterization of dwarf mice who were deficient in anterior pituitary hormones (i.e. GH, prolactin, thyroxine). These mice were reported to have a severely hypoplastic thymus and some reports described susceptibility to infection and early death [20,21]. We had also detected thymic abnormalities in these mice but no outwardly deleterious effects on overall health [22].

We reasoned that the differences in health of the mice may have been attributable to the differences in mouse housing conditions over the years. Most mouse colonies are now specific-pathogen-free (SPF), and thus the mice are simply not exposed to the extent they were when the earlier studies were performed. We found the CD4/CD8 pre-T cell numbers in these mice were significantly reduced and administration of GH could restore thymic cellularity [22]. Importantly, amounts of GH were used that did not result in significant weight gain suggesting that the thymopoietic effects were independent of the anabolic effects. However, other groups reported that thymic cellularity in these mice was relatively normal [23–25]. Careful examination on any potential differences in experimental protocols was then performed as the dwarf mice were obtained from the same source. One condition that stood out was the differences in housing. The study that reported no defect in thymic cellularity had the mice housed alone, whereas we had our dwarf mice housed with their normal-sized littermates. The mice that were used were all female and fighting among female mice is not a usual occurrence. However, it was possible that the dwarf mice were stressed by the presence of their normal-sized littermates. An experiment was performed altering the housing conditions, placing some dwarf mice with their normal littermates and others alone. The results reconciled the data from the two laboratories in that dwarf mice housed alone had a relatively normal thymus whereas the dwarf mice housed with their littermates had a hypoplastic thymus (Figure 2). Interestingly, the administration of GH could

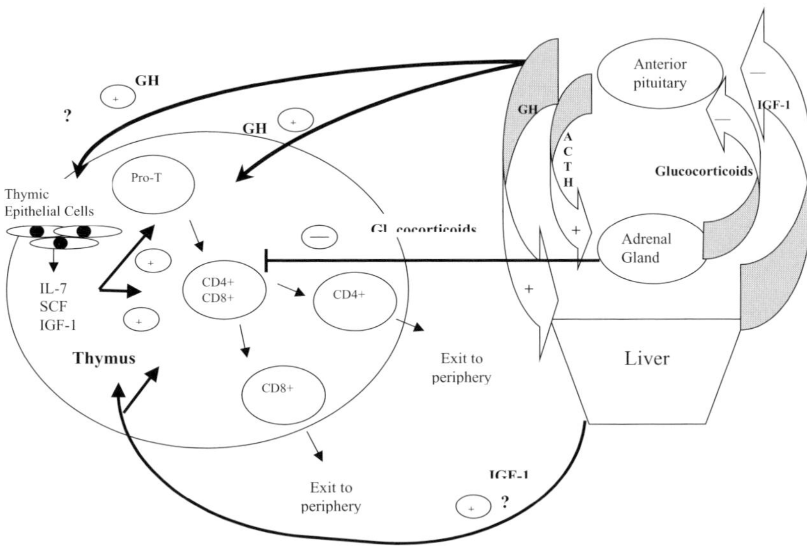

Figure 3. Potential role of GH on thymopoiesis. While GH and IGF-1 are produced in lymphoid tissues, the primary source of circulating GH is the anterior pituitary which is controlled through a negative feedback loop following IGF-1 production in the liver. Possible targets for the pro-thymopoietic effects are illustrated. Glucocorticoids can induce apoptosis in CD4+CD8+ thymocytes. The mechanism by which GH reverses the effect of glucocorticoids on thymopoiesis is not known.

reverse this sensitivity to stress within the thymus (W.J.Murphy and K. Dorshkind, manuscript in preparation).

These results would then suggest that the thymopoietic effects attributed to GH may solely be in the ability to make the thymus resistant to stress. This may reconcile the previously contradictory results observed from different laboratories over the years when neuroendocrine hormone deficient dwarf mice were assessed for their T cell status. The housing conditions, and most likely, conditions in the colony, markedly affected the T cell compartment due to their sensitivity to stress. Experiments involving the adrenalectomy of dwarf mice would provide definitive data as to the role of stress in this situation. How GH is protecting the thymus from stress is unclear. Until direct assessment of glucocorticoid production is ascertained it is possible, although unlikely, that GH may affect secretion of these immunosuppressive agents during a stress response. More likely, GH may affect pre-T cell survival. This could occur directly or through secondary mediators such as IGF-1 which has also been shown to affect thymus cellularity [26] or through the production of cytokines (i.e. IL7) by thymic epithelial cells (Figure 3). As BMT and HIV are conditions where stress can occur, it is possible that significant thymopoietic effects can be obtained if GH is applied clinically. Additionally, the previous studies were performed in normal, resting (i.e. untreated), and young dwarf mice. It is also possible that GH can affect thymopoiesis directly in the aged individual.

More work needs to be performed to determine if the affects of GH may depend on the age of the recipient. For example, younger individuals may only show thymopoietic effects of GH when under stress conditions whereas older individuals, due to the atrophied nature of the thymus, may show direct thymopoietic effects of GH. It will be interesting to determine the differences of expression of GH and IGF-1 receptors in the thymus with age. It is also of interest to note that the susceptibility of the thymus to stress was most noticeable in mice deficient in GH, not normal mice. This would imply that the paramount role of GH in T cell development would be in the resistance of the thymus to stress and that GH is not an obligatory growth factor in T cell differentiation.

4. ADDITIONAL IMMUNOMODULATORY EFFECTS OF GROWTH HORMONE

Increasing evidence shows that administration of recombinant human growth hormone (rhGH) systemically alter immunologic function in animal models and in the clinical setting. This work has been supported by the finding that GH receptors and IGF receptors are present on murine thymocytes, human B cells and on subsets of human T and NK cells [27–29]. Initially, the major role for the immunomodulatory function of GH administration was thought to occur through improvement of thymic function, but there is evidence for activity on peripheral immune function. GH was able to stimulate the in vitro proliferation of ConA or anti-CD3-activated murine T lymphocytes, confirming the biological significance of the receptors present on these cells. The proliferative effect of GH is shown exclusively on activated T lymphocytes [30]. Additional studies demonstrate augmentation of peripheral T cell function by GH by stimulation of human Th0 and Th2 clones in the presence of antigen, neutralizing antibodies to GH suggest endogenous GH activity in Th1 clone proliferation [31] and enhanced Th1 function was observed in an in vivo murine burn model following rhGH treatment [32].

GH receptors have been detected on murine B lymphocytes and lipopolysaccharide (LPS) treatment increases GH binding, however, GH did not alter B lymphocyte proliferation [33].
GH-deficiency in humans and rodents results in reduced NK cell number and activity compared to normal cohorts [34]. Although GH or GH-releasing hormone treatment has not always resulted in restoration of NK cell function [34,35]. rhGH may reverse alterations in immunologic function in animals receiving chemotherapy. In these studies, NK cell activity was significantly elevated, when compared to the same population in the chemotherapy group [36].
In GH deficient patients, markers of monocyte activation such as tumor necrosis factor (TNF) -α and interleukin-6 production are increased and GH replacement reverses some but not all of the observed abnormalities [37]. Despite the observed suppression of monocyte activity in the previous clinical situation, in a porcine sepsis model, administration of rhGH did not diminish granulocyte function but it did lower serum TNF-α [38]. GH enhances macrophage function as it is a potent chemoattractant [39], increases interferon gamma secretion and macrophage function [40] in a mouse viral infection model and augments anti-bacterial activity in vitro [41] and in vivo [42]. Overall, these studies demonstrate that GH administration directly or indirectly impacts macrophage function. The mechanism of GH action of myeloid cells can occur through induction of IGF-1 or GH engagement of prolactin receptors [43].

5. FUTURE DIRECTIONS

It appears that GH and possibly IGF-I can be used to promote thymopoiesis during states in which the peripheral T cell compartment has been ablated. The mechanism(s) by GH can exert thymopoietic effects is not clear, although providing resistance during stress responses appears to be at least one mechanism. It is also possible that GH can act directly on the thymus, particularly during aging, although more work needs to be performed addressing the mechanisms of thymic involution during aging and if the normal loss of GH with age contributes to this. GH appears to affect multiple stages in hematopoietic development as well as peripheral immune cell function. While these properties would also be advantageous after BMT, it is the potential thymopoietic effects that make GH particularly attractive, not only after BMT but also in AIDS. There are still unanswered questions as to how GH administration may also affect tumor growth if it is used after BMT in cancer. Additionally, due to the immuno-stimulatory properties of GH, it is most likely imperative that it be given in conjunction with HAART in HIV although this may also provide a means to remove the virus reservoir that often exists. The relatively low toxicities observed after chronic GH administration also make it attractive, particularly when compared with the often severe toxicities observed after systemic administration of cytokines. Thus, GH may be of considerable use to augment T cell recovery in BMT and HIV, as well as possibly to promote thymic function in the elderly but more work needs to be performed to adequately assess the real potential of this approach.

REFERENCES

1. Haynes BF, Markert ML, Sempowski GD, Patel DD, Hale LP. The role of the thymus in immune reconstitution in aging, bone marrow transplantation, and HIV-1 infection. Annu Rev Immunol 2000;529–60.
2. Haynes BF, Hale LP, Weinhold KJ, Patel DD, Liao HX, Bressler PB, Jones DM, Demarest JF, Gebhard-Mitchell K, Haase AT, Bartlett JA. Analysis of the adult thymus in reconstitution of T lymphocytes in HIV- 1 infection. J Clin Invest 1999;453–60.
3. Schuurman HJ, Krone WJ, Broekhuizen R, van Baarlen J, van Veen P, Golstein AL, Huber J, Goudsmit J. The thymus in acquired immune deficiency syndrome. Comparison with other types of immunodeficiency diseases, and presence of components of human immunodeficiency virus type 1. Am J Pathol 1989;1329–38.
4. Pakker NG, Notermans DW, de Boer RJ, Roos MT, de Wolf F, Hill A, Leonard JM, Danner SA, Miedema F, Schellekens PT. Biphasic kinetics of peripheral blood T cells after triple combination therapy in HIV-1 infection: a composite of redistribution and proliferation. Nat Med 1998;208–14.
5. Bucy RP, Hockett RD, Derdeyn CA, Saag MS, Squires K, Sillers M, Mitsuyasu RT, Kilby JM. Initial increase in blood CD4(+) lymphocytes after HIV antiretroviral therapy reflects redistribution from lymphoid tissues. J Clin Invest 1999;1391–8.
6. Zhang L, Lewin SR, Markowitz M, Lin HH, Skulsky E, Karanicolas R, He Y, Jin X, Tuttleton S, Vesanen M, Spiegel H, Kost R, van Lunzen J, Stellbrink HJ, Wolinsky S, Borkowsky W, Palumbo P, Kostrikis LG, Ho DD. Measuring recent thymic emigrants in blood of normal and HIV-1-infected individuals before and after effective therapy. J Exp Med 1999;725–32.
7. Walker RE, Carter CS, Muul L, Natarajan V, Herpin BR, Leitman SF, Klein HG, Mullen CA, Metcalf JA, Baseler M, Falloon J, Davey RT, Jr., Kovacs JA, Polis MA, Masur H, Blaese RM, Lane HC. Peripheral expansion of pre-existing mature T cells is an important means of CD4+ T-cell regeneration HIV-infected adults. Nat Med 1998;852–6.
8. Small TN, Papadopoulos EB, Boulad F, Black P, Castro-Malaspina H, Childs BH, Collins N, Gillio A, George D, Jakubowski A, Heller G, Fazzari M, Kernan N, MacKinnon S, Szabolcs P, Young JW, O'Reilly RJ. Comparison of immune reconstitution after unrelated

and related T-cell- depleted bone marrow transplantation: effect of patient age and donor leukocyte infusions. Blood 1999;467–80.

9. Weinberg K, Blazar BR, Wagner JE, Agura E, Hill BJ, Smogorzewska M, Koup RA, Betts MR, Collins RH, Douek DC. Factors affecting thymic function after allogeneic hematopoietic stem cell transplantation. Blood 2001;1458–66.
10. Douek DC, Vescio RA, Betts MR, Brenchley JM, Hill BJ, Zhang L, Berenson JR, Collins RH, Koup RA. Assessment of thymic output in adults after haematopoietic stem-cell transplantation and prediction of T-cell reconstitution. Lancet 2000;1875–81.
11. Murphy W, Durum S, Anver M, Longo D. Immunologic and hematologic effects of neuroendocrine hormones. Studies on DW/J dwarf mice. J Immunol 1992;3799–805.
12. Merchav S, Tatarsky I, Hochberg Z. Enhancement of human granulopoiesis in vitro by biosynthetic insulin- like growth factor I/somatomedin C and human growth hormone. J Clin Invest 1988;791–7.
13. Tsarfaty G, Longo D, Murphy W. Human insulin-like growth factor I exerts hematopoietic growth-promoting effects after in vivo administration. Exp Hematol 1994;1273–7.
14. Tian ZG, Woody MA, Sun R, Welniak LA, Raziuddin A, Funakoshi S, Tsarfaty G, Longo DL, Murphy WJ. Recombinant human growth hormone promotes hematopoietic reconstitution after syngeneic bone marrow transplantation in mice. Stem Cells 1998;193–9.
15. Taub D, Tsarfaty G, Lloyd A, Durum S, Longo D, Murphy W. Growth hormone promotes human T cell adhesion and migration to both human and murine matrix proteins in vitro and directly promotes xenogeneic engraftment. J Clin Invest 1994;293–300.
16. Montecino-Rodriguez E, Clark R, Dorshkind K. Effects of insulin-like growth factor administration and bone marrow transplantation on thymopoiesis in aged mice. Endocrinology 1998;4120–6.
17. Wyllie AH. Glucocorticoid-induced thymocyte apoptosis is associated with endogenous endonuclease activation. Nature 1980;555–6.
18. Ashwell JD, Lu FW, Vacchio MS. Glucocorticoids in T cell development and function. Annu Rev Immunol 2000;309–45.
19. Kim K, Lee CK, Sayers TJ, Muegge K, Durum SK. The trophic action of IL-7 on pro-T cells: inhibition of apoptosis of pro-T1, -T2, and -T3 cells correlates with Bcl-2 and Bax levels and is independent of Fas and p53 pathways. J Immunol 1998;5735–41.
20. Baroni C. Thymus, peripheral lymphoid tissues and immunological responsiveness of the pituitary dwarf mouse. Experientia 1967;282–3.
21. Duquesnoy RJ. The pituitary dwarf mouse: a model for study of endocrine immunodeficiency disease. Birth Defects Orig Artic Ser 1975;536–43.
22. Murphy W, Durum S, Longo D. Role of neuroendocrine hormones in murine T cell development. Growth hormone exerts thymopoietic effects in vivo. J Immunol 1992;3851–7.
23. Montecino-Rodriguez E, Clark RG, Powell-Braxton L, Dorshkind K. Primary B cell development is impaired in mice with defects of the pituitary/thyroid axis. J Immunol 1997;2712–9.
24. Dumont F, Robert F, Bischoff P. T and B lymphocytes in pituitary dwarf Snell-Bagg mice. Immunology 1979;23–31.
25. Schneider GB. Immunological competence in Snell-Bagg pituitary dwarf mice: response to the contact-sensitizing agent oxazolone. Am J Anat 1976;371–93.
26. Jardieu P, Clark R, Mortensen D, Dorshkind K. In vivo administration of insulin-like

growth factor-I stimulates primary B lymphopoiesis and enhances lymphocyte recovery after bone marrow transplantation. J Immunol 1994;4320–7.

27. Arrenbrecht S. Specific binding of growth hormone to thymocytes. Nature 1974;255–7.
28. Badolato R, Bond HM, Valerio G, Petrella A, Morrone G, Waters MJ, Venuta S, Tenore A. Differential expression of surface membrane growth hormone receptor on human peripheral blood lymphocytes detected by dual fluorochrome flow cytometry. J Clin Endocrinol Metab 1994;984–90.
29. Tapson VF, Boni-Schnetzler M, Pilch PF, Center DM, Berman JS. Structural and functional characterization of the human T lymphocyte receptor for insulin-like growth factor I in vitro. J Clin Invest 1988;950–7.
30. Gagnerault M, Postel-Vinay M, Dardenne M. Expression of growth hormone receptors in murine lymphoid cells analyzed by flow cytofluorometry. Endocrinology 1996;1719–1726.
31. Yamashita N, Hashimoto Y, Honjo M. The effect of growth hormone on the proliferation of human Th cell clones. Life Sci 2000;1929–35.
32. Takagi K, Suzuki F, Barrow RE, Wolf SE, Herndon DN. Recombinant human growth hormone modulates Th1 and Th2 cytokine response in burned mice. Ann Surg 1998;106–11.
33. Postel-Vinay MC, de Mello Coelho V, Gagnerault MC, Dardenne M. Growth hormone stimulates the proliferation of activated mouse T lymphocytes. Endocrinology 1997;1816–20.
34. Kiess W, Malozowski S, Gelato M, Butenand O, Doerr H, Crisp B, Eisl E, Maluish A, Belohradsky BH. Lymphocyte subset distribution and natural killer activity in growth hormone deficiency before and during short-term treatment with growth hormone releasing hormone. Clin Immunol Immunopathol 1988;85–94.
35. Span JP, Pieters GF, Smals AG, Koopmans PP, Kloppenborg PW. Number and percentage of NK-cells are decreased in growth hormone- deficient adults. Clin Immunol Immunopathol 1996;90–2.
36. Harrison LE, Port JL, Hochwald S, Blumberg D, Burt M. Perioperative growth hormone improves wound healing and immunologic function in rats receiving adriamycin. J Surg Res 1995;646–50.
37. Serri O, St-Jacques P, Sartippour M, Renier G. Alterations of monocyte function in patients with growth hormone (GH) deficiency: effect of substitutive GH therapy. J Clin Endocrinol Metab 1999;58–63.
38. Balteskard L, Unneberg K, Halvorsen D, Ytrebo LM, Waage A, Sjursen H, Revhaug A. The influence of growth hormone on tumour necrosis factor and neutrophil leukocyte function in sepsis. Scand J Infect Dis 1997;393–9.
39. Wiedermann CJ, Reinisch N, Braunsteiner H. Stimulation of monocyte chemotaxis by human growth hormone and its deactivation by somatostatin. Blood 1993;954–60.
40. Takagi K, Suzuki F, Barrow RE, Wolf SE, Kobayashi M, Herndon DN. Growth hormone improves immune function and survival in burned mice infected with herpes simplex virus type 1. J Surg Res 1997;166–70.
41. Inoue T, Saito H, Hashiguchi Y, Fukatsu K, Inaba T, Lin MT, Han I, Furukawa S, Muto T. Growth hormone and insulin-like growth factor I augment Escherichia coli-killing activity of murine peritoneal exudative cells. Shock 1996;345–50.
42. Edwards CK, 3rd, Yunger LM, Lorence RM, Dantzer R, Kelley KW. The pituitary gland is required for protection against lethal effects of Salmonella typhimurium. Proc Natl Acad

Sci USA 1991;2274–7.
43. Wiedermann CJ, Reinisch N, Kahler C, Braunsteiner H. Regulation of myeloid phagocyte development and function by growth hormone: a review. J Pediatr Endocrinol 1993;85–91.

Growth and Lactogenic Hormones
Edited by L. Matera and R. Rapaport

Signal Transduction by Prolactin Receptors

LI-YUAN YU-LEE

Departments of Medicine, Molecular & Cellular Biology, and Immunology, Baylor College of Medicine, Houston, Texas, USA

ABSTRACT

The pleiotropic actions of prolactin (PRL) are mediated by its receptor, PRL receptor (PRL-R), which is a member of the hematopoietin cytokine receptor superfamily. PRL can act either as an endocrine hormone secreted by the pituitary or as a local cytokine produced in many extra-pituitary sites. Signaling from the PRL-R mediates the numerous biological activities of PRL, including proliferation, differentiation, apoptosis and cell survival. The focus of this review is on PRL-R signaling as part of a proliferation pathway to regulate the expression of a target gene, the transcription factor interferon regulatory factor-1 (IRF-1). PRL is involved in not only positive but also negative regulatory signaling to the IRF-1 gene. In understanding this process, we obtained insights into how PRL signaling cross talks with TNFα signaling in regulating target gene expression.

1. INTRODUCTION

The pituitary hormone prolactin (PRL) is a systemic hormone as well as an autocrine/paracrine cytokine, which mediates its diverse biological functions through the PRL receptor (PRL-R). The PRL-R belongs to the Class I cytokine receptor or hematopoietin/ cytokine receptor superfamily, and shares many structural and functional similarities with other cytokine receptors. This review focuses on PRL-R signal transduction, first on the proximal signaling pathways emanating from the PRL-R, then on the signaling molecules that mediate PRL actions, and finally on the transcription of a PRL responsive target gene. We will examine both positive and negative signaling by PRL in regulating the expression of the interferon regulatory factor 1 (IRF-1) gene as an example. From our molecular analysis, we provide a possible explanation of how PRL signaling cross-talks with the TNFα signaling at the IRF-1 promoter.

2. PRL TARGET TISSUES

PRL is a peptide hormone that is synthesized and secreted primarily by lactotrophic cells in the anterior pituitary. PRL is also expressed in a number of extra-pituitary sites [1], ranging from neurons in the brain to epithelial cells of secretory glands to cells of the immune system [2,3]. PRL acts on a wide range of tissues, with over 300 effects described in vertebrates [3]. PRL regulates *differentiation* of the mammary gland, ovary, male sex accessory organs, submaxillary and lacrimal glands, pancreas and liver [4]. PRL regulates *proliferation* in the pigeon crop sac epithelium, pancreatic beta cells, astrocytes, anterior pituitary cells and T lymphocytes [4,5]. PRL also exerts *anti-apoptotic* effects in lymphocytes undergoing glucocorticoid-induced programmed cell death. Additional PRL target tissues were confirmed in mice in which either the PRL or the PRL-R gene was ablated, including the brain (maternal behavior), uterus (implantation), bone (osteoblasts) and adipocytes [5,6]. The apparent lack of overt immune phenotype in the PRL and PRL-R knockout (KO) animals could be due to cytokine/cytokine receptor redundancy. These KO animals have led to a renewed interest in understanding PRL as a stress-adaptation hormone/cytokine [7] whose immunoregulatory properties will be manifested under conditions of stress (refer to the accompanying articles in this issue). How pituitary or extra-pituitary PRL modulates the growth, differentiation and function of target tissues also depends on the cell type and its stage of differentiation.

3. PRL-R SIGNAL TRANSDUCTION

3.1. PRL-R structure

The diverse activities of PRL are mediated by the PRL-R which is expressed on many cell types. The PRL-R is encoded by a single gene [8]. However, several receptor forms exist, including the long (90 kDa) and short (42 kDa) form PRL-Rs, which result from differential splicing of 3' end exons encoding the cytoplasmic domain (Figure 1) [8]. A naturally occurring intermediate PRL-R form (65 kDa), found in rat Nb2 T lymphoma cells, results from an in-frame truncation in the intracellular domain (ICD), generating a shortened receptor tail. A similar intermediate PRL-R form is found in human mammary tumors [9]. Several motifs in the PRL-R ICD are important for signal transduction. A proline-rich sequence (I-F-P-P-V-P-X-P) proximal to the transmembrane domain is important for interacting with receptor-associated protein tyrosine kinase (PTK) JAK2 [8]. Several tyrosine residues, presumably phosphorylated by JAK2, are critical for receptor signaling. The last tyrosine Y382 in the Nb2 PRL-R, or its equivalent Y580 in the long PRL-R, is important for mediating differentiated functions [8], while both Y309 and Y382 in the Nb2 PRL-R are needed for signaling to an immediate early response gene IRF-1 [10] (Figure 1). One function of phosphorylated receptor tyrosine residues is to provide a "docking site" for the recruitment of SH2-containing proteins, including Stats, phosphatases and adaptor molecules [8]. Other as yet undefined PRL-R ICD regions interact with cytoplasmic factors that modulate PRL-R signaling [11,12]. The intermediate Nb2 PRL-R is more potent than the long PRL-R in both mitogenic [13] and lactogenic [6] signaling. The short PRL-R is suggested to modulate the activity of the long or Nb2 PRL-R by engaging them in heterodimer complex formation and thereby reducing their signaling capacity [5,6]. Most tissues that express the PRL-R contain about 300–400 PRL-R/cell. In the Nb2 T cells, where PRL-R is abundant (12,000 PRL-R/cell), only 30% occupancy of surface PRL-R is needed to elicit maximal proliferative response [14].

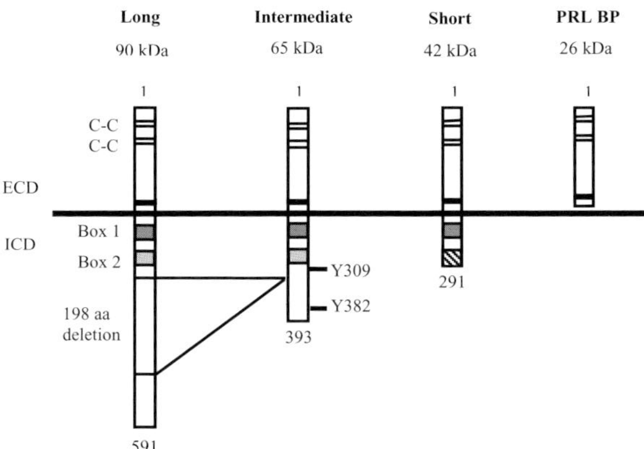

Figure 1. Schematic representation of the rat PRL-R isoforms. The extracellular domain (ECD) (1–210 aa) of the PRL-R contains two cysteine doublets (C-C) and a WS motif, which are characteristic of other Type I cytokine receptors. The intracellular domain (ICD) (235–591 aa) of the PRL-R contains a conserved Box 1 or proline-rich motif that interacts with JAK2 and an acidic Box 2 motif. An in-frame deletion (323–520 aa) generates a truncated intermediate form. The last exon of the short PRL-R is unique, resulting from alternative splicing of 3' exons. A membrane truncation generates a soluble PRL binding protein (PRLBP). Only the two critical tyrosine residues Y309 and Y382 in the intermediate Nb2 receptors are highlighted (see text for discussion). The equivalent Y residues in the long PRL-R isoform are not shown.

3.2. PRL-inducible JAK/Stat pathway

The primary signaling pathway activated by PRL binding to PRL-R is the "JAK/Stat pathway" [15], which is used by all hematopoietin/cytokine receptors (Figure 2). One of the first molecules to be activated in the PRL-R signaling pathway is the PTK JAK2. It is interesting to note that JAK2 is prebound to the inactive PRL-R monomer, in contrast to other cytokine receptors where JAK PTKs are recruited into the receptor complex. Briefly, upon ligand binding, the PRL-R homodimerizes, bringing together receptor-associated JAK2 [8]. Activated JAK2 then phosphorylates downstream targets, including tyrosine residues on the PRL-R ICD and a family of latent transcription factors called Stats (signal transducers and activators of transcription) [3,13]. Stats 1, 3 and 5 become activated by tyrosine phosphorylation and form homo- (Stat1/1, Stat3/3, Stat5/5) or heteromeric (Stat1/3) complexes, translocate into the nucleus, bind to conserved DNA elements called interferon (IFN) gamma activated sequence (GAS) and regulate target gene transcription. One unique feature of the JAK/Stat pathway is that all of the signaling components are already pre-existing in the cytoplasm such that signaling is initiated within minutes (1–5 min) by a series of tyrosine phosphorylation events, and transcription of nuclear target genes are detected within minutes (5–10 min) of PRL stimulation.

3.3. JAK2 and parallel kinase pathways

In addition to the JAK/Stat signaling pathway, other PTKs are also activated by PRL stimulation in different target cells. These include Fyn, Src, Ras, and Raf, as well as serine/threonine kinases such as ZAP-70, PI3 kinase, MAPK, JNK and protein kinase C [5] (Figure 3). Activation of these parallel kinase cascades in coordination with the JAK/Stat signaling pathway elicits

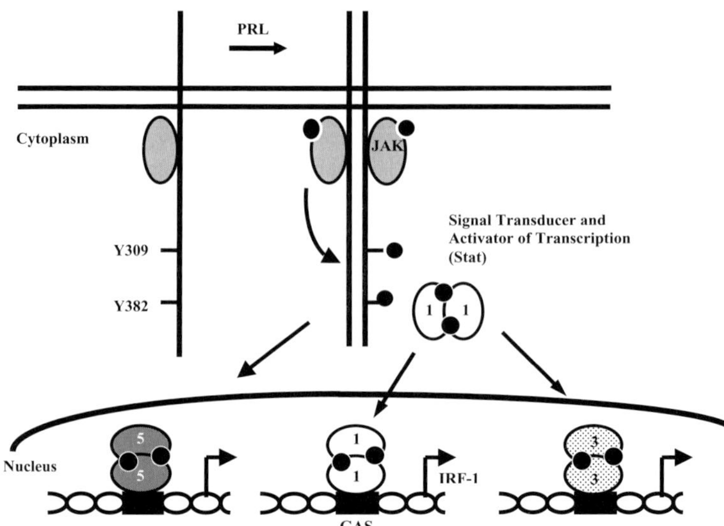

Figure 2. PRL-R signaling pathway. Rapid signaling through the JAK/Stat pathway is possible as all of the signaling components are pre-existing, and signaling is initiated by a series of tyrosine phosphorylation events. See text for details. JAK, Janus kinase; IRF-1, interferon regulatory factor 1; GAS, Interferon gamma activation sequence.

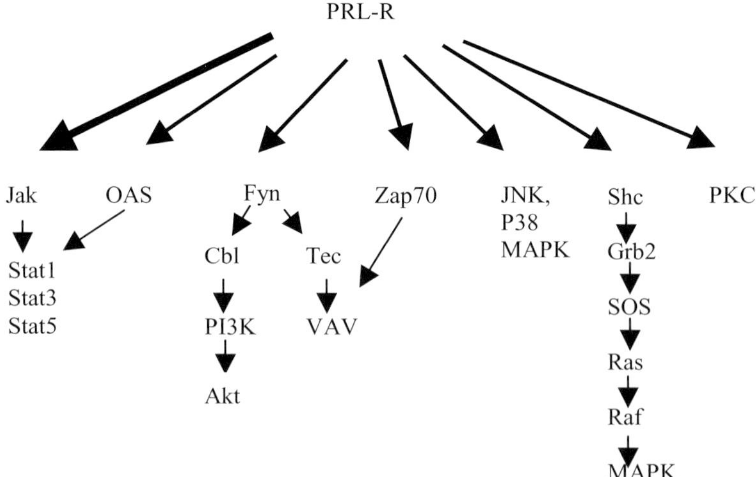

Figure 3. PRL inducible JAK/Stat and parallel kinase signaling pathways. One of the most well-characterized signaling pathways activated by PRL stimulation is the JAK/Stat pathway (bold arrow). Depending on the cell type, other kinase cascades are also activated in parallel. JAK2, Fyn and OAS appear to be constitutively associated with the PRL-R, while the other molecules are recruited into the PRL-R complex upon PRL stimulation. How cells respond to PRL is likely to be determined by a combination of the various kinase pathways activated in a cell type specific manner.

specific patterns of gene expression in various PRL responsive cells and tissues. The pleiotropic actions of PRL on cellular proliferation, differentiation or apoptosis will likely be determined by the interactions amongst these parallel kinase cascades.

4. PRL-R SIGNALING MOLECULES

4.1. Stat factors

Stat factors are one of the earliest mediators of signaling from cytokine receptors [15,18]. Seven mammalian Stat genes, Stat1–4, 5a, 5b and 6, have been identified. Stat factors are in general 750–800 amino acids in size and contain distinct functional domains. These include a coiled-coiled domain, DNA binding domain, linker domain, SH2 domain, a critical tyrosine residue that is important for dimerization, nuclear translocation and DNA binding, and a carboxyl terminus transactivation domain [18]. Additional serine [16] and tyrosine [19] residues in the carboxyl terminus of Stat1, Stat3 and Stat5 further contribute to the ability of these factors to regulate gene transcription. Stat1, 3, 5a, 5b and 6 contain naturally occurring splice variants with truncations in the carboxyl terminus, generating dominant negative β isoforms [20,21]. Stat factors utilize various domains to interact/cross talk with a diverse set of proteins, both in the cytoplasm and nucleus, to mediate target gene transcription.

4.2. Stat and cytoplasmic protein interactions

Stat factors reside basally in the cytoplasm of unstimulated cells. In addition to being recruited into the receptor complex, Stats may also directly interact with JAK PTK. Stat dimers can undergo tetramer formation through their amino terminus to bind to tandemly-occurring weak affinity sites [22]. The coiled-coil domain of Stats (except Stat2) can interact with the cytoplasmic N-myc interacting protein (Nmi) [15], forming a Stat/Nmi complex which enhances Stat transactivation potentials. Other Stat interacting proteins include StIP1 (***St***at3 ***i***nteracting ***p***rotein) which interacts with both JAK2 and Stat3 [23] and PIAS (***p***rotein ***i***nhibitors of ***a***ctivated ***S***tats) [17], which downregulates Stat transcriptional activity. Stat1 also interacts in the cytoplasm with the nuclear transport importin α/β complex for transport into the nucleus [24]. Thus, in the cytoplasm, Stats acquire *signal-transducing* capability via interactions with cytokine receptors, JAK and other factors to mediate transcriptional responses.

4.3. Stat and nuclear protein interactions

Activated Stat complexes translocate within minutes into the nucleus [15]. Once in the nucleus, Stats can interact with other nuclear proteins, bind to cognate DNA elements (ISRE or GAS), and participate in gene transcription (*activators of transcription*). The transactivation potentials of Stats are modulated by interactions with nuclear proteins, such as p48 (a member of the IRF family), IRF-1, c-jun, Sp1, Src, nuclear hormone receptors, MCM5 and BRCA1 [17,18,25,26]. The activities of Stat proteins are further enhanced by their interaction with a class of nuclear factors called coactivators, which by themselves do not bind DNA but can integrate the activities of multiple DNA binding proteins by protein-protein interactions [27]. Co-activators not only facilitate interactions with other co-activators as well as with components of the basal transcription machinery, but many co-activators also exhibit intrinsic histone acetyltransferase (HAT) activities, which acetylate histones and participate in remodeling of chromatin and thereby enhance transcriptional activation of genes. Stat1 interacts with three regions within the co-activator protein CBP/p300 [18,28]. One of the Stat1 interacting regions in CBP/p300 also interacts with Stat5 [29], leading to the speculation that Stat5 competition with Stat1 for binding to CBP/p300 forms one basis for competitive action between these two Stats at target promoters (see below). Thus, coactivators can integrate the activities of Stats with other factors in regulating gene transcription. Conversely, coactivators are targets of competitive binding between different Stats or between Stats and other proteins which can lead to inhibition of gene transcription [30]. Such competitive interaction of different transcription factors for a limiting pool of co-activator proteins results in "squelching" of CBP/p300 and inhibition of gene transcription [27,31].

4.4. PRL-inducible stats

PRL stimulates the rapid tyrosine phosphorylation of Stat1, Stat3 and Stat5 in many cell types (Figure 2) [32,34]. Stat5a was first cloned as a PRL-inducible mammary gland factor (MGF), and is followed by the cloning of the closely-related Stat5b [35,36]. PRL stimulates primarily Stat5 to regulate transcription of the milk protein and other genes, which are markers of differentiation [26,34,37]. PRL stimulates Stat1 to regulate transcription of the IRF-1 gene, an immediate early response gene in T cells [10]. Unexpectedly, PRL inducible Stat5 inhibits IRF-1 promoter activity [31,36].

4.5. PRL stimulates IRF-1 gene transcription

The IRF-1 gene is a PRL-inducible immediate-early response gene that is activated during mitogenic stimulation in Nb2 T cells [2,38,39]. IRF-1, one of the most PRL-responsive genes, is itself a multifunctional transcription factor that regulates the expression of numerous genes involved in mediating T helper and innate immune responses [2]. PRL also stimulates IRF-1 expression in normal rat and human leukocytes [40,41]. Additionally, mutations and/or deletions in IRF-1 are correlated with a high incidence of leukemias and myelodysplasia, suggesting a role of IRF-1 in tumor suppression [42]. In Nb2 T cells, PRL stimulates the biphasic transcription of the IRF-1 gene, with a transient but dramatic 25-fold induction during early G1 (1 hr) and a second peak of induction at the G1/S transition phase (8–10 hr) of the cell cycle [38].

4.5.1. *Positive mediators: Stat1, CBP/p300 and Sp1*

PRL stimulation of IRF-1 gene transcription during early G1 is positively mediated by at least three factors interacting at the IRF-1 promoter: Stat1 (binding to GAS at –110/–120 bp) [38], Sp1 (–200 bp) [43], and protein-protein interaction between Stat1 and the co-activator CBP/p300 [31] (Figure 4A). Our working hypothesis is that upon PRL stimulation, activated Stat1 binds to the IRF-1 GAS and together with the pre-bound Sp1 forms an enhanceosome (assembly of transcription factors) [44], which recruits coactivators such as CBP/p300 and CRSP (cofactor required for Sp1) [45], as well as the general transcription machinery, for transcriptional activation of the IRF-1 gene. Using chromatin immunoprecipitation (ChIP) assays to identify DNA that are found in "active" chromatin, we show that the IRF-1 promoter is associated with acetylated H4 histones in response to PRL stimulation [45]. Thus, HAT activities of coactivators and chromatin remodeling at the IRF-1 promoter play a role in coordinating PRL stimulation of IRF-1 gene transcription in vivo.

4.5.2. *Negative mediators: Stat5 and corepressors*

Interestingly, Stat5 is detected as a minor component in the G1 PRL-inducible IRF-1 GAS complex in PRL stimulated Nb2 T cells [33]. Surprisingly, Stat5 interaction at the IRF-1 promoter results in transcriptional repression [31,36]. Our data suggest that Stat5b is not compet-

A. Positive signaling **B. Negative signaling**

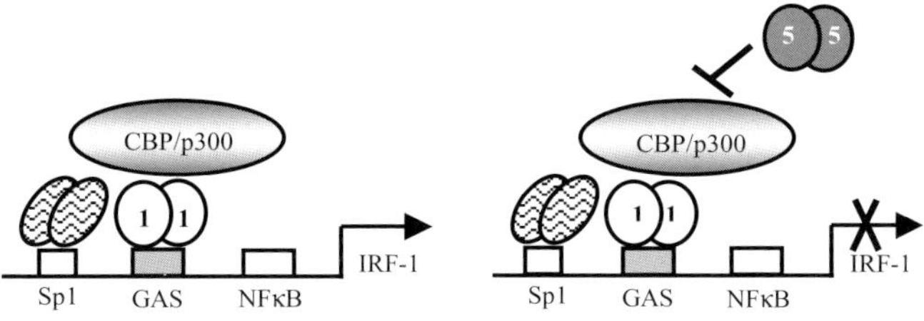

C. Cytokine signal cross talk

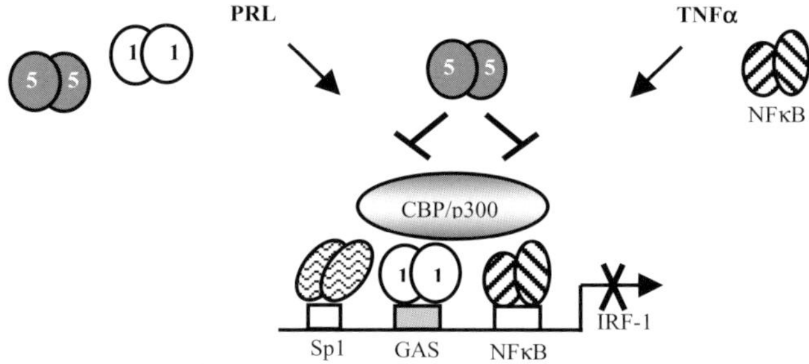

Figure 4. PRL regulation of IRF-1 gene transcription. A. Positive signaling to the IRF-1 promoter involves PRL-inducible Stat1, constitutively bound Sp1 and recruitment of CBP/p300 coactivators to integrate the activities of these factors. B. Negative signaling is mediated by PRL-inducible Stat5, which "squelches" limiting amounts of the co-activators. C. Stat5 inhibits NFkB signaling at the IRF-1 promoter by "squelching" limiting amounts of coactivators, as one mechanism underlying negative signal cross talk. This type of functional antagonism between signaling molecules may provide one mechanism by which PRL antagonizes TNFα signaling. See text for details.

ing with Stat1 for binding to the IRF-1 promoter, but that Stat5b is competing for a factor via protein/protein interactions ("squelching") to inhibit PRL signaling to the IRF-1 promoter [13] (Figure 4B). One target of Stat5b inhibition at the IRF-1 promoter is the coactivator p300 [31]. Thus, PRL signaling to the IRF-1 promoter involves Stat1 binding to the IRF-1 GAS and cooperative interactions between Stat1, Sp1 and coactivators to promote IRF-1 transcription. On the other hand, PRL signaling to Stat5 appears to squelch limiting amounts of coactivators, leading to IRF-1 promoter inactivation and IRF-1 gene downregulation. That Stat5 can act as transcriptional repressors in vivo was confirmed in the Stat5a/5b KO mice [47] in which the "increased expression" or "derepression" of certain genes have been observed. As another mechanism for mediating transcriptional repression, Stat5b can interact directly with corepressor proteins (data not shown) which in turn recruit histone deacetylase to turn off gene transcription [48]. Thus, two distinct mechanisms may mediate transcriptional repression by Stat5: Stat5 can squelch coactivators and/or recruit corepressors at target promoters to shut down gene transcription.

4.5.3. *Opposite functions of Stat5*

Interestingly, the functional activity of PRL-inducible Stat5 interaction with CBP/p300 depends on the target promoter. For example, the Stat5/p300 complex is involved in activation of the β-casein promoter but repression of the IRF-1 promoter. While the mechanistic details of how Stat5 functions in such opposite ways at different promoters are still unclear, these studies show that transcriptional regulation by Stats is a complex process. Stats can act as transcriptional activators or transcriptional repressors, depending on the target promoter, the complement of coactivators, corepressors and other DNA binding proteins, which are recruited into the specific promoter, and the stage of differentiation of the target cell.

4.6. Signals that downregulate PRL-R signaling

How cytokine signaling is turned off is also a highly regulated process involving multiple factors. Two types of SH2-containing protein tyrosine phosphatases (PTP) [6,49] have been implicated in turning off cytokine signaling. SHP-1 is found primarily in hematopoietic cells and SHP-2 is ubiquitous. PTPs are recruited into the receptor complex via their SH2 domains, become activated by tyrosine phosphorylation, and appears to dephosphorylate JAKs and/or the cytokine receptor, thereby shutting down the signaling pathway [6]. Interestingly, not all PTPs are involved in turning off signaling, as SHP-2 is actually needed for initiating PRL-R signaling [6]. Further, nuclear PTPs appears to be involved in dephosphorylating and thereby downregulating Stat activity in the nucleus [49]. Using the PRL-R cytoplasmic domains as bait, we isolated 2'5'-oligoadenylate synthetase (OAS) as a PRL-R interacting protein [11], and showed that OAS attenuates PRL-R signaling by reducing Stat1 tyrosine phosphorylation and DNA binding. Finally, a group of proteins, which are induced by PRL stimulation feedback inhibit signaling from the PRL-R. These include SOCS (suppressors of cytokine signaling) which bind to and inhibit JAKS [51 and see 41], and PIAS which bind to and inhibit Stats [17]. Again, not all SOCS proteins inhibit receptor signaling as SOCS2, which binds directly to the PRL-R, potentiates PRL-R signaling [52].

5. SIGNAL CROSS TALK

Stat factors are involved in protein/protein interactions with numerous cytoplasmic as well as nuclear proteins (see above and [17]). Consequently, Stats can cross talk positively as well as negatively with other factors to regulate target gene expression. For example, Stats interact physically as well as functionally with NFκB (nuclear factor of κB) [53]. Stat1 synergizes with NFκB while Stat5b antagonizes NFκB signaling to the IRF-1 promoter [31] (Figure 4C). Interestingly, Stat5b appears to antagonize TNFα mediated NFκB signaling again through squelching of limiting co-activators [31]. These observations provide a mechanistic understanding as well as a testable hypothesis in which the PRL+TNFα-inducible Stat1/NFκB complex would mediate proinflammatory responses while the PRL+TNFα-inducible Stat5/NFκB complex would block inflammatory responses. Similarly, Stat/glucocorticoid receptor (GR) interactions may be involved in PRL protection against glucocorticoid-mediated apoptosis while Stat/Smad interactions may be involved in PRL protection against TGFβ mediated myelosuppression in various disease states.

6. CONCLUSIONS AND PERSPECTIVES

PRL is a highly versatile hormone/cytokine, which mediates a variety of cellular responses through the PRL-R. Our studies focus primarily on PRL-R signaling through the JAK/Stat pathway and highlight novel features of PRL-inducible Stat functions: 1) Stats can function as either transcriptional activators or repressors depending on target promoter and cell type; 2) Stats work in concert with non-Stat proteins to regulate gene transcription; and 3) Cytokine signals cross talk via protein/protein interactions at the level of target genes. Understanding signal cross talk may elucidate how PRL antagonizes TNFα, TGFβ and GR mediated signaling in various inflammatory and immunosuppressive diseases. It is interesting to note that PRL and the closely related growth hormone (GH) both stimulate similar receptor proximal signaling events, for example, activation of JAK2, Stat1, Stat3 and Stat5. However, the biological consequence of PRL versus GH signaling is often different in target tissues. Thus, one unresolved issue is how specificity is established in the PRL versus GH signaling pathways. In particular, does PRL versus GH elicit similar, overlapping or distinct biological responses in cells of the immune system? Some of these questions are being addressed in the accompanying articles. The challenge for future studies is to sort out which PRL-inducible signaling pathways are involved in regulating specific patterns of gene expression, and how the resulting PRL signal affects immune cell functions. These studies will help elucidate the immunoregulatory properties of PRL under normal versus disease states.

ACKNOWLEDGEMENT

L.-y. Yu-Lee is supported by grants from the NIH, Linda and Ronald Finger Lupus Research Center and the Women's Fund for Health, Education and Research.

REFERENCES

1. Ben-Jonathan N, Mershon J, Allen D, Steinmetz R. Extrapituitary prolactin: Distribution, regulation, functions and clinical aspects. Endocrine Rev 1996;17:639–669.
2. Yu-Lee L-y. Molecular actions of prolactin in the immune system. Proc Soc Exp Biol Med 1997;215:35–52.
3. Bole-Feysot C, Goffin V, Edery M, Binart N, Kelly PA. Prolactin (PRL) and its receptor: actions, signal transduction pathways and phenotypes observed in PRL receptor knockout mice. Endocr Rev 1998;19:225–268.
4. Horseman ND, Yu-Lee L-y. Transcriptional regulation by the helix bundle peptide hormones: Growth hormone, prolactin, and hematopoietic cytokines. Endocr Rev 1994;15:627–649.
5. Horseman ND. Prolactin. Boston: Kluwer Academic Publishers, 2001.
6. Goffin V, Binart N, Clement-Lacroix P, et al. From the molecular biology of prolactin and its receptor to the lessons learned from knockout mice models. Genet Anal Biomol Engin 1999;15:189–201.
7. Dorshkind K, Horseman ND. The roles of prolactin, growth hormone, insulin-like growth factor-I, and thyroid hormones in lymphocyte development and function: insights from genetic models of hormone and hormone receptor deficiency. Endocr Rev

2000;21:292–312.
8. Goffin V, Kelly PA. The prolactin/growth hormone receptor family: Structure/function relationships. J Mamm Gland Biol Neoplasia 1997;2:7–17.
9. Clevenger CV, Chang WP, Ngo W, Pasha TLM, Montone KT, Tomaszweski JE. Expression of prolactin and prolactin receptor in human breast carcinoma. Am J Pathol 1995;146:695–705.
10. Wang Y, O'Neal KD, Yu-Lee L-y. Multiple prolactin receptor cytoplasmic residues and Stat1 mediate prolactin signaling to the IRF-1 promoter. Mol Endocrinol 1997;11:1353–1364.
11. McAveney KM, Book ML, Ling P, Horvath G, Chebath J, Yu-Lee L-y. Association of 2',5'-oligoadenylate synthetase with the prolactin receptor: Alteration in prolactin-inducible Stat1 signaling to the IRF-1 promoter. Mol Endocrinol 2000;295–306.
12. Pezet A, Favre H, Kelly PA, Edery M. Inhibition and restoration of prolactin signal transduction by suppressors of cytokine signling. J Biol Chem 1999;274:24497–24502.
13. Yu-Lee L-y, Luo G, Moutoussamy S, Finidori J. Prolactin and growth hormone signal transduction in lymphohemopoietic cells. Cell Mol Life Sci 1998;54:1067–1075.
14. Gertler A. Recombinant analogues of prolactin, growth hormone, and placental lactogen: correlations between physical structure, binding characteristics, and activity. J Mamm Gland Biol Neoplasia 1997;2:69–80.
15. Schindler C. Cytokine and Jak-Stat signaling. ExpCell Res 1999;253:7–14.
16. Bromberg J, Darnell JE, Jr. The role of STATs in transcriptional control and their impact on cellular function. Oncogene 2000;19:2468–2473.
17. Shuai K. Modulation of Stat signaling by Stat-interacting proteins. Oncogene 2000;19:2683–2644.
18. Horvath CM. Stat proteins and transcriptional responses to extracellular signals. Trends Biochem Sci 2000;25:496–502.
19. Kazansky AV, Kabotyanski E, Wyszomierski SL, Yel J, Rosen JM. Differential effects of prolactin and src/abl kinase on the nuclear translocation of Stat5b and Stat5a. J Biol Chem 1999;274:22484–22492.
20. Moriggl R, Fouilleux-Gruart V, Jahne R, et al. Deletion of the carboxyl-terminal transactivation domain of MGF-Stat5 results in sustained DNA binding and a dominant negative phenotype. Mol Cell Biol 1996;16:5691–5700.
21. Shaefer TS, Sanders LK, Park OK, Nathans D. Functional differences between Stat3a and Stat3b. Mol Cell Biol 1997;17:5307–5316.
22. John S, Vinkemeier U, Idaini E, Darnell JE, Jr., Leonard WJ. The significance of tetramerization in promoter recruitment by Stat5. Mol Cell Biol 1999;19:1910–1918.
23. Collum RG, Brutsaer S, Lee G, Schindler C. A Stat3 interacting protein (StIP1) regulates cytokine signal transduction. Proc Natl Acad Sci 2000;97:10120–10125.
24. Sekimoto T, Imamoto N, Makajima K, Hirano T, Yoneda Y. Extracellular signal-dependent nuclear import of Stat1 is mediated by nuclear pore targeting complex formation with NPI-1, but not Rch1. EMBO J 1997;16:7067–7077.
25. Chatterjee-Kishore M, Wright KL, Ting JPY, Stark GR. How Stat1 mediates constitutive gene expression: a complex of unphosphorylated Stat1 and IRF1 supports transcription of the LMP2 gene. EMBO J 2000;19:4111–4122.
26. Rosen JM, Wyszomierski SL, Hadsell D. Regulation of milk protein gene expression. Annu Rev Nutr 1999;19:407–436.
27. Glass CK, Rose DW, Rosenfeld MG. Nuclear receptor coactivators. Curr Op Cell Biol 1997;9:222–232.

28. Kurokawa R, Kalafus D, Ogliastro MH, et al. Differential use of CREB binding protein-coactivator complexes. Science 1998;279:700–703.
29. Pfitzner E, Jahne R, Wissler M, Stoecklin E, Groner B. p300/CREB-binding protein enhances the prolactin-mediated transcriptional induction through direct interaction with the transactivation domain of Stat5, but does not participate in the Stat5-mediated suppression of the glucocorticoid response. Mol Endocrinol 1998;12:1582–1593.
30. Horvai AE, Xu L, Korzus E, et al. Nuclear integration of Jak/Stat and Ras/AP-1 signaling by CBP and p300. Proc Natl Acad Sci 1997; 94:1074–1079.
31. Luo G, Yu-Lee L-y. Stat5 inhibits NFκB-mediating signaling. Mol Endocrinol 2000; 14:114–123.
32. DaSilva L, Rui H, Erwin RA, et al. Prolactin recruits Stat1, Stat3, and Stat5 independent of conserved receptor tyrosines Tyr402, Tyr479,Tyr515 and Tyr 580. Mol Cell Endocrinol 1996;117:131–140.
33. Wang Y, Yu-Lee L-y. Multiple Stat complexes interact at the IRF-1 GAS in prolactin-stimulated Nb2 T cells. Mol Cell Endocrinol 1996;121:19–28.
34. Gouilleux F, Wakao H, Mundt M, Groner B. Prolactin induces phosphorylation of Tyr694 of Stat5 (MGF), a prerequisite for DNA binding and induction of transcription. EMBO J 1994;13:4361–4369.
35. Wakao H, Gouilleux F, Groner B. Mammary gland factor (MGF) is a novel member of the cytokine regulated transcription factor gene family and confers the prolactin response. EMBO J 1994;13:2182–2191.
36. Luo G, Yu-Lee L-y. Transcriptional inhibition by Stat5: Differential activities at growth-related versus differentiation-specific promoters. J Biol Chem 1997;272:26841–26849.
37. Richards JS, Fitzpatrick SL, Clemens JW, Morris JK, Alliston T, Sirois J. Ovarian cell differentiation: A cascade of multiple hormones, cellular signals, and regulated genes. Rec Prog Horm Res 1995;50:223–254.
38. Stevens AM, Wang Y, Sieger KA, Lu H, Yu-Lee L-y. Biphasic transcriptional regulation of the interferon regulatory factor-1 gene by prolactin: Involvement of gamma-interferon activated sequence and Stat-related proteins. Mol Endocrinol 1995;9:513–525.
39. Bole-Feysot C, Perret E, Roustan P, Bouchard B, Kelly PA. Analysis of prolactin-modulated gene expression profiles during the Nb2 cell cycle using differential screening techniques. Genome Biol 2000;1:research0008.1–0008.15
40. Dogusan Z, Book ML, Verdood P, Yu-Lee L-y, Hooghe-Peters E. Prolactin activates interferon regulatory factor-1 expression in normal lympho-hemopoietic cells . Eur Cyto Netw 2000;11:435–442.
41. Hooghe R, Devos S, Dogusan Z, Hooghe-Peters L. Singal transduction and modulation of gene expression by prolactin in human leukocytes. In: Matera L, Rapaport R, editors. Growth and Lactogenic Hormones. Neuroimmune Biology, Vol. 2. Elsevier, 2002; 123–136 (this volume).
42. Tanaka N, Taniguchi T. The interferon regulatory factors and oncogenesis. Semin Cancer Biol 2000;10:73–81.
43. McAlexander MB, Yu-Lee L-y. Sp1 is required for prolactin activation of the interferon regulatory factor 1 gene. 2000, submitted.
44. Carey M. The enhanceosome and transcriptional synergy. Cell 1998;92:5–8.
45. Ryu S, Zhou S, Ladurner AG, Tijan R. The transcriptional cofactor complex CRSP is required for activity of the enhancer-binding protein Sp1. Nature 1999;397:446–450.
46. McAlexander MB, Yu-Lee L-y. Prolactin activation of IRF-1 transcription involves changes

in histone acetylation. FEBS Lett 2001;488:91–94.
47. Teglund S, McKay C, Schuetz E, et al. Stat5a and Stat5b proteins have essential and nonessential, or redundant, roles in cytokine responses. Cell 1998;93:841–850.
48. Ayer DE. Histone deacetylases: transriptional repression with SINers and NuRDs. Trends Cell Biol 1999;9:193–198.
49. Hunter T. Protein kinases and phosphatases: The yin and yang of protein phosphorylation and signaling. Cell 1995;80:225–236.
50. Haspel RL, Darnell JE, Jr. A nuclear tyrosine phosphatase is required for the inactivation of Stat1. Proc Natl Acad Sci 1999;96:10188–10193.
51. Naka T, Fujimoto M, Kishimoto T. Negative regulation of cytokine signaling: Stat-induced Stat inhibitor. Trends Biochem Sci 1999;24:394–398.
52. Finidori J. Regulators of growth hormone signaling. Vitam Horm 2000;59:71–97.
53. Perkins ND. The Rel/NFκB family: friend and foe. Trends Biochem Sci 2000;25:434–440.

Growth and Lactogenic Hormones
Edited by L. Matera and R. Rapaport

Signal transduction and modulation of gene expression by prolactin in human leukocytes

R. HOOGHE [1,2], S. DEVOS[1], Z. DOGUSAN[1], E.L. HOOGHE-PETERS[1]

[1]*Pharmacology Department, Medical School, Free University of Brussels (VUB), B-1090 Brussels, Belgium*
[2]*Environmental Toxicology, Flemish Institute for Technological Research (VITO), B-2400 Mol, Belgium*

ABSTRACT

Leukocytes have receptors for prolactin (PRL) and the immune system is indeed a direct target of this pleiotropic hormone that originates from the pituitary gland but also, to a minor extent, from other sites such as the lymphoid system. PRL thus could act on leukocytes in an endocrine, a paracrine or an autocrine fashion. Seminal studies were performed in the PRL-responsive Nb2 rat T-cell lymphoma line. PRL signals mainly through Janus kinase (Jak)/signal transducers and activators of transcription (Stat) and by the mitogen-activated protein-kinase (MAP-K) pathways. PRL induces the expression of genes that are involved in innate and acquired immune responses. Our own studies were done with normal rat and human leukocytes. The expression of receptors for PRL was detected with PCR in rat splenocytes and bone marrow cells and in human peripheral blood mononuclear cells (PBMC). In human granulocytes, however, receptor expression was below detection level. Exposure to physiological concentrations of PRL led to the activation of the Jak-2 and Stat-5 in rat cells and in human PBMC. In human granulocytes, PRL activated Stat-1 but not Jak-2 nor Stat-5. PRL stimulated the expression of the interferon regulatory factor (IRF)-1 gene in rat spleen and bone marrow cells. In man, genes induced by PRL include several members of the SOCS-family (suppressors of cytokine signaling), inducible nitric oxide synthase (iNOS) and IRF-1, which are all highly relevant to immune responses. Further work should identify the functional consequences of these biochemical events at the level of survival, proliferation, differentiation and functional activity. For instance, the sharing of signaling pathways accounts for synergy between PRL and cytokines such as IL-2 and IL-12. Also, PRL induces SOCS factors thereby modulating signal transduction by cytokines.

1. INTRODUCTION

The immune system is one of the many direct targets of prolactin (PRL). PRL acts on all types of leukocytes and is considered to have a globally immunostimulatory activity on both innate and acquired immune responses [1–5]. In addition, indirect effects of PRL on the immune system are likely, for instance through other target tissues. The role of PRL in the immune system was first demonstrated in hypophysectomized rats and in rodents treated with the PRL-lowering drug bromocriptine. Detailed biochemical studies were performed in PRL-dependent rat Nb2 T-cell lymphoma cells. However, the whole concept of prolactin as a lympho-hemopoietic growth and differentiation factor nearly collapsed in 1998 after it appeared that the development of the immune system was unaffected in PRL knockout and in PRL-receptor knockout mice. In addition, these mice respond normally to various antigenic challenges [3,4]. The compensation of inborn deficits as a result of redundancy in the cytokine network has been advocated to account for the apparent contradiction between data from knockout mice and earlier in vivo experiments. This view could not be tested so far, as conditional knockout mice (to test the effect of abrupt PRL depletion in adults) or double knockout mice (to test for redundancy between e.g. IL-2 or IL-3 and PRL) have not been generated yet. A role for PRL is indeed very hard to demonstrate in the normal lympho-hemopoietic system whereas immunostimulatory effects of PRL are clear, not only after hypophysectomy, but also after e.g. ovariectomy, acute bleeding, or treatment with glucocorticoids or azathioprine [1–7]. Many in vitro experiments also confirmed the immunomodulatory activity of PRL on rodent and human leukocytes [1–7,29–32,35]. The availability of recombinant hPRL and its very low toxicity would make its use attractive and clinical trials are indeed under way to evaluate to which extent immunocompromised patients can benefit from treatment with PRL.

2. PROLACTIN

Whereas the bulk of PRL is produced by the pituitary gland, many non-endocrine cells produce PRL. In particular, leukocytes express PRL [1,2,8,9]. We have detected PRL mRNA and protein in rat and human bone marrow (1% of the cells), spleen (red pulp and marginal zone), thymus (very few cells in the medulla, the cortico-medullary junction, and in the subcapsular zone) [1,2]. In rat bone marrow and in human peripheral blood cells, granulocytes express PRL mRNA, as judged by in situ hybridization. The total number of PRL-producing cells in the immune system is far from negligible but the actual production has not been properly estimated. The expression of pituitary and extra-pituitary PRL is controlled by different factors: in the pituitary, estrogens and dopamine are respectively positive and negative regulatory factors. In extra-pituitary sites, PRL expression can be initiated from the “pituitary” promoter or from a distal promoter [8,10]. A paracrine or autocrine role for PRL has been proposed but will not be discussed here. We have not addressed the origin (pituitary versus locally-expressed) of PRL that stimulates leukocytes in vivo in our recent work. Rather, our objective was to elucidate signaling pathways used by PRL and to analyze effects of PRL in leukocytes after the addition of exogenous PRL.

2.1. PRL-receptor

The PRL-R is a member of the hemopoietin-cytokine receptor family. It shares homology in particular with the receptors for growth hormone, for erythropoietin and for thrombopoietin. It should be stressed that the PRL-R binds all lactogenic hormones, namely PRL, placental lactogen and growth hormone (from primates only) [11]. In all species, there are several variants of the receptor resulting from alternative splicing or post-translational modifications. There is, however, only one gene coding for the PRL-R. Binding of the ligand can be reduced when part of the extracellular domain is missing, whereas differences in the length of the intracellular domain result in the activation of different signaling pathways. In man, until recently, only one receptor had been identified, corresponding to the long form (full-length receptor) described in other species [11]. The group of Clevenger has identified 2 splice variants of the human PRL-R: the first one is an "intermediate" form, with a deletion and a frame shift in the intra-cellular domain. This receptor is still able to signal through Jak but not through Fyn [12]. The second one is the ΔS1 form, lacking about one half of the extracellular domain and thus able to bind PRL to a much lower extent than the full-length receptor [13].

By immunocytochemistry, PRL-R was found on all mononuclear leukocyte populations. In mouse and man, among leukocytes, quiescent T cells express fewer receptors per cell than macrophages. B cells and NK cells express the highest levels of PRL-R. Stimulated T cells express more receptors than unstimulated T cells [1–5,14–17]. Recently, studies on human leukocytes have been repeated with biotin-labeled PRL and a much lower percentage of PRL-R positive cells were found than in the earlier studies (done with monoclonal antibodies) [18]. This may reflect as well technical problems as variable reactivity of the various probes with the PRL-R variants. The number of PRL-R per cell is low, not only on leukocytes, but also on classical target cells, such as the mammary gland epithelium [4].

Our studies on PRL-R expression and effects of PRL in normal leukocytes from rat and man are summarized in the next paragraphs.

3. OUR OWN DATA

3.1. Signaling by PRL in the rat immune system [19]

A physiological concentration of PRL stimulates the phosphorylation of Jak-2 and Stat-5 in rat bone marrow and spleen cells and activates the IRF-1 gene.

Signaling studies were performed on normal rat bone marrow and spleen cells. PRL-R mRNA expression was monitored by RT-PCR. Stronger signals were obtained in normal spleen and thymus than in bone marrow cells. Biosynthesis of PRL-R was monitored with ^{35}S-methionine labeling followed by immunoprecipitation, SDS-PAGE and autoradiography. The rate of protein synthesis paralleled mRNA expression levels in spleen cells. In bone marrow, mRNA expression and protein biosynthesis were low, whereas PRL-R protein levels as estimated by Western blotting and immunocytochemistry were high. In bone marrow and spleen cells, PRL treatment promoted tyrosine phosphorylation of Jak-2 after 15 min and activated Stat-5 factor to bind a gamma-interferon-activated DNA sequence (GAS) from the IRF-1 promoter after 30 min. One of the targets in the PRL-R signaling pathway also in normal leukocytes is indeed the IRF-1 gene, as demonstrated by RT-PCR. In summary, a physiological concentration of PRL was sufficient to activate the Jak-Stat-GAS pathway and this probably led to the expression of the IRF-1 gene.

3.2. Signaling of PRL in the human immune system

Studies were done on leukocytes from healthy donors, after separation of peripheral blood mononuclear cells (PBMC) and granulocytes.

3.2.1. *PRL-R expression is detected in PBMC but not in granulocytes*
PCR analysis confirmed PRL-R expression in the mononuclear fraction from peripheral blood cells (PBMC), in bone marrow cells and in tonsillar B and T cells, with higher expression on B cells. Purified granulocytes from peripheral blood, however, were consistently negative although several primer sets were used, including primers that should allow the detection of the full-length, intermediate and ΔS1 forms of the human PRL-R [12,13].

3.2.2. *PRL induces tyrosine phosphorylation of Jak-2 in PBMC, but not in granulocytes*
Jak-2 was constitutively phosphorylated albeit at a low level in unstimulated PBMC and after 20 min PRL treatment (100 ng/ml), the level of tyrosine phosphorylation was increased. In granulocytes, there was no constitutive phosphorylation and no phosphorylation of Jak-2 was detected after stimulation with PRL. However, tyrosine phosphorylation of Jak-2 was observed in GM-CSF treated granulocytes.

3.2.3. *PRL induces the phosphorylation of Stat-5 in PBMC and that of Stat-1 in granulocytes*
After 30 min. treatment with as little as 10 ng/ml PRL, Stat-5 was phosphorylated in PBMC but not in granulocytes. With 100 ng/ml PRL, Stat-5 was already phosphorylated after 15 min. in PBMC but again, not in granulocytes. Phosphorylation of Stat-5, however, was induced in granulocytes by treatment with GM-CSF. In contrast, no phosphorylation of Stat-1 could be detected in PBMC upon PRL stimulation though IFN-γ induced Stat-1 phosphorylation. In granulocytes, both PRL and IFN-γ induced Stat-1 phosphorylation.

3.2.4. *PRL activates p38 MAP-K in PBMC and in granulocytes*
PRL also caused a significant increase in the phosphorylation level of p38 MAP-K in PBMC and granulocytes.

3.3. Gene expression

Modulation of gene expression by PRL has been studied in PBMC, granulocytes, and cells from tonsils and bone marrow. Particular attention was paid to members of the recently described SOCS-CIS-family (Suppressors Of Cytokine Signaling, Cytokine-Inducible SH-2 proteins) [15,16].

3.3.1. *PRL increased SOCS-3 and iNOS gene expression in PBMC, IRF-1, CIS, SOCS-2 and iNOS gene expression in granulocytes, CIS and SOCS-2 in bone marrow cells and SOCS-2 and SOCS-7 in tonsillar cells*
In PBMC, SOCS-3 and inducible nitric oxide synthase (iNOS) gene expression levels were significantly enhanced upon PRL treatment (10 ng/ml). In granulocytes, this physiological concentration of PRL increased IRF-1 and SOCS-2 expression and induced CIS and iNOS gene expression. In bone marrow cells, CIS and SOCS-2 were also induced by PRL. In tonsillar cells, the expression of CIS was increased and SOCS-2 and SOCS-7 were induced by PRL [20].

Recently, we have reproduced several of the findings reported above (rapid activation of Stat, induction of gene expression) in human T cell clones. This suggests that although T cells have the lowest numbers of PRL-R among mononuclear cells, some T cell subpopulations are also responsive to PRL [manuscript in preparation].

3.4. Effect of PRL on leukemic cell lines

R. Kooijman discusses the production of PRL by leukemic cells in a separate contribution [21]. Many leukemic cells express PRL-R [2,22,23]. In some cases, an autocrine loop is thus possible (and has indeed been demonstrated in one cell line [23]). There is no known human equivalent of the PRL-dependent rat Nb2 T-cell lymphoma, or the 2779 rat lymphoma where the PRL-R acts as an oncogene [24,25]. The significance of PRL-R expression in leukemic cells has received little attention so far. We have found PRL-R expression in myeloid lines (THP-1, MEG-01) and in myeloma lines (MMS-1, RPMI-8226 and EJM). We also monitored the expression of SOCS genes in several lines: PRL increased the expression of SOCS-2 in Jurkat cells [21]. The significance of increased SOCS-2 in these cells is not known.

4. DISCUSSION

Pituitary PRL mediates signaling from the brain to the periphery and there has been much speculation about possible effects of PRL on the immune system.

Not only do the different pituitary hormones have direct or indirect effects on the immune system, the hypothalamo-pituitary axis also responds to signals from leukocytes. In addition, it was proposed that PRL produced in lymphoid tissue acts as a cytokine [1–5]. Among leukocytes, T cells express the highest levels of PRL and the lowest levels of PRL-R. There is evidence suggesting an autocrine or, more likely, a paracrine role for PRL in the lympho-hemopoietic system [26]: according to this scheme, leukocytes, mainly T cells, produce PRL that acts mainly on target cells having higher number of receptors, such as B cells, monocytes and NK cells. This hypothesis, based in part on data from Pellegrini et al. [15] was already summarized in the cover picture of Molecular Endocrinology in July 1992.

Functional studies have shown that PRL stimulates NO production by granulocytes [27], antibody production by B cells [28], the maturation of dendritic cells [29, Garman et al., quoted in 6], NK activity and the generation of LAK cells [30,31]. More limited effects have been observed on T-cells [32].

4.1. PRL Signal transduction

A wealth of information is now available on signaling by PRL. The bulk of the data has been obtained in the rat Nb2 T-cell lymphoma line (discussed in this issue by L.-y. Yu-Lee [34]). Other data were obtained mainly in mammary cells, in liver cells and in various cell types -including leukocytes- after transfection with the PRL-R cDNA. The main signaling pathways used by the PRL-R are the Jak-2 - Stat-5a/b (and to a lesser extent Stat-1 and Stat-3) and the MAP-K pathways [11–13]. In addition, other pathways can be activated (see below). The binding of PRL to its receptor first activates a protein kinase, either Jak-2 (that is promiscuous and can phosphorylate various substrates in addition to Stats), ZAP 70 or a member of the Src or Tec family. Downstream from these kinases, many other kinases, adaptor molecules and transcription factors can be activated, which leads to gene expression or metabolic responses [11–13]. Our data show that several conclusions reached with Nb2 cells can be extrapolated to normal rat and human leukocytes but they also point to interesting differences. Clearly, a short treatment with physiological concentrations of PRL rapidly activates signaling molecules, transcription factors and stimulates gene expression in the different populations of leukocytes tested.

In rat leukocytes, the Jak-2 - Stat-5 - GAS pathway was activated by PRL, as is the case in Nb2 cells [19, 34].

In human PBMC, PRL stimulated the phosphorylation of Jak-2, Stat-5 and p38 MAP-K. In granulocytes, no PRL-R expression could be detected, Jak-2 was not phosphorylated but Stat-1 and p38 MAP-K were activated and different target genes were induced. Our studies and these by Fu et al. clearly suggest that PRL is able to signal in granulocytes [35]. Stat activation does not rely on Jak only. Receptor tyrosine kinases and some non-receptor tyrosine kinases such as Src and Fyn may also phosphorylate Stat [36]. The PRL-R is apparently expressed below the detection limit of our PCR system. In addition, we speculate that Stat-1 is activated through a kinase of the Src family, such as Fyn. We also show that p38 MAP-K was markedly activated by PRL in both PBMC and granulocytes. This is the first report of p38 activation by PRL. It is not known whether p38 MAP-K alone is responsible for the activation of some transcription factors (see below) or rather has a permissive action on gene expression by inducing chromatin relaxation [37].

It should be recalled that in Nb2 cells or in other systems, PRL signals also through Shc- Grb2 - Sos -Ras- Raf - MEK- ERK-1 and ERK-2; JNK; Vav - rho - rac; Tec; PKC; FAK; IRS-1; calcium mobilisation; Vav and Sos are activated either through Jak-2 or through Src or Tec kinases [11–13,34]. We have not explored these pathways in leukocytes.

4.2. Gene expression

Our data confirm that IRF-1, iNOS, CIS, SOCS-2 and SOCS-3 are targets of PRL. The induction of SOCS-7 expression had not been reported before. As for signal transduction, extensive studies on gene induction by PRL were done in the Nb2 line [34,38]. In addition, the expression of milk proteins was studied in mammary epithelium cells. The Nb2 lymphoma is especially useful for studies on proliferation, cell cycle progression or apoptosis. It is less relevant for lymphocyte differentiation and function.

4.2.1. IRF-1

PRL is able to induce IRF-1 expression in normal rat leukocytes [20] and in human granulocytes. Although modulation of IRF-1 expression by PRL in human PBMC was not detected by PRL, increased expression in a subpopulation of PBMC cannot be ruled out. In two human T cell clones, we observed that PRL stimulated the expression of IRF-1 [manuscript in preparation]. The rapid induction by PRL of IRF-1 expression in the Nb2 line led to intense speculation about the immunomodulatory role of PRL [33]. Indeed, IRF-1 is a key transcription factor in leukocyte biology, as shown e.g. by the abnormal immune responses in IRF-1 knockout mice. As IRF-1 knockout mice have a Th2 dominance [39], it was proposed that PRL favored Th1 versus Th2 responses. This hypothesis has not received experimental confirmation.

The expression of IRF-1 in granulocytes from normal donors has not been reported before.

4.2.2. iNOS

PRL also stimulated iNOS gene expression in human granulocytes and PBMC.

iNOS is of great importance for innate immune responses and is probably a key factor in immunoprotection afforded by PRL in vivo [40]. Di Carlo et al. have shown that PRL stimulates NO production in rat neutrophils [27] and induces the production of iNOS and the release of NO in the rat C6 astrocytic cell line [41].

4.2.3. SOCS

4.2.3.1. Introduction to SOCS factors

In human leukocytes, PRL, as do cytokines and other hormones, induces the expression of suppressors of cytokine signaling (CIS/SOCS), mainly known as negative feedback regulators of the Jak-Stat signaling pathway [42]. Four of these factors (CIS, SOCS-1, -2 and -3) have received much attention but the study of other members of this family has just started.

The first cloned member of this family, CIS binds to cytokine receptors that recruit Stat-5 and inhibits Stat-5 activation [42]. Also, MAP-K activation after stimulation of the T cell receptor (TCR) is greater in T cells from CIS transgenic mice than from control mice [43]. SOCS-1 and SOCS-3 were shown to reduce Jak activity [42]. Although SOCS factors are not specific for a given cytokine, the phenotypes of SOCS transgenic and knockout mice indicate that they have a preferential impact on one or a few transduction pathways. In CIS transgenics, for instance, the body weight is lower than in wild-type mice, suggesting a defect in growth hormone signaling. Female CIS transgenic mice fail to lactate after parturition because of incomplete differentiation of the mammary gland, compatible with a defect in PRL signaling. The IL-2-dependent up-regulation of the IL-2 receptor α chain and proliferation are also partially suppressed in the T cells from CIS transgenic mice. These signs fit within the concept that CIS interferes with Stat-5 activation [45]. SOCS-1 knockout mice in contrast, are hypersensitive to IFN-γ, indicating that SOCS-1 reduces in particular IFN-γ signaling [46]. SOCS-1 transgenic mice have also been generated, using the lck proximal promoter that drives transgene expression in the T cell lineage. These mice strikingly resemble mice lacking the common γ-chain or Jak-3, indicating that in T cells, SOCS-1 inhibits the functions of common γ-chain-using cytokines [47]. Gigantism is the main feature of SOCS-2 knockout mice, which suggests that SOCS-2 negatively modulates GH and/or IGF-1 signal transduction [44]. SOCS-3 knockout mice die early, with marked erythrocytosis that results from hypersensitivity to erythropoietin [47]. Taken together, the available information indicates that SOCS knockout mice over-react to one or several cytokines whereas SOCS transgenics show signs of cytokine depletion. Very little is known about the regulation and function of SOCS-4 to -7. Ours is actually the first study showing SOCS-7 induction by any factor [21]. The function of SOCS-7 is not known but from its tissue distribution and interacting protein partners, it can be speculated that SOCS-7 interferes with cytokine and growth factor signaling [49]. Clearly, the identification of interacting proteins is only one step in the elucidation of functional properties. SOCS-1, for instance, interacts with both Jaks and Tec, but this interaction leads to a strong inhibition of Jak activity and only a minimal inhibition of Tec kinase activity [42,50].

4.2.3.2. SOCS factors and PRL

The induction by PRL of CIS, SOCS-2, SOCS-3 and also SOCS-1 (which we have not investigated) in other cell types had been reported [51,52]. Overexpression of SOCS-1 and SOCS-3 inhibits signal transduction through the PRL-R [51–53]. These SOCS factors thus terminate PRL signaling. In addition, recent data show that SOCS factors induced by a given hormone or cytokine also act on signal transduction through other receptors (inhibition of heterologous signaling). For instance, the immunosuppressive activity of IL-10 is explained in part by the induction of SOCS-3, which reduces signal transduction by interferons [54]. Also, SOCS-3, induced by IL-3, interferes with signal transduction by IL-11; SOCS-1, induced by IL-6 or IFN-γ limits the activation of respectively Stat-1 (by IFN-γ) or Stat-6 (by IL-4) [55,56,72]. In different systems, PRL induced at least 5 SOCS factors of which only SOCS-1 and -3 have been shown to actually inhibit PRL signaling through Stat in vitro [21,51,52]. In addition, abnormal development of the mammary gland in CIS transgenic mice also suggested that CIS interferes with PRL signaling, although this was not the case in vitro [53]. Overexpression of only SOCS-2 had limited effects but when both SOCS-1 and SOCS-2 were overexpressed together, SOCS-2 counteracted the effect of SOCS-1 by restoring Jak activity [51]. The functional relevance of SOCS-7 induction by PRL is unknown. The case of PRL is certainly not unique: IL-9 too induces at least 3 SOCS factors (CIS, SOCS-1 and -3), of which only SOCS-3 interferes with IL-9 signal transduction [57].

In different systems, SOCS genes are targets of PRL. We suggest that PRL, through the induction of SOCS factors, acts as a modulator of signal transduction by various cytokines and other agonists and in this way exerts a "buffering" or homeostatic effect in the immune system. Current descriptions of SOCS factors stress mainly their inhibitory activity [42]. Positive effects can however result from inhibition of suppression. Also, SOCS factors could shift the balance of differentiation versus proliferation, survival versus apoptosis, Th2 versus Th1 and, in various ways, favor a more robust immune response [72]. A positive effect of CIS has also been shown on TCR signaling [43]. Interestingly, in the developing nervous system, Polizzoto et al. speculate that SOCS factors favor neuronal differentiation by shifting the balance from predominantly Jak-Stat towards increased MAP-K signaling [58].

4.3. Signaling pathways: cross talk?

Undoubtedly, huge gaps remain in our understanding of the role of PRL in the immune system. There is still much uncertainty about the type and the number of PRL-R on the different subtypes of leukocytes. To what extent PRL signaling results in metabolic changes and in gene expression under physiological and pathological conditions is not known. Although no essential role for PRL has been identified in leukocytes so far, PRL shares signaling pathways with a large number of cytokine receptors as well as the B- or T cell receptor and could, in principle, modulate their signals.

Stat-5 is the final common pathway used by several cytokines. In the immune system, Stat-5 activation follows stimulation through a large number of cytokines, such as IL-2, IL-3, IL-5, IL-7, IL-9, IL-15, GM-CSF [59]. Synergy of PRL with IL-2 has indeed been demonstrated in at least 2 different systems [31,60]. Stat-5 alone or in combination with other factors, allows the transcription of many genes including IRF-1 (in rat), CIS, oncostatin M, egr-1, p21 waf/cip1, the serine protease inhibitor (Spi) 2.1 and the IL-2-R α chain [59]. As mentioned above, we observed increased expression of irf-1 and cis after stimulation of normal leukocytes with PRL. Increased expression of IL-2 receptor has also been reported in normal leukocytes and that of egr-1 in Nb2 cells treated with PRL [38,60]. Expression of egr-1, however, is also stimulated by p38 (see below).

The activation of p38 follows stimulation with cytokines such as IL-1, IL-12 and TNF-α, but also stimulation through the B-cell receptor or CD40, through the TCR or CD28, through CD49 by integrins, or stimulation of neutrophils by FMLP, or of monocytes by LPS [61,62]. Synergy of PRL with IL-12 has been demonstrated [63].

p38 activates the transcription factors ATF2, CHOP, MEF2C and SAP-1 and it can also modulate the transactivation capacity of NFκB [64–66]. Finally, it induces the relaxation of chromatin and in this way, has a permissive role for the activity of various transcription factors [37]. Among genes induced by LPS through p38 in human monocytes are interferon-induced gene 15, neuroleukin, radiation-inducible immediate-early gene-1, the zinc finger protein A20, IL-1β, IL-8 and superoxide dismutase [64]. Genes induced by anisomycin through p38 in Jurkat cells include the transcription factors c-jun, fra-1 and egr-1; the c-src kinase csk, the nucleotide exchange factor ras-GRF and the growth arrest gene gadd153. The insulin receptor, grb2 and myc are down regulated through p38 [65].

If the activation of Vav by PRL can be confirmed in normal leukocytes, this would be another important signal possibly enhanced by PRL. The physiologic consequences of Vav activation have been discussed in detail in a recent review [13]. As a result of defective T cell signaling, knockout mice lacking the vav1 locus have reduced T cell proliferation and differentiation and are immunosuppressed [67].

As PRL-R shares signaling pathways with both the B and the T cell receptor, there could be interference, either synergy or competition for signaling molecules. The latter was indeed observed in Nb2 cells and in normal human T cells, where stimulation of the TCR with anti CD3 reduced PRL signal transduction [68]. PRL induced the rapid phosphorylation of multiple, TCR/CD3 complex proteins, an event required for lymphocyte activation. Two of these phosphorylated proteins were identified to be CD3ε and ZAP-70.

Whether we are dealing with Stat-5, Stat-1, IRF-1, or Vav, all these signaling molecules are used by quite a few different receptors. Depending e.g. on time and concentration, PRL could synergize or compete with other agonists. Redundancy in the cytokine network probably explains why effects of PRL on the immune system are easier to study in immunocompromised hosts.

4.4. Is PRL immunostimulatory?

Earlier work suggested that PRL had a globally positive (immunostimulatory, immunoprotective) action [1–5,40]. The dramatic effects of PRL in Nb2 cells suggested in particular effects at the T cell level, possibly favoring Th1 over Th2-type responses [33]. Our observations, in particular on granulocytes, together with many data from the literature, rather indicate that PRL has positive effects on innate immunity [1–5,40]. Through stimulating e.g. the maturation of dendritic cells, PRL also contributes to antigen-specific responses [29,30]. A recent study suggests that PRL favors the rupture of tolerance, which may result in autoimmune disease [69]. An aggravating role for PRL in autoimmune diseases, in particular systemic lupus erythematosus, has been advocated [70]. Soon, comparisons of gene expression profiling on microarray will be available for leukocytes treated with PRL versus not treated. Comparison with available data on genes expressed in quiescent versus activated versus tolerant lymphocytes will give better clues and indicate whether PRL does indeed stimulate/inhibit certain types of immune response and contribute to e.g. the gender differences in normal and pathologic immune responses [71]. Also, with clinical trials of PRL in leukopenic patients now in progress, the immunological and hematological effects of treatment with PRL will soon be known. In an effort to reconcile experiments showing clear-cut effects of PRL in rodents with the lack of immunological phenotype in PRL- and PRL-R- knockout mice, Dorshkind and Horseman proposed that PRL has immunostimulatory/immunoprotective activity mainly apparent after some insult has been inflicted to the immune system [4,5]. Indeed, clear-cut effects of PRL have been reported in animals manipulated in various ways e.g. hypophysectomy, ovariectomy, bleeding or treatment with glucocorticoids, bromocriptine or azathioprine [1–5,30]. Whether this can be extrapolated to man is currently not known. The tide, however, is turning again. In the late eighties, there was little doubt that PRL had immunostimulatory activity. From PRL- and PRL-R knockout animals generated in the late nineties, we learned that PRL is dispensable for the developing immune system. More recent work confirms that PRL is active in the rodent and human immune systems. Can this have clinical relevance? The answer will soon be known.

ACKNOWLEDGEMENTS

We thank R. Kooijman and N. Martens for critical reading of the manuscript. Authors' research is supported by grants from the Ministry of Research of the Brussels-Capital Region, from the Flemish Government (GOA 97-02-4) and the V.U.B. (OZR).

REFERENCES

1. Kooijman R, Hooghe-Peters EL, Hooghe R. Adv Immunol 1996;63:377–454.
2. Hooghe-Peters EL and Hooghe R. Growth hormone, prolactin and IGF-I as lymphohematopoietic cytokines. RG Landes, New York: Austin/Springer-Verlag, 1995;256.
3. Hooghe-Peters EL, Dogusan Z, Hooghe R. In vitro effects of prolactin on the lympho-hemopoietic system In: Horseman ND, editor. Prolactin. Boston: Kluwer Academic Publishers, 2001;317–339.
4. Dorshkind K, Horseman ND. The roles of prolactin, growth hormone, insulin-like growth factor-I, and thyroid hormones in lymphocyte development and function: insights from genetic models of hormone and hormone receptor deficiency. Endocr Rev 2000;21:292–312.
5. Dorshkind K, Horseman ND. Anterior pituitary hormones, stress, and immune system homeostasis. Bioessays 2001;23:288–294.
6. Richards SM, Murphy WJ. Use of human prolactin as a therapeutic protein to potentiate immunohematopoietic function. J Neuroimmunol 2000;109:56–62.
7. Woody MA, Welniak LA, Sun R, et al. Prolactin exerts hematopoietic growth-promoting effects in vivo and partially counteracts myelosuppression by azidothymidine. Exp Hematol 1999;27:811–816.
8. Ben-Jonathan N, Mershon JL, Allen DL, Steinmetz RW. Extrapituitary prolactin: distribution, regulation, functions, and clinical aspects. Endocr Rev 1996;17:639–669.
9. Freeman ME, Kanyicska B, Lerant A, Nagy G. Prolactin: structure, function, and regulation of secretion. Physiol Rev 2000;80:1523–1631.
10. Reem GH, Ray DW, Davis JR. The human prolactin gene upstream promoter is regulated in lymphoid cells by activators of T-cells and by cAMP. J Mol Endocrinol. 1999;22:285–292.
11. Goffin V, Binart N, Clement-Lacroix P, et al. From the molecular biology of prolactin and its receptor to the lessons learned from knockout mice models. Genet Anal 1999;15:189–201.
12. Kline JB, Roehrs H, Clevenger CV. Functional characterization of the intermediate isoform of the human prolactin receptor. J Biol Chem 1999;274:35461–35468.
13. Clevenger CV, Rycyzyn MA, Syed F, Kline JB. Prolactin receptor signal transduction. In: Horseman ND (Ed). Prolactin. Boston: Kluwer Academic Publishers, 2001;355–379.
14. Pellegrini I, Lebrun JJ, Ali S, Kelly PA. Expression of prolactin and its receptor in human lymphoid cells. Mol Endocrinol 1992;6:1023–1031.

15. Dardenne M, de Moraes M, Kelly PA, Gagnerault MC. Prolactin receptor expression in human hematopoietic tissues analyzed by flow cytofluorometry. Endocrinology 1994;134:2108–2114.
16. Matera L, Muccioli G, Cesano A, Bellussi G, Genazzani E. Prolactin receptors on large granular lymphocytes: dual regulation by cyclosporin A. Brain Behav Immun. 1988;2:1–10.
17. Gagnerault MC, Touraine P, Savino W, Kelly PA, Dardenne M. Expression of prolactin receptors in murine lymphoid cells in normal and autoimmune situations. J Immunol 1993;150:5673–5681.
18. Bresson JL, Jeay S, Gagnerault MC, et al. Growth hormone (GH) and prolactin receptors in human peripheral blood mononuclear cells: relation with age and GH-binding protein. Endocrinology 1999;140:3203–3209.
19. Dogusan Z, Book ML, Verdood P, Yu-Lee L-y, Hooghe-Peters EL. Prolactin activates interferon regulatory factor-1 expression in normal lympho-hemopoietic cells. Eur Cytokine Netw 2000;11:435–442.
20. Dogusan Z, Hooghe-Peters EL, Berus D, Velkeniers B, Hooghe R. Expression of SOCS genes in normal and leukemic human leukocytes stimulated by prolactin, growth hormone and cytokines. J Neuroimmunol 2000;109:34–39.
21. Kooijman R, Gerlo S. Prolactin expresssion in the immune system. In: Matera L, Rapaport R, editors. Growth and Lactogenic Hormones. Neuroimmune Biology, Vol. 2. Elsevier, 2002; 147–159 (this volume).
22. Hooghe R, Merchav S, Gaidano G, Naessens F, Matera L. A role for growth hormone and prolactin in leukaemia and lymphoma? Cell Mol Life Sci 1998;54:1095–1101.
23. Matera L, Cutufia M, Geuna M, et al. Prolactin is an autocrine growth factor for the Jurkat human T-leukemic cell line. J Neuroimmunol 1997;79:12–21.
24. Tanaka T, Shiu RP, Gout PW, Beer CT, Noble RL, Friesen HG. A new sensitive and specific bioassay for lactogenic hormones: measurement of prolactin and growth hormone in human serum. J Clin Endocrinol Metab1980;51:1058–1063.
25. Barker CS, Bear SE, Keler T, et al. Activation of the prolactin receptor gene by promoter insertion in a Moloney murine leukemia virus-induced rat thymoma. J Virol 1992;66:6763–6768.
26. Hartmann DP, Holaday JW, Bernton EW. Inhibition of lymphocyte proliferation by antibodies to prolactin. FASEB J 1989;3:2194–2202.
27. Meli R, Raso GM, Gualillo O, Pacilio M, Di Carlo R. Prolactin modulation of nitric oxide and TNF-alpha production by peripheral neutrophils in rats. Life Sci 1997;61:1395–1403.
28. Lahat N, Miller A, Shtiller R, Touby E. Differential effects of prolactin upon activation and differentiation of human B lymphocytes. J Neuroimmunol 1993;47:35–40.
29. Matera L, Galetto A, Geuna M, et al. Individual and combined effect of granulocyte-macrophage colony-stimulating factor and prolactin on maturation of dendritic cells from blood monocytes under serum-free conditions. Immunology 2000;100:29–36.
30. Matera L, Contarini M, Bellone G, Forno B, Biglino A. Up-modulation of interferon-gamma mediates the enhancement of spontanous cytotoxicity in prolactin-activated natural killer cells. Immunology 1999;98:386–392.
31. Matera L, Bellone G, Lebrun JJ, et al. Role of prolactin in the in vitro development of interleukin-2-driven anti-tumoural lymphokine-activated killer cells. Immunology 1996;89:619–626.
32. Gala RR, Shevach EM. Influence of prolactin and growth hormone on the activation of

dwarf mouse lymphocytes in vivo. Proc Soc Exp Biol Med 1993;204:224–230.
33. Yu-Lee L-y. Molecular actions of prolactin in the immune system. Proc Soc Exp Biol Med. 1997;215:35–52.
34. Yu-Lee L-y. Signal transduction by prolactin receptors. In: Matera L, Rapaport R, editors. Growth and Lactogenic Hormones. Neuroimmune Biology, Vol. 2. Elsevier, 2002; 111–122 (this volume).
35. Fu YK, Arkins S, Fuh G, et al. Growth hormone augments superoxide anion secretion of human neutrophils by binding to the prolactin receptor. J Clin Invest 1992;89:451–457.
36. Rane SG, Reddy EP. Janus kinases: components of multiple signaling pathways. Oncogene 2000;19:5662–5679.
37. Clayton AL, Rose S, Barratt MJ, Mahadevan LC. Phosphoacetylation of histone H3 on c-fos- and c-jun-associated nucleosomes upon gene activation. EMBO J 2000;19:3714–3726.
38. Bole-Feysot C, Perret E, Roustan P, Bouchard B, Kelly PA. Analysis of prolactin-modulated gene expression profiles during the Nb2 cell cycle using differential screening techniques. Genome Biol 2000;1:Research0008.1–0008.15.
39. McElligott DL, Phillips JA, Stillman CA, Koch RJ, Mosier DE, Hobbs MV. CD4+ T cells from IRF-1-deficient mice exhibit altered patterns of cytokine expression and cell subset homeostasis. J Immunol 1997;159:4180–4186.
40. Di Carlo R, Meli R, Galdiero M, Nuzzo I, Bentivoglio C, Carratelli CR. Prolactin protection against lethal effects of Salmonella typhimurium. Life Sci 1993;53:981–989.
41. Raso GM, Meli R, Gualillo O, Pacilio M, Di Carlo R. Prolactin induction of nitric oxide synthase in rat C6 glioma cells. J Neurochem 1999;73:2272–2277.
42. Yasukawa H, Sasaki A, Yoshimura A. Negative regulation of cytokine signaling pathways. Annu Rev Immunol 2000;18,143–164.
43. Li S, Chen S, Xu X, et al. Cytokine-induced Src homology 2 protein (CIS) promotes T cell receptor-mediated proliferation and prolongs survival of activated T cells. J Exp Med 2000;191:985–994.
44. Metcalf D, Greenhalgh CJ, Viney E, et al. Gigantism in mice lacking suppressor of cytokine signalling-2. Nature 2000;405:1069–1073.
45. Matsumoto A, Seki Y, Kubo M, et al. Suppression of STAT5 functions in liver, mammary glands, and T cells in cytokine-inducible SH2-containing protein 1 transgenic mice. Mol Cell Biol 1999;19:6396–6407.
46. Marine JC, Topham DJ, McKay C, et al. SOCS1 deficiency causes a lymphocyte-dependent perinatal lethality. Cell 1999;98:609–616.
47. Fujimoto M, Naka T, Nakagawa R, et al. Defective thymocyte development and perturbed homeostasis of T cells in STAT-induced STAT inhibitor-1/suppressors of cytokine signaling-1 transgenic mice. J Immunol 2000;165:1799–1806.
48. Marine JC, McKay C, Wang D, et al. SOCS3 is essential in the regulation of fetal liver erythropoiesis. Cell 1999;98:617–627.
49. Matuoka K, Miki H, Takahashi K, Takenawa T. A novel ligand for an SH3 domain of the adaptor protein Nck bears an SH2 domain and nuclear signaling motifs Biochem Biophys Res Commun 1997;239:488–492.
50. Ohya Ki, Kajigaya S, Yamashita Y, et al. SOCS-1/JAB/SSI-1 can bind to and suppress Tec protein-tyrosine kinase. J Biol Chem 1997;272:27178–27182.
51. Pezet A., Favre H, Kelly PA, Edery M. Inhibition and restoration of prolactin signal transduction by suppressors of cytokine signaling. J Biol Chem 1999;274:24497–24502.

52. Tomic S, Chughtai N, Ali S. SOCS-1, -2, -3: selective targets and functions downstream of the prolactin receptor. Mol Cell Endocrinol 1999;158:45–54.
53. Helman D, Sandowski Y, Cohen Y, Matsumoto A, Yoshimura A, Merchav S, Gertler A. Cytokine-inducible SH2 protein (CIS3) and JAK2 binding protein (JAB) abolish prolactin receptor-mediated STAT5 signaling. FEBS Lett 1998;441:287–291.
54. Ito S, Ansari P, Sakatsume M, et al. Interleukin-10 inhibits expression of both interferon alpha- and interferon gamma- induced genes by suppressing tyrosine phosphorylation of STAT1. Blood 1999;93:1456–1463.
55. Magrangeas F, Boisteau O, Denis S, Jacques Y, Minvielle S. Negative cross-talk between interleukin-3 and interleukin-11 is mediated by suppressor of cytokine signalling-3 (SOCS-3). Biochem J 2001;353:223–230.
56. Dickensheets HL, Venkataraman C, Schindler U, Donnelly RP. Interferons inhibit activation of STAT6 by interleukin 4 in human monocytes by inducing SOCS-1 gene expression. Proc Natl Acad Sci USA 1999;96:10800–10805.
57. Lejeune D, Demoulin JB, Renauld JC. Interleukin 9 induces expression of three cytokine signal inhibitors: cytokine-inducible SH2-containing protein, suppressor of cytokine signalling (SOCS)-2 and SOCS-3, but only SOCS-3 overexpression suppresses interleukin 9 signalling. Biochem J 2001;353:109–116.
58. Polizzotto MN, Bartlett PF, Turnley AM. Expression of "suppressor of cytokine signaling" (SOCS) genes in the developing and adult mouse nervous system. J Comp Neurol 2000;423:348–358.
59. Heim MH. The Jak-STAT pathway: cytokine signalling from the receptor to the nucleus. J Recept Signal Transduct Res 1999;19:75–120.
60. Viselli SM, Stanek EM, Mukherjee P, Hymer WC, Mastro AM. Prolactin-induced mitogenesis of lymphocytes from ovariectomized rats. Endocrinology 1991;129:983–990.
61. Lin J, Weiss A. T cell receptor signalling. J Cell Sci 2001;114:243–244.
62. Hehner SP, Hofmann TG, Dienz O, Droge W, Schmitz ML. Tyrosine-phosphorylated Vav1 as a point of integration for T-cell receptor- and CD28-mediatedactivation of JNK, p38, and interleukin-2 transcription. J Biol Chem 2000;275:18160–18171.
63. Matera L, Mori M. Cooperation of pituitary hormone prolactin with interleukin-2 and interleukin 12 on production of interferon-g by Natural Killer and T cells. Ann NY Acad Sci 2001;917:505–513.
64. Manthey CL, Wang SW, Kinney SD, Yao Z. SB202190, a selective inhibitor of p38 mitogen-activated protein kinase, is a powerful regulator of LPS-induced mRNAs in monocytes. J Leukoc Biol 1998;64:409–417.
65. Rolli-Derkinderen M, Gaestel M. p38/SAPK2-dependent gene expression in Jurkat T cells. Biol Chem 2000;381:193–198.
66. Vanden Berghe W, Plaisance S, Boone E, et al. p38 and extracellular signal-regulated kinase mitogen-activated protein kinase pathways are required for nuclear factor-kB p65 transactivation mediated by tumor necrosis factor. J Biol Chem 1998;273:3285–3290.
67. Zhang R, Alt FW, Davidson L, Orkin SH, Swat W. Defective signalling through the T- and B-cell antigen receptors in lymphoid cells lacking the vav proto-oncogene. Nature 1995;374:470–473.
68. Krumenacker JS, Montgomery DW, Buckley DJ, Gout PW, Buckley AR. Prolactin receptor signaling: shared components with the T-cell antigen receptor in Nb2 lymphoma cells. Endocrine 1998;9:313–320.

69. Peeva E, Grimaldi C, Spatz L, Diamond B. Bromocriptine restores tolerance in estrogen-treated mice. J Clin Invest 2000;106:1373–1379.
70. Walker SE, Jacobson JD. Roles of prolactin and gonadotropin-releasing hormone in rheumatic diseases. Rheum Dis Clin North Am 2000;26:713–736.
71. Glynne R, Ghandour G, Rayner J, Mack DH, Goodnow CC. B-lymphocyte quiescence, tolerance and activation as viewed by global gene expression profiling on microarrays. Immunol Rev 2000;176:216–246.
72. Diehl S, Anguita J, Hoffmeyer A, Zapton T, Ihle JN, Fikrig E, Rincon M. Inhibition of Th1 differentiation by IL-6 is mediated by SOCS1. Immunity 2000;13:805–815.

Growth and Lactogenic Hormones
Edited by L. Matera and R. Rapaport

Regulation of PRL Release by Cytokines and Immunomodifiers: Interrelationships between Leptin and Prolactin Secretion. Functional Implications

ORESTE GUALILLO[1], EDUARDO CAMINOS[2], RUBEN NOGUEIRAS[2], CELIA POMBO[2], FRANCISCA LAGO[1], FELIPE F. CASANUEVA[1] and CARLOS DIÉGUEZ[2]

[1]*Santiago de Compostela University Clinical Hospital (CHUS), University of Santiago de Compostela, Spain*
[2]*Department of Physiology, School of Medicine, University of Santiago de Compostela, Spain*

ABSTRACT

The discovery of leptin in the mid 1990s has focused attention on the role of proteins secreted by fat cells, giving a new vision of the adipose tissue as an endocrine organ. Leptin is an adipocyte-derived hormone, that belongs structurally to the long-chain helical cytokine family which also comprises the hormones GH and Prolactin, and signals by a class I cytokine receptor (Ob-R). Leptin represents an important link between fat mass and other endocrine systems, and it has been shown to be involved in immunoregulation. Specifically, leptin has been suggested to function as a prominent regulator of immune system activity, linking the function of T-lymphocytes to nutritional status. Besides its role as regulator of food intake and energy expenditure, new and previously unsuspected neuroendocrine roles have emerged for leptin. This hormone plays also a role in the neuroendocrine control of all the pituitary hormones and their axis. In general terms, leptin reports the state of fat stores to the hypothalamus regulating the activity of several neuroendocrine systems, so that they adapt their function to the current status of energy homeostasis and fat stores.

Prolactin is a pituitary hormone that affects more physiological process than all other pituitary hormones combined. Among these are the regulation of mammary gland development, initiation and maintenance of lactation, immune modulation, osmoregulation, and behavioral modification. Moreover, several data as well as similarities in the biochemical structure of both hormones and their cognate receptors, indicate the existence of a reciprocal regulation betwen leptin and prolactin. In this review, after a short introduction summarizing the general characteristics of these hormones, we will present the current knowledge on the relationships between leptin and prolactin, two hormones that have distinct metabolic roles but share the property of being a humoral modulators of the immune system.

1. LEPTIN, A SHORT OVERVIEW

Leptin is a 16 kDa protein mainly synthesized by the adipose tissue although low levels have been detected in the placenta, skeletal muscle, gastric and mammary epithelium and brain [1–3]. Its structure is similar to the helical structure of cytokines and it is highly conserved among mammals. Leptin circulates in the bloodstream as bound hormone and plasmatic cleareance is prevalently renal. Leptin secretion and expression is modulated by a host of factors, including glucocorticoids, acute infections and inflammation and pro-inflammatory cytokines [4]. In contrast, cold exposure, adrenergic stimulation, GH and thiazolidindiones decrease leptin. Leptin levels are higher in females than in males, as a possible consequence of the inhibitory action of androgens and the higher proportion of subcutaneous fat in females [5].

Leptin is secreted in a pulsatile manner [6] (Figure 1). The main role of leptin is to decrease appetite and increase energy expenditure through action in the brain mediated by the cognate receptor (OB-R). The long isoform of the leptin receptor is localized prevalently in the hypothalamus and its activation mediates the biological effects of leptin through a cascade involving Janus kinase and signal transducers and activator of the transcription (JAK/STAT pathway). It has been considered that the prevalent physiological role of leptin is to serve as a hormone of adaptation between fed and fasted states [7]. Leptin decreases during starvation triggering important metabolic and neuroendocrine responses in rodents such as suppression of GH, thyroid and reproductive hormones and activation of the hypothalamic-pituitary-adrenal axis. Starvation is also associated with marked abnormalities of the immune response [8]. Leptin and its receptor share structural and functional similarities with members of the long chain helical cytokines [9], including prolactin. Leptin displays proliferative and antiapoptotic effects on a variety of cell types, particularly on the cells of the immune system such as T-lymphocytes and macrophages. Leptin has also been implicated in other roles including glucose metabolism, lipid oxidation, substrate partitioning and adipocyte apoptosis [7]. A critical role of leptin in reproduction is suggested by the failure of pubertal maturation in humans and rodents with total leptin deficiency or insensitivity [10]. Leptin treatment restores puberty and fertility in ob/ob mice and accelerates puberty when administered to wild type rodents as well as in patients with mutations of the leptin gene [11,12]. Other actions of leptin on the endocrine system include regulation of insulin production, steroid secretion by ovarian granulosa cells, and modulation of pituitary hormones secretion [1–3]. As for all newly discovered proteins, the original view of leptin as a metabolic hormone has been rapidly replaced by a more complex one. Leptin clearly shows multisystemic actions and although recent important contributions have been made to this fields, future studies should address the potential role of leptin in the regulation of physiopathologic conditions.

2. PROLACTIN, A SHORT OVERVIEW

Prolactin is a 23 kDa peptide synthesized and secreted by the lactotrophic cells of the anterior pituitary of all vertebrates, and by various extrapituitary tissues including decidual cells of the placenta [13], lymphocytes [14] and breast cancer cells of epithelial origin [15,16]. Besides the classical role in development of mammary gland during pregnancy and initiation of lactation in the post partum, a wide variety of biological actions have been ascribed to prolactin. These include osmoregulation, regulation of secretory glands such as prostate and lacrimal gland [17], regulation of gonadal functions through steroidogenesis and corpus luteus formation and

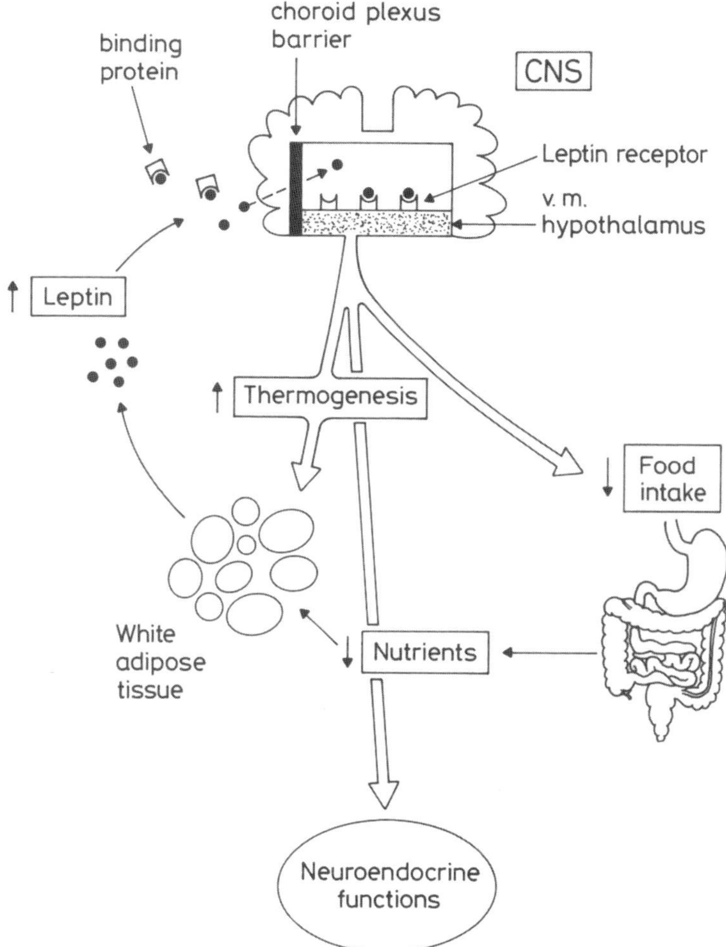

Figure 1. General scheme of leptin physiology. Leptin is secreted from the adipocytes and circulates as free and bound forms. At both the choroid plexus and the blood-brain barrier, leptin is transported by a saturable system into the central nervous system (CNS), where it binds to specific receptors in the ventro-medial hypothalamus. The three actions modulated by a rise in leptin are a reduction in food intake, an increase in thermogenesis, and several neuroendocrine functions over different systems.

regulation of luteinizing receptors [18]. Prolactin may also exert multiple effects in the immune system, some of them shared with leptin [19]. Prolactin biological actions are mediated by specific receptors belonging to the cytokine receptor superfamily [20], which are expressed as short and long forms, differing in the lenght and sequence of their cytoplasmic domain, because of alternative splicing of a single prolactin receptor gene [21]. Prolactin receptors are expressed at widely varying levels in virtually all tissues, both adult and fetal [22–24]. Although both forms of the prolactin receptor are dimerized by the binding of a single molecule of prolactin to activate the Jak2, Fyn, and mitogen associated protein MAP kinase system , only the long form of the receptor can activate the Stat5 transcription factor and initiate milk protein gene transcription [25].

3. LEPTIN REGULATION OF PROLACTIN SECRETION

It is now widely accepted that leptin plays an important neuroendocrine role as shown by its ability to activate hypothalamus-pituitary-gonadal axis at puberty. It up-modulates GH and TSH secretion in vivo and regulates the hypothalamus-pituitary-adrenal axis [26]. Conflicting data exist about a possible regulatory role of leptin on PRL secretion. Prolactin secretion is altered in states of high leptin levels such as obesity and leptin restores lactation in ob/ob mice [4]. Moreover, Yu et al. [27] showed that leptin could significantly stimulate PRL release in vitro from the anterior pituitary of male rats after a 3-h incubation. However, this effect was observed only at extremely high concentrations of leptin such as 10^{-7}–10^{-5} M , which are 10^{3}–10^{5} times higher than the circulating level of leptin in normally-fed male rats (about 10^{-10}M). Furthermore, data reported by other authors [28] suggest that a dose of 10^{-7} M of leptin did not modulate PRL release from incubate anterior pituitary of fasted rats within 2 hours. In addition, it was recently reported that i.c.v. administration of the fragment 116–130 of leptin caused a significant rise in serum PRL levels in food deprived male rats [29]. Finally, it was demonstrated [30] that chronic but not acute administration of leptin stimulates PRL secretion in male rats in a dose dependent manner. Taken together, these data indicate that leptin is needed for lactation to proceed adequately. The finding that leptin antibodies delayed the preovulatory surge of prolactin, while the blunted PRL surge of starved rats was reversed towards normalization by i.c.v. leptin administration indicated that leptin could play a physiological role in the regulation of PRL secretion in adult animals, this was not confirmed in other studies [31].

Whether the main action of leptin are exerted at hypothalamic levels or directly at the lactotrophs needs to be established. Leptin effects could be possibly due to an intermediatory role of hypothalamic peptides as wells as to an increased release of PRL releasing factors, such as thyrotropin–releasing hormone or vasoactive intestinal peptide (VIP). On the other hand, it is well established that the arcuate nucleus of the hypothalamus is the site that show the most abundant expression of leptin receptors as well as high density of neurons that produce NPY, alpha-MSH, beta-endorphin and other related appetite regulating factors. Although NPY does not seem to play a significant role in the physiological regulation of PRL secretion [32], both alpha-MSH and beta-endorphin are considered as exerting an excitatory input on PRL release [33,34]. Thus, it is conceivable that leptin stimulatory action on PRL secretion could be mediated by these proopiomelanocortin (POMC) gene products.

4. EFFECT OF PROLACTIN ON LEPTIN SECRETION

It has been reported that hyperprolactinemia in humans may be associated with a relatively high rate of obesity, and weight is lost after normalization of serum prolactin levels [35–37]. Therefore, it was of interest to assess whether the effects of PRL on body weight homeostasis could influence leptin gene expression. Data obtained in rats showed that PRL increased leptin gene expression and the evidence is as follows. Prolactin has been demostrated to stimulate leptin secretion by the white adipose tissue in rats (38). It has been observed that PRL is able to significantly increase leptin serum levels in both pituitary grafted ovariectomized female rats

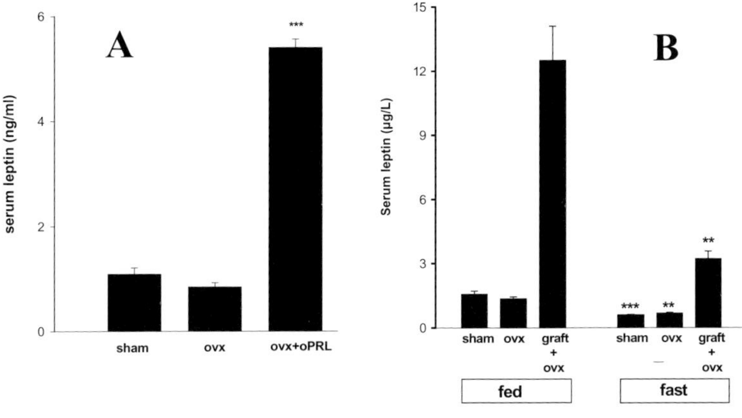

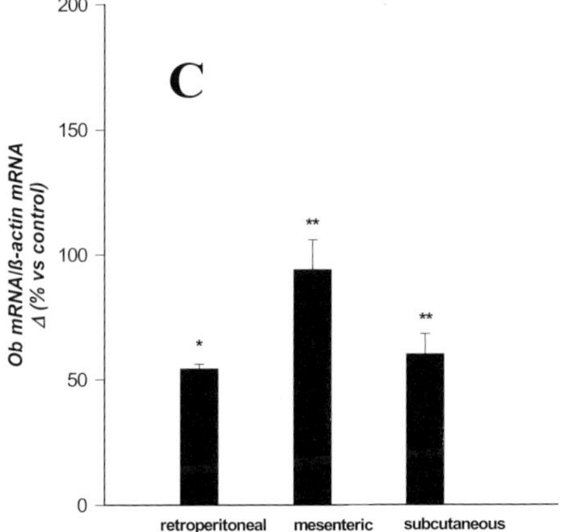

Figure 2. **Panel A:** Effect of PRL administration (5 mg/kg sc every 8 hours, for 4 days) on serum leptin levels of bilaterally ovaryectmised rats. *** p<0.001. **Panel B:** Food deprivation (48 h) reverts serum leptin increase driven by pituitary graft in feeding animals. ***p<0.001; **p<0.01. **Panel C:** Effect of pituitary graft on in vivo leptin mRNA expression of white adipose tissue areas, in comparison to control ovariectomized rats. *p<0.05; ** p<0.01.

(hyperprolactinemic) and PRL treated rats at the dose of 5 mg/kg s.c. every 8 hours for 4 days (Figure 2). This stimulatory effect of PRL on serum leptin levels was significantly reduced by food deprivation. This PRL-induced leptin release appears to be mediated by an increase in leptin mRNA in retroperitoneal, mesenteric and subcutaneous white adipose tissue. The mechanisms by which prolactin exerts its stimulatory effects on leptin secretion and mRNA synthesis is still unclear. On a theoretical basis, it could be possible through a direct effect on PRL receptor on adipocytes. The facts that the density of PRL receptors in these cells is relatively low [39] and the lack of effect of PRL on in vitro leptin secretion, argue against the possibility of the effect being mediated by PRL-R on adipocytes. One possibility is that PRL exerts its effects through an indirect mechanism, connected to the PRL-driven induction of serum factors as well as pro-inflammatory cytokines such as TNF-alpha and interleukin-1, notwithstanding that the

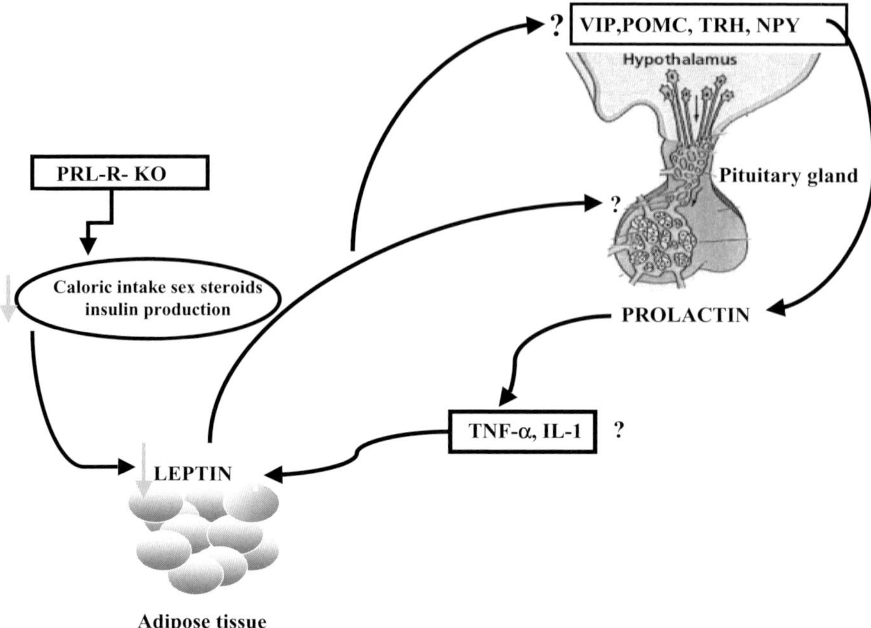

Figure 3. An integrated view of the interrelationships between leptin and prolactin.

biochemical events that interplay between these cytokines and leptin levels are not fully clarified.

Data regarding the possible correlation between fat mass and increased PRL levels are also conflicting. In fact it has been reported that obesity per se is not usually associated with increased PRL serum levels [40]. However, in a large population study, a weak, but significant, correlation was revealed between serum PRL levels and body weight [41]. Therefore, the higher incidence of obesity in hyperprolactinemic patients is unlikely to be mediated by a leptin deficient state, but rather supports the existence of a leptin resistant state.

Finally, a very recent report by Freemark et al. [42] attempted to clarify the roles of prolactin on homeostasis by studying plasma leptin levels in a unique model of lactogen resistance present in the prolactin receptor knockout mouse (PRLR-KO) (Figure 3). This experimental model was created by targeted deletion of the gene encoding the mouse PRLR. PRLR knockout mice are resistant to the action of mouse PRL and mouse placental lactogen, which bind only to the mouse PRLR. Female homozygous PRLR-KO mice are sterile as a consequence of progesterone deficiency and defects in egg transport and implantation. On the other hand, male homozigous mutants appear to have near-normal reproductive activity and normal serum testosterone concentrations [43]. The results obtained by this group showed that plasma leptin concentrations were reduced in females but not in males when compared to the levels of the wild type littermate. The absence of PRLRs in KO mice was accompanied by a small, but progressive, reduction in the rate of weight gain after a 16 weeks of age and a reduction in abdominal fat mass. As serum leptin concentrations in humans and rodents correlate strongly with adipocytes mass and percent body fat, the reduction in fat mass in PRLR deficient females probably contributed to the reduction in plasma leptin levels. In addition, leptin levels correlate more strongly with the amount of subcutaneous fat than with visceral fat stores [44] and are higher in females than in males.

In conclusion, it is possible that the effects of PRLR deficiency on the distribution of fat in females may differ from its effects on fat distribution in males. What are the explanations for the changes in body weight, abdominal fat, and serum leptin in adult PRLR deficient mice? According to Freemark et al., there are at least four possible contributing factors. First: the reduction in body weight and fat content may reflect in part a reduction in caloric intake. Second, the reduction in abdominal fat content may reflect a reduction in insulin production. Third, PRLR deficiency in female mice is accompanied by a state of progesterone deficiency and hypoestrogenemia, so the reduction in abdominal fat content and serum leptin level in PRLR deficient females may derive in part from a deficiency of sex steroids. Interestingly, the testosterone levels in PRLR-KO male mice are normal; thus the small reductions in abdominal fat content in male mice cannot be explained by a deficiency or an excess of testosterone. Finally, the detection of PRLRs in normal mouse adipose tissue suggests that the reduction in abdominal fat in PRLR deficient mice might reflect the failure of lactogens, and particularly prolactin, to exert direct lipogenic effects on adipose cell development and/or metabolism.

5. LEPTIN, PROLACTIN, PREGNANCY AND LACTATION

Elevated concentrations of leptin are reported during pregnancy and lactation [45]. The effects of leptin in vivo seems to be opposite to those of prolactin, at least from a metabolic point of view, since prolactin treatment in mature female rats induces an increase of food intake and fat deposition [46,47]. Moreover, during pregnancy and lactation in both rats and hamsters, prolactin results in suppression of brown fat thermogenesis thus increasing the efficiency of energy utilization [48,49].

Both prolactin and leptin concentrations are significantly higher early in pregnancy when compared to controls suggesting that these early hormonal changes may be actively promoting shifts in substrate used for energy metabolism. The results reported by several authors support a role for leptin in conjunction with prolactin on energy storage during pregnancy and lactation. Novel roles for lactogens in fetal and maternal adipose tissue development and functions are suggested by the patterns of hormone production, adipose tissue accumulation, and leptin expression during pregnancy. The mass of adipose tissue and the serum concentrations of leptin increase during the first 26–32 weeks of gestation in humans and during the third trimester in the human fetus. The accumulation of fat mass and the rise in serum leptin coincide with striking increase in the concentrations of lactogenic hormones in maternal and fetal blood. Thus, increases in prolactin levels together with high levels of progesterone and estrogen may contribute to the accumulation of adipose tissue stores and induction of serum leptin in the mother and in the fetus. In conclusion, all these observations suggest novel roles for prolactin as well as other lactogenic hormones in adipose tissue development and function during gestation and possibly in the postnatal life.

6. CONCLUSIONS

Current evidence indicates that both prolactin and leptin serve a regulatory role in the loop between nutritional status and neuroendocrine function. Despite this progress, many questions about the possible interrelationship between leptin and prolactin, and other accessory factors remain to be answered. This will hopefully lead to a greater knowledge of the regulatory mechanisms involved in pituitary hormone secretion, and adaptative responses to various environmental situations, as well as physiopathological mechanisms of different disease states.

REFERENCES

1. Friedman JM, Halaas JL. Leptin and the regulation of body weight in mammals. Nature 1998;395:763–770.
2. Flier JS. In search of leptin's physiologic role. J Clin Endocrinol Metab 1998;83:1407–13.
3. Ahima RS, Flier JS. Leptin. Annu Rev Physiol 2000;62:413–37.
4. Gualillo O, Eiras S, Lago F, Dieguez C, Casanueva FF. Elevated serum leptin concentrations induced by experimental acute inflammation.Life Sci 2000;67(20):2433–41.
5. Rosenbaum M, Leibel RL. Role of gonadal steroids in the sexual dimorphism in body composition and circulating concentration of leptin. J Clin Endocrinol Metab 1999;84:1784–1789.
6. Licinio J, Mantzoros C, Negrao AB, et al. Human leptin levels are pulsatile and invesely related to pituitary –adrenal function. Nat Med 1997;3:575–579.
7. Ahima RS, Flier JS. Adipose Tissue as an endocrine. TEM 2000;11:327-332.
8. Matarese G. Leptin and immune system: how nutritional status influences the immune response. Eur Cytokine Netw 2000;11:1–7.
9. Fantuzzi G, Faggioni R. Leptin in the regulation of immunity, inflammation, and hematopoiesis. J Leucocyte Biol 2000;68:437–445.
10. Chehab FF, Lim ME, Lu R. Correction of the sterility defect in homozigous obese female mice by treatment with the human recombinant leptin. Nat Gen 1996;12:318–320.
11. Chehab FF, Mounzih K, Lu R, Lim ME. Early onset of reproductive function in normal female mice treated with leptin. Science 1997;275:88–90.
12. Farooki IS, Jebb SA, Lanmack G, et al. Effects of recombinant leptin therapy in a child with congenital leptin deficiency. N Eng J Med 1999;341:879–884.
13. Lee DW, Markoff E. Synthesis and release of glycosilated prolactin by human decidua in vitro. J Clin Endocrinol Metab 1986;62:990–994.
14. Montgomery DW, LeFevre JA, Ulrich ED, Adamson CR, Zukoski CF. Identification of prolactin-like proteins synthesized by normal murine lymphocytes. Endocrinology 1990;127:2601–2603.
15. Clevenger CV, Chang WP, Ngo W, Pasha TLM, Montone KT, Tomaszewski JE. Expression of prolactin and prolactin receptor in human breast carcinoma. Am J Pathol 1995;146:695–705.
16. Ginsberg E, Vonderhaar BK. Prolactin synthesis and secretion by human breast cancer cells. Cancer Res 1995;55:2591–2595.
17. Costello LC, Franklin RB. Effect of prolactin on the prostate. Prostate 1994;24:162–166.
18. Kelly PA, Djiane J, Postel-Vinay MC, Edery M. The prolactin/growth hormone receptor family. Endocrinol Rev 1991;12:235–251.
19. Gala RR. Prolactin and growth hormone in the regulation of immune system. Proc Soc Exp Biol Med 1991;198:513–527.
20. Bazan JP. A novel family of growth factor receptors: A common binding domainin the growth hormone, prolactin, erythropoietin and IL-6 receptor, and the p75 IL-2 receptor B chain. Biochem Biophys Res Comm 1989;164:788–795.

21. Arden KC Boutin JM, Djiane J, Kelly PA, Cavanee WK. The receptor for prolactin and growth hormone colocalize the same region of human chromosome 5. Cytogen Cell Genet 1990;53:161–165.
22. Nagano M, Kelly PA. Tissue distribution and regulation of rat prolactin receptor gene expression. Quantitative analysis by polymerase chain reaction. J Biol Chem 1994;269:13337–13345.
23. Freemark M, Nagano M, Edery M, Kelly PA. Prolactin receptor gene expression in the fetal rat. J Endocrinol 1995;144:285–292.
24. Royster M, Driscoll P, Kelly PA Freemark M. The prolactin receptor in the fetal rat: cellular localization of messenger ribonucleic acid, immunoreactive protein and ligand binding activity and induction of expression in late gestation. Endocrinology 1995;136:3892–3900.
25. Goilleux FH, Wakao H, Mundt M, Groner B. Prolactin induces phosphorylation of Tyr 694 of Stat5 (MGF), a prerequisite for DNA binding and induction of transcription. Embo J 1994;13:4361–4369.
26. Casanueva FF, Dieguez C. Neuroendocrine regulations and actions of leptin. Frontiers in Neuroendocrinology 1999;20:317–363.
27. Yu WH, Kimura M, Walczewska A, Karanth S, McCann SM. Proc Natl Acad Sci USA 1997;94:1023–1028.
28. Tena Sampere M, Pinilla M, Gonzalez LC, Dieguez C, Casanueva FF, Aguilar E. Leptin inhibits testorene secretion from adult rat testis in vitro. J Endocrinol 1999;161:211–218.
29. Gonzalez LC, Pinilla L, Tena-Sampere M, Aguilar E. Leptin (116–130) stimulates prolactin and luteinizing hormone secretion in fasted adult male rats. Neuroendocrinology1999;70:213–220.
30. Watanobe H, Schiot HB, Suda T. Stimulation of prolactin secretion by chronic, but not acute, administration of leptin in the rat. Brain Res 2000;887:426–431.
31. Gonzalez LC, Pinilla L, Tena-Sampere M, Dieguez C, Casanueva FF, Aguilar E. Effect of acute immunoneutralization of endogenous leptin on prolactin and LH secretion during the afternoon of pro-oestrus or in steroid-treated ovariectomized female rats. J Reprod Fertil 2000;118:39–45.
32. Kalra SP, Crowley WR. Neuropeptide Y: a novel neuroendocrine peptide in the control of pituitary hormone secretion, and its relation to luteinizing hormone. Front Neuroendocrinol 1992;13:1–46.
33. Nunez L, Frawley LS. Alpha MSH potentiate the responsivness of mammotropes by increasing Ca^{++}entry. Am J Physiol 1998;274:E971–977.
34. Kehoe L, Parman R, Janik J, Callahan P. Opiate receptor subtype involvement in the stimulation of prolactin release by beta-endorphin in female rats. Neuroendocrinology 1993;57:875–883.
35. Greenman Y, Tordjman K, Stern N. Increased body weight associated with prolactin secreting pituitary adenomas: Weight loss with normalization of prolactin levels. Clin Endocrinol 1998;48:547–553.
36. Lachelin G, Abu-Fadil S, Yen S. Functional delineation of hyperprolactinemic amenorrhea. J Clin Endocrinol Metab 1997;44:1163–1174.
37. Nunes S, Sobrinho L, Calhaz-Jorge C, Santos M, Mauricio J, Sousa M. Psychosomatic factors in patients with hyperprolactinemia and/or galactorhea. Obestet Gynecol 1980;55:591–595.
38. Gualillo O, Lago F, Garcia M, et al. Prolactin stimulates leptin secretion by rat white

adipose tissue. Endocrinology 1999;140:5149–5153.

39. McAveney MK, Gimble GM, YuLee LY, Prolactin receptor expression during adipocytes differentiation of bone marrow stroma. Endocrinology 1996;137:5273–5276.
40. Scaglione R, Averna MR, Dichiara MA, et al. Thyroid function and release of thyroid-stimulating hormone and prolactin from the pituitary in obesity. J Int Med Res 1991;19:389–394.
41. Wang DY, De Stavola BL, Bulbrook RD, et al. The relationship between blood prolactin levels and risk of breast cancer in pre menopausal women. Eur J Cancer Clin Oncol 1987;23:1541–1548.
42. Freemark M, Fleenor D, Driscoll P, Binart N, Kelly PA. Body weight and fat deposition in prolactin receptor deficient mice. Endocrinology 2001;142:532–537.
43. Ormandy CY, Camus A, Barra J, Damotte D, Lucas B, Buteau H, Edery M, et al. Null mutation of the prolactin receptor gene produces multiple reproductive defects in the mouse. Genes Dev 1997;11:167–178.
44. Montague CT, O'Rahilly S. The perils of portliness: causes and consequences of visceral adiposity.Diabetes,2000;49:883–888.
45. Mukherjea R, Castonguay TW, Douglass LW, Moser-Veillon P. Elevated leptin concentrations in pregnancy and lactation: Possible role as a modulator of substrate utilization. Life Sci 1999;65:1183–1193.
46. Byatt JC, Staten NR, Salgiver WJ, Kostelc JG, Collier RJ. Stimulation of food intake and weight gain in mature female rats by bovine prolactin and bovine growth hormone. Am J Physiol 1993;264(6 Pt 1):E986–92.
47. Noel MB, Woodside B. Effects of systemic and central prolactin injections on food intake, weight gain, and estrous cyclicity in female rats. Physiol Behav 1993;54(1):151–4.
48. Chan E, Swaminathan R. Role of prolactin in lactation-induced changes in brown adipose tissue. Am J Physiol 1990;258:R51–6.
49. Wade GN, Jennings G, Trayhurn P. Energy balance and brown adipose tissue thermogenesis during pregnancy in Syrian hamsters.Am J Physiol 1986;250:R845–50.

Growth and Lactogenic Hormones
Edited by L. Matera and R. Rapaport

Prolactin Expression in the Immune System

RON KOOIJMAN and SARAH GERLO

Department of Pharmacology, Medical School, Free University of Brussels (V.U.B.), Laarbeeklaan 103, B-1090 Brussels, Belgium

ABSTRACT

Prolactin (PRL) is not only synthesized by the pituitary, as originally described, but also by several other tissues. We will review the literature on the production and regulation of PRL in the immune system, and discuss the possible implications of locally produced PRL for immune responses and disorders of the immune system. It is evident that a different molecular regulation mechanism for PRL exists in leukocytes, and that cells of the immune system produce several molecular weight variants in addition to the major form of pituitary PRL (23 kDa). Bioassays and other functional assays suggest that these molecules exert autocrine or paracrine immunomodulatory effects. However, further work on the regulation of PRL, and the identification and functional analysis of PRL variants is required to elucidate the role of PRL produced by cells of the immune system.

1. INTRODUCTION

PRL is a pleiotropic hormone which is mainly produced in the pituitary. Although the main functions of PRL are mammary gland development, and initiation and maintenance of lactation, PRL receptors are widely distributed throughout many different tissues. The fact that PRL subserves many different functions is in accordance with the expression of PRL in extra-pituitary tissues, the existence of several molecular variants of PRL with different activities, and the tissue-specific regulation of PRL [1–3]. There is compelling evidence for local production of PRL in the immune system and for the ability of PRL to modulate humoral and cellular immune responses in several species [4]. Although PRL is not an obligatory factor in the immune system [5,6], it may act as an autocrine, paracrine or endocrine mediator of immunomodulatory actions [7]. In addition, PRL may be involved in the pathophysiology of immune disorders. In this review we will provide a survey of the production and regulation of PRL in leukocytes, lymphoid organs and sites of inflammation. We will also discuss the possible implications of locally produced PRL for immune responses and disorders of the immune system.

2. REGULATION OF PRL EXPRESSION IN THE IMMUNE SYSTEM

The human PRL gene is composed of six exons and is more than 15 kb long [8]. The PRL mRNA that is expressed in the decidua, the myometrium and in immune cells, is approximately 150 bp longer than its pituitary counterpart [9,10]. This difference is due to the fact that in extrapituitary tissues transcription starts at exon 1a, which is located 5.8 kb upstream of the pituitary start site at exon 1b. Exon 1a is then spliced to exon 1b, giving rise to the longer mRNA. Since exon 1a is non-coding, extrapituitary PRL is indistinguishable from 23 kDa pituitary PRL. Whereas pituitary PRL expression is under the control of a promoter located immediately 5' to exon 1b [11], extrapituitary PRL transcription is under the control of an alternative upstream promoter, located 5' to exon 1a [12,13]. Interestingly, the presence of an alternative promoter governing extrapituitary PRL expression has only been reported in humans and was suggested in the rhesus macaque where the decidual PRL message also contains a unique 5' sequence [14].

Regulation of PRL expression in the pituitary depends mainly on the transcription factor Pit-1 [15]. Prolactin-releasing peptide and thyrotropin releasing hormone (TRH) have been identified as the main factors stimulating PRL expression, whereas dopamine is the main inhibitor [3]. The stimulatory effect of TRH on PRL secretion and synthesis appears to be mediated by protein kinase C (PKC) activation [16]. Inhibition by dopamine is linked to a decreased activity of adenylate cyclase, resulting in a fall of intracellular cAMP levels [3]. Other factors shown to induce PRL transcription in the pituitary are endothelial growth factor, fibroblast growth factor, triiodothyronine, Ca_2^+ channel agonists, and the vitamin D receptor [17,18], whereas glucocorticoids inhibit PRL transcription [17].

Whereas regulation of PRL expression in the pituitary has been extensively studied, regulation of ectopic PRL expression has remained relatively unexplored. Of the few studies that have been carried out, most have focused on the control of PRL expression in the decidua. In endometrial stroma cells, PRL expression is a marker of decidualization and PRL transcription is independent of the regulators orchestrating pituitary PRL expression. The regulation of decidual PRL expression has been reviewed earlier [1].

One of the first reports on PRL expression in immune cells described PRL expression in the B-lymphoblastoid cell line IM-9-P33 [10]. The same authors investigated how PRL expression was regulated in these cells. At the time this article was presented, the alternative upstream promoter had not yet been identified, which prompted the authors to focus on agents that are known to affect pituitary PRL expression. They found that, as in decidua, the most important regulators of pituitary PRL expression were all without effect on IM-9-P33 PRL expression. Only the synthetic glucocorticoid dexamethasone inhibited PRL expression in IM-9-P33 cells as it does in lactotrope cells [19]. The observed inhibition however appeared not to result from modification of transcriptional activity, but from a negative effect on PRL mRNA stability. The same authors showed that PRL expression can also be positively regulated at the posttranscriptional level by retinoic acid which stabilizes the PRL message [20].

Since the unresponsiveness of IM-9-P33 cells to modulators of pituitary PRL expression could be due to the fact that the appropriate receptors were not present on these cells, the authors tried to circumvent this problem by directly activating the second messengers involved in the TRH and dopamine effects in the pituitary, being PKC and cAMP. When IM-9-P33 cells were stimulated with TPA, which activates PKC, PRL expression was affected in a biphasic manner with an early stimulation of secretion, followed by a later inhibition in synthesis [21]. No clear effect of cAMP could be detected in IM-9-P33 cells. Only a minor increase of 35% in PRL secretion was observed after one hour exposure to dibutyryl-cAMP. It could be however that in

IM-9-P33 basal adenylate cyclase activity was already high and the system was saturated [21].

In these initial studies, the differences in PRL expression in pituitary versus ectopic sources were explained by the use of different signalling pathways in one cell type versus the other. Studies that have been done after the upstream promoter had been described, addressed alternative promoter usage. As mentioned earlier, extrapituitary PRL expression is controlled by promoter and enhancer regions 5' to exon 1a. Several studies focused on a molecular analysis of this upstream promoter. In the T-leukemic cell line Jurkat, the activity of the upstream promoter was almost completely dependent upon the sequence between – 453 and –67 bp relative to exon 1a. When the upstream promoter was transfected into GH3 or HeLa cells, no transcriptional activity was measured. In contrast, in Jurkat, the pituitary promoter was inactive. Footprinting experiments showed the presence of a Jurkat specific enhancer located between –375 and –212 that confers up to 50% of the activity of the upstream promoter [12]. It was furthermore shown that upstream promoter activity was not restricted to lymphoid cells expressing endogenous PRL. In contrast, in human decidualized endometrial cells promoter activity did reflect the endogenous PRL status [13]. In a computer search for consensus sequences, two Pit-1 binding sites and seven half sites for glucocorticoid receptor/progesterone receptor binding motifs were detected in the upstream PRL promoter. The presence of the Pit-1 transcription factor had also been shown in cells of the immune system [22]. However, coexpression studies showed that Pit-1 is not involved in regulation of PRL transcription in lymphocytes nor endometrial stroma. Also, the activated progesterone receptor could not activate the upstream promoter [13].

Recently, it was shown that cAMP induces the upstream PRL promoter in the T-leukemic Jurkat cell line, with an early response that maximizes after 6 hours. Mutation of the CRE at –25 to –12 leads to loss of half of the response to cAMP. The factors that account for the other half of the response remain to be established. PHA and TPA also activated PRL transcription but in a very modest manner. However, a combination of TPA and cAMP synergistically induced PRL promoter activation. The basis of this synergism is not elucidated [23]. These findings suggest that PRL expression is enhanced in activated T-cells. However, it would be desirable to confirm these data using more physiologic T-cell activators (e.g. anti CD-3, anti CD28) in normal T cells.

The effect of cAMP on the upstream promoter had been studied earlier in human endometrial cells, where the response to cAMP was biphasic with an initial weak induction within 12 hours that is followed by a late, much more intense, induction after 12 hours. As in Jurkat, the CRE at position –25 to –12 seemed to be involved in the early induction of transcription [24], whereas the delayed response, that is absent in Jurkat, is regulated by C/EBP binding to two overlapping consensus sequences between –332 and –270 [25].

Although PRL has been implicated as a cytokine, its regulation by other cytokines in the immune system has not yet been explored. However, several publications report cytokine modulation of PRL expression in other tissues using the upstream promoter. It was shown that interleukin (IL)-1 and tumor necrosis factor-α inhibit the induction of the PRL gene in decidual cells [26,27] whereas in myometrial cells, interferon-γ and IL-4 inhibit PRL expression [28]. None of these studies have focused on the molecular basis of regulation of transcription though.

The latest publications provide evidence that, besides the pituitary and upstream PRL promoters, the region separating the two promoters might also be involved in regulation of extrapituitary PRL transcription. DNaseI protection experiments of an activating region in intron A-1, separating the pituitary from the extrapituitary transcription start site, revealed binding sites for Pit-1, several ubiquitous factors and one Jurkat-specific factor. The Jurkat factor binding region was furthermore able to activate a heterologous (TK) promoter when transfected in Jurkat

[29]. By DNaseI mapping of chromatin from PRL producing (IM-9-P33) and non-producing (IM-9-P6) human lymphoblastoid cell lines, hypersensitive sites specific to the IM-9-P33 cells were revealed in intron A-1. In transfection experiments it was shown that this region can only confer transcriptional activation to the rat minimal PRL promoter in the antisense orientation and that surprisingly it was active independently of the PRL-producing status of the cells [30].

It is clear that the selection between alternate promoters in pituitary versus extrapituitary tissues requires a complex interplay between various transcription factors. The molecular mechanisms at the basis of this selection however remain to be elucidated. Furthermore, cis-acting regions involved in transcriptional activation in extrapituitary tissues might not be limited to the region 5' to exon 1a as suggested by the latest data. Also there is evidence that extrapituitary PRL expression can be regulated at the posttranscriptional level, by effects on mRNA stability. Finally certain factors modulate the secretion of stored PRL. However, regulation of secretion is probably far more important in the lactotrope, where the hormone is stored intracellularly in secretory granules, than in immune cells where intracellular storage capacity is low.

3. PRL EXPRESSION IN LEUKOCYTES AND LYMPHOID TISSUES

It was demonstrated by Wu et al. [31] that in all lymphoid tissues tested leukocytes as well as some epithelial and vascular endothelial cells expressed PRL mRNA. In this section we will summarize PRL expression in lymphoid organs, blood and sites of inflammation.

3.1. Bone marrow

Bellone et al. [32] demonstrated by Western blot analysis that cultured stroma cells from human bone marrow secreted 23 and 25 kDa immunoreactive PRL. Both expression of PRL transcripts and secretion of 23 kDa PRL were increased by platelet-activating factor (PAF). PAF increased the colony formation of $CD34^+$ hematopoietic progenitors that were cultured on a monolayer of bone marrow stroma cells via secretion of PRL as evidenced by the finding that anti-PRL antibodies completely abrogated the effect of PAF. The biological activity of bone marrow stroma-derived PRL was confirmed in the Nb2 bioassay [33]. Using in situ hybridization Delhase et al. [22] detected PRL mRNA in rat bone marrow cells.

The idea that bone marrow stroma-derived PRL exerts paracrine effects is compatible with the presence of PRL receptors on bone marrow cells in rats [34], mice [35] and humans [36,37]. Moreover, human $CD34^+$ stem cells also express PRL receptors and PRL augmented granulopoiesis and erythropoiesis from purified $CD34^+$ progenitors [37].

3.2. Thymocytes

One of the first studies on PRL production by leukocytes showed that Con A induced the production of 22 kDa PRL in murine thymocytes, whereas 33 and 35 kDa immunoreactive PRL variants were constitutively expressed [38]. PRL cDNA was cloned from human thymocytes and sequence analysis revealed that this cDNA encodes a 23 kDa PRL, which is identical to pituitary PRL [39]. Northern blot analysis showed that thymocytes express the long PRL mRNA, whereas the pituitary-sized PRL transcripts were not detectable [39,40]. The messengers were constitutively expressed in freshly isolated thymocytes. Synthesis and secretion of biologically active 23 kDa PRL was detected in Con-A stimulated human thymocytes [40]. Pellegrini et al.

[41] demonstrated the expression of the long PRL messenger by RT-PCR using primers for exon 1a and exon 5, and showed that thymocytes express higher levels of PRL transcripts than T cells, B cells and monocytes from peripheral blood. Wu et al. [31] confirmed PRL expression in human thymocytes by RT-PCR and revealed by in situ hybridization the presence of PRL mRNA in the subcapsular cortex of the thymus.

3.3. Spleen

Expression of PRL in human splenocytes was shown by RT-PCR. In situ hybridization studies demonstrated the presence of PRL transcripts in the periarterial lymphatic sheet and in the marginal zone of the human spleen [31]. In rat spleen, PRL positive cells were detected by immunocytochemistry and in situ hybridization in the red pulp and in the marginal zone [22,42]. Double labelling experiments on dissociated splenocytes showed that the protein and the mRNA was generally expressed in the same cells, although some cells only contained detectable amounts of the protein. It was also shown that PRL positive splenocytes also expressed the transcription factor Pit-1 [42]. Although PRL expression in the human immune system is Pit-1 independent, the role of Pit-1 in rodent leukocytes remains to be elucidated. Montgomery et al. [43] were the first to show the secretion of bioactive and immunoreactive PRL-like material from Con A-stimulated murine splenocytes [43]. Shah et al. [38] extended these observations and detected secretion of a 46 kDa PRL-like molecule and expression of PRL mRNA. In another study, it was shown that both non-stimulated and Con A-stimulated murine splenocytes contain a similar (48 kDa) immunoreactive PRL-like molecule [44].

3.4. Lymph nodes and tonsils

PRL mRNA in lymph nodes and tonsils was detected by RT-PCR and according to in situ hybridization studies localized in the paracortex [31].

3.5. Peripheral blood leukocytes

RT-PCR analysis by Pellegrini et al. [41] on freshly isolated non-stimulated peripheral blood mononuclear cells (PBMC) subsets showed that PRL was expressed by T cells, to a lesser extent by B cells and not by monocytes. The relatively high PRL expression in T cells as compared with B cells is in accordance with the high levels of PRL transcripts in T cell areas in spleen, lymph nodes and tonsils [31]. PRL mRNA in human PBMC was also found by Sabharwal et al. [45]. Western blot analysis revealed the presence of 60 kDa PRL in non-stimulated and Con A-stimulated PBMC. Autocrine effects of PBMC-derived PRL were indicated by the inhibition of Con-A-induced proliferation of PBMC by anti-PRL antibodies.

The presence of 60 kDa immunoreactive PRL in Con A-stimulated PBMC and the secretion of biologically active PRL was confirmed in one other study [46]. In contrast, other investigators detected the synthesis and secretion of 23 and 27 kDa PRL [40] by metabolic labelling and immune precipitation studies. In addition, Matera et al. [47] detected the secretion of metabolically labelled 27 kDa PRL in IL-2-stimulated but not in non-stimulated PBMC. The same group also showed that the natural killer (NK) cell line YT and purified peripheral T cells express PRL mRNA, whereas purified NK cells were negative. However, the presence of immunoreactive PRL in NK cells was demonstrated by immunocytochemistry, and it was postulated that this was due to uptake of external PRL. Since the development of IL-2-induced lymphokine-activated killer cells was inhibited by an anti-PRL antibody, it was hypothesized that T cell-derived PRL served as a paracrine factor in the differentiation process.

Recently, PRL expression has also been demonstrated in myeloid cells from peripheral blood. Gingras et al. [48] showed by RT-PCR that purified human monocytes expressed PRL. They also observed a 60-fold increase in PRL mRNA levels during the differentiation of the myelomonocytic cell line U937. Interestingly, PRL was identified by suppressive substraction hybridization as one of the six new markers of U937 differentiation to monocytes. We detected PRL mRNA in freshly isolated human peripheral blood granulocytes by in situ hybridization [42] and RT-PCR [33]. Although granulocytes contain 43 kDa immunoreactive PRL, we never detected secretion of biologically active material as assessed by the Nb2 bioassay [33].

3.6. PRL in neoplastic cells

A role for PRL in leukaemia and lymphoma has been suggested by several investigators [49,50]. RT-PCR analysis and in situ hybridization studies by Wu et al. [31] revealed the expression of PRL in neoplastic spleen and lymph nodes. PRL expression in these tissues was more diffuse than in normal lymphoid organs. Pellegrini et al. [41] showed expression of long, extra-pituitary type, PRL transcripts in the Hut78 and U937 cell lines, but not in the Molt4 cell line. Furthermore, 60 kDa immunoreactive PRL in the Burkitt lymphoma cell line AG-876 was detected by Western blotting [45]. In a more detailed study on non-Hodgkin lymphoma cell lines, Matera et al. [51] showed by RT-PCR that all 6 AIDS-related non-Hodgkin lymphoma cell lines tested expressed lymphocytic PRL mRNA and secreted 23 kDa and 25 kDa (glycosylated) PRL. Two out of six non-AIDS-related non-Hodgkin lymphomas expressed extra-pituitary type PRL mRNA, and two cell lines were negative. In the other two cell lines PRL mRNA was only detected using a primer set in the coding region, which amplifies both pituitary and lymphocytic PRL cDNA, but not by a primer set in the non-coding regions which amplifies only lymphocytic PRL cDNA. This result indicates that these cells expressed the pituitary type of PRL mRNA. PRL receptors were expressed by three cell lines, but autocrine stimulation of proliferation is unlikely, because addition of inhibiting antibodies to PRL or exogenous PRL did not affect proliferation.

Several leukemic cell lines have been shown to express PRL. The T-leukemic cell line Jurkat and the leukemic NK-like cell line YT both express the long mRNA for PRL, whereas the promyelocytic cell line HL60 and the erythroleukemia cell line K562 were negative [41]. Metabolic labelling studies followed by immunoprecipitation revealed that 23 kDa PRL was synthesized and secreted by Jurkat cells [41]. Secretion of biologically active PRL was confirmed by Matera et al. [50] using immunoassays and Nb2 bioassays. This group also showed that autocrine PRL served as a growth factor for Jurkat [50]. The eosinophilic leukaemia cell line Eol-1 expresses the long extrapituitary type mRNA for PRL and synthesizes immunoreactive PRL with a molecular weight of 23 kDa. The biological activity in Eol-1 culture medium, as determined by the Nb2 bioassay, co-eluted with recombinant human PRL on an S-200 Sephacryl gel filtration column and could be blocked by anti-PRL antiserum [52]. However, autocrine effects of PRL are unlikely, because PRL receptors were undetectable and Eol-1 cells did not respond to exogenous PRL. Hatfill et al. [53] found that 16 out of 28 patients diagnosed with acute myeloid leukaemia (AML) showed increased levels of serum PRL, and that leukemic blasts from one patient contained immunoreactive PRL. The authors proposed that the increased serum levels of PRL were due to secretion by leukemic blasts. We confirmed the presence of 23 kDa immunoreactive PRL in leukemic blasts from another AML patient by Western blot analysis using two different anti-PRL antibodies. In addition, we showed that leukemic blasts released 23 kDa PRL into the culture medium. The amount of released PRL (0.02 ng/10^6 cells in 18 h) could account for increased serum levels in AML patients which may have 10^7–10^8 blasts/ml [52].

Taken together, it is well established that some neoplastic leukocytes produce PRL, but PRL as a growth factor for neoplastic leukocytes has only been implicated in the human Jurkat cell line [50] and in the rat Nb2 lymphoma cell line [54]. Possibly, the putative autocrine effects of PRL may be evident only at certain stages of the disease. More studies using freshly isolated transformed cells from patients are needed to address the possible role of PRL as an autocrine growth factor. Also the putative effects of PRL on immune responses against transformed leukocytes [55] remain to be elucidated.

3.7. PRL and autoimmunity

Systemic lupus erythematosus (SLE) is a multisystem autoimmune disease characterized by production of autoantibodies against many different autoantigens, typically including nucleic acid and associated proteins. The relation between PRL and systemic lupus erythematosus (SLE) has recently been discussed by Hooghe et al. [56]. Some SLE patients show increased serum PRL levels, and disease activity in a group of seven patients was decreased during bromocriptine treatment, indicating that serum PRL levels can modulate disease activity. A possible role for autocrine or paracrine PRL in SLE is indicated by a more than 100% increase of PRL secretion by PBMC from SLE patients [57]. Moreover, Larrea et al. [46] observed increased levels of PRL mRNA in Con-A-stimulated PBMC from SLE patients as compared with controls. Furthermore, Con-A-stimulated PBMC from patients contained higher levels of 60 kDa immunoreactive PRL and secreted higher levels of biologically active PRL-like material. Also non-stimulated cells from SLE patients secreted more biological activity. It remains to be established whether 60 kDa PRL is identical to the biologically active material secreted by PBMC and whether PRL secretion by PBMC contributes to hyperprolactenemia in some patients with SLE. It has been postulated that the close genomic linkage between the PRL gene and HLA alleles is responsible for the association between SLE and increased PRL production [58]. Stevens et al. [59] found that a single nucleotide polymorphism in the upstream promoter region of the PRL gene was increased in a cohort of SLE patients. Since one allele was responsible for increased activation of the upstream promoter in transfection studies using the Jurkat cell line, it is tempting to speculate that increased PRL levels in SLE patients may be due to increased transcription of extra-pituitary PRL.

An intriguing question is whether polymorphisms in the PRL promoter region are also linked with other autoimmune diseases such as Sjögren's syndrome. Local production of PRL was also observed in patients with Sjögren's syndrome [60]. This is a chronic autoimmune disease characterized by lymphocytic infiltration of the salivary and lacrimal glands leading to loss of secretory functions. The percentage of PRL expressing acinar epithelial cells in salivary glands was five times higher in patients with extraglandular manifestations (sensory neuropathy, Raynaud or pulmonary involvement) as compared with controls or patients without extraglandular manifestations. Furthermore, the fraction of PRL positive cells correlated well with the presence of autoantibodies. Immunoprecipitation studies revealed that salivary glands of patients produced much more 60 kDa and slightly more 16 kDa PRL-like molecules. Remarkably, PRL was not detected in infiltrating mononuclear cells. The expression of PRL receptors on ductal epithelial cells opens up the possibility for paracrine effects of PRL. Indeed, patients showed increased cathepsin B and D levels in minor salivary glands, and PRL augmented these levels in both patients and controls [61]. Since cathepsins are known to be involved in antigen processing, it was postulated that these proteinases could be involved in the processing of autoantigens in salivary glands.

The production of PRL by synovium infiltrating T cells and to a lesser extent by synovial fibroblasts was found in patients with rheumatoid arthritis (RA). The presence of PRL receptors on fibroblast-like synovial cells and the effects of PRL on proliferation and cytokine production strongly suggest an autocrine or paracrine function for PRL in the synovial membrane of patients with RA [62]. On the other hand, PRL levels in the synovial fluid do not seem to be affected by the disease as shown by Hedman et al. [63]. They found that PRL levels in the synovial fluid from patients with RA were identical to those from patients with other pathology (mainly psoriasis, ankylosing spondylitis and reactive arthritis). They also showed that the mean PRL levels in the synovial fluid were identical to those in serum. We also observed that PRL concentrations in the synovial fluid of six patients with RA (6.6 ± 2.0 ng/ml) were not different from normal concentrations in serum (unpublished observation).

4. CONCLUDING REMARKS

Autocrine and paracrine effects of PRL in the immune system are conceivable, because leukocytes as well as epithelial and endothelial cells in lymphoid organs, and sites of inflammation express PRL. The synthesis and secretion of bioactive 23 kDa pituitary PRL is well established in human bone marrow stroma cells, thymocytes and PBMC [32,40,41] and in a few leukemic cell lines [41,50,52]. Taken together, the production of bioactive PRL by normal leukocytes and the induction by PRL of genes in normal rat and human leukocytes that are relevant for immune responses, as discussed by Hooghe et al. in this issue [56], suggest an autocrine or paracrine function for PRL. However, size-heterogeneity in immunoreactive PRL from leukocytes is evident. Heterogeneity of human PRL has been excellently reviewed by Sinha [2]. Pituitary-derived PRL variants are glycosylated, phosphorylated or proteolytically modified. Unique functions or modified specific activities have been assigned to some of these variants. The biological activity of glycosylated and phosphorylated PRL is slightly decreased, whereas a 16 kDa proteolytic fragment, which hardly binds to the PRL receptor, appears to exert anti-angiogenic actions through a receptor of its own [64].

Although the identity and biological significance of most PRL-like molecules from the immune system are not clear, there are indications that they are biologically active. For instance, Sabharwal et al. [45] demonstrated that Con A-activated PBMC secreted biologically active PRL-like material, while only one PRL-like molecule of 60 kDa could be detected by Western blotting. Montgomery et al. [40] showed that human thymocyte-derived 11 kDa PRL excised from a polyacrylamide gel exhibited biological activity in an Nb2 bioassay. However, to the best of our knowledge, PRL gene products larger than 30 kDa resulting from either alternative splicing or glycosylation have not been described. PRL dimerization or aggregation with other proteins by noncovalent or disulfide bonds are unlikely, because the high molecular weight variants have been detected by Western blotting after SDS-polyacrylamide gel electrophoresis under reducing conditions. Interestingly, Walker et al. [65] showed that aggregation of PRL with immunoglobulins by covalent bonds may result in high molecular weight forms. Remarkably, these complexes were not active in the classical Nb2 bioassay but proliferation of lymphocytes from patients with chronic lymphocytic leukaemia was markedly enhanced, possibly by engaging both PRL and immunoglobulin receptors. The characterization PRL-like molecules has been precluded by lack of material. The use of new techniques in mass spectrometry [66] allowing the identification and analysis of very small amounts of proteins may contribute to delineate the significance of PRL-like molecules in the immune system.

The relations between extrapituitary PRL production and autoimmune diseases or proliferative disorders are not clear. Although there is no evidence that elevated PRL levels can cause immune disorders, increased levels of PRL production may be secondary to immune disorders and in turn, through their immunostimulatory actions, influence the course of the disease. Future research on the factors that regulate local PRL production in the immune system and functional analysis of PRL variants will contribute to the elucidation of the relation between PRL and immune disorders.

ACKNOWLEDGEMENTS

We would like to thank Dr. L. Verbruggen for providing synovial fluid, J. Schietecatte for measuring PRL concentrations, and Drs. R. Hooghe and E.L. Hooghe-Peters for their comments. This research in our group has been funded by the Flemish Government (GOA 97-02-4), the Fund for Scientific Research-Flanders, Belgium (F.W.O.) and institutional grants from the V.U.B.

REFERENCES

1. Ben-Jonathan N, Mershon JL, Allen DL, Steinmetz RW. Extrapituitary prolactin: distribution, regulation, functions, and clinical aspects. Endocr Rev 1996;17:639–669.
2. Sinha YN. Structural variants of prolactin: occurrence and physiological significance. Endocr Rev 1995;16:354–369.
3. Freeman ME, Kanyicska B, Lerant A, Nagy G. Prolactin: structure, function, and regulation of secretion. Physiol Rev 2000;80:1523–1631.
4. Kooijman R, Hooghe-Peters EL, Hooghe R. Prolactin, growth hormone and insulin-like growth factor-I in the immune system. Adv Immunol 1996;63:377–453.
5. Bouchard B, Ormandy CJ, Di Santo JP, Kelly PA. Immune system development and function in prolactin receptor- deficient mice. J Immunol 1999;163:576–582.
6. Horseman ND, Zhao W, Montecino-Rodriguez E, et al. Defective mammopoiesis, but normal hematopoiesis, in mice with a targeted disruption of the prolactin gene. EMBO J 1997;16:6926–6935.
7. Dorshkind K, Horseman ND. The roles of prolactin, growth hormone, insulin-like growth factor-I, and thyroid hormones in lymphocyte development and function: insights from genetic models of hormone and hormone receptor deficiency. Endocr Rev 2000;21:292–312.
8. Truong AT, Duez C, Belayew A, et al. Isolation and characterization of the human prolactin gene. EMBO J 1984;3:429–437.
9. Gellersen B, DiMattia GE, Friesen HG, Bohnet HG. Prolactin (PRL) mRNA from human decidua differs from pituitary PRL mRNA but resembles the IM-9-P3 lymphoblast PRL transcript. Mol Cell Endocrinol 1989;64:127–130.
10. DiMattia GE, Gellersen B, Bohnet HG, Friesen HG. A human B-lymphoblastoid cell line produces prolactin. Endocrinology 1988;122:2508–2517.
11. Peers B, Voz ML, Monget P, et al. Regulatory elements controlling pituitary-specific

expression of the human prolactin gene. Mol Cell Biol 1990;10:4690–4700.

12. Berwaer M, Martial JA, Davis JRE. Characterization of an up-stream promoter directing extrapituitary expression of the human prolactin gene. Mol Endocrinol 1994;8:635–642.
13. Gellersen B, Kempf R, Telgmann R, DiMattia GE. Nonpituitary human prolactin gene transcription is independent of Pit-1 and differentially controlled in lymphocytes and in endometrial stroma. Mol Endocrinol 1994;8:356–373.
14. Brown NA, Bethea CL. Cloning of decidual prolactin from rhesus macaque. Biol Reprod 1994;50:543–552.
15. Peers B, Monget P, Nalda MA, et al. Transcriptional induction of the human prolactin gene by cAMP requires two cis-acting elements and at least the pituitary-specific factor Pit-1. J Biol Chem 1991;266:18127–18134.
16. Sutton CA, Martin TF. Thyrotropin-releasing hormone (TRH) selectively and rapidly stimulates phosphatidylinositol turnover in GH pituitary cells: a possible second step of TRH action. Endocrinology 1982;110:1273–1280.
17. Berwaer M, Monget P, Peers B, et al. Multihormonal regulation of the human prolactin gene expression from 5000 bp of its upstream sequence. Mol Cell Endocrinol 1991;80:53–64.
18. Castillo AI, Jimenez-Lara AM, Tolon RM, Aranda A. Synergistic activation of the prolactin promoter by vitamin D receptor and GHF-1: role of the coactivators, CREB-binding protein and steroid hormone receptor coactivator-1 (SRC-1). Mol Endocrinol 1999;13:1141–1154.
19. Gellersen B, DiMattia GE, Friesen HG, Bohnet HG. Regulation of prolactin secretion in the human B-lymphoblastoid cell line IM-9-P3 by dexamethasone but not other regulators of pituitary prolactin secretion. Endocrinology 1989;125:2853–2861.
20. Gellersen B, Kempf R, Hartung S, Bonhoff A, DiMattia GE. Posttranscriptional regulation of the human prolactin gene in IM-9-P3 cells by retinoic acid. Endocrinology 1992;131:1017–1025.
21. Gellersen B, DiMattia GE, Friesen HG, Bohnet HG. Phorbol ester stimulates prolactin release but reduces prolactin mRNA in the human B-lymphoblastoid cell line IM-9-P3. Mol Cell Endocrinol 1989;66:153–161.
22. Delhase M, Vergani P, Malur A, Hooghe R, Hooghe-Peters EL. The transcription factor Pit-1/GHF-1 is expressed in hematopoietic and lymphoid tissues. Eur J Immunol 1993;23:951–955.
23. Reem GH, Ray DW, Davis JR. The human prolactin gene upstream promoter is regulated in lymphoid cells by activators of T-cells and by cAMP. J Mol Endocrinol 1999;22:285–292.
24. Telgmann R, Gellersen B. Marker genes of decidualization: activation of the decidual prolactin gene. Hum Reprod Update 1998;4:472–479.
25. Pohnke Y, Kempf R, Gellersen B. CCAAT/enhancer-binding proteins are mediators in the protein kinase A- dependent activation of the decidual prolactin promoter. J Biol Chem 1999;274:24808–24818.
26. Frank GR, Brar AK, Jikihara H, Cedars MI, Handwerger S. Interleukin-1 beta and the endometrium: an inhibitor of stromal cell differentiation and possible autoregulator of decidualization in humans. Biol Reprod 1995;52:184–191.
27. Jikihara H, Handwerger S. Tumor necrosis factor-alpha inhibits the synthesis and release of human decidual prolactin. Endocrinology 1994;134:353–357.
28. Bonhoff A, Gellersen B. Modulation of prolactin secretion in human myometrium by cytokines. Eur J Obstetrics & Gynecology and Reproductive Biology 1994;54:55.
29. Van De Weerdt C., Peers B, Belayew A, Martial JA, Muller M. Far upstream

sequences regulate the human prolactin promoter transcription. Neuroendocrinology 2000;71:124–137.
30. Gaasenbeek M, Gellersen B, DiMattia GE. DNase I hypersensitivity analysis of non-pituitary human prolactin gene expression. Mol Cell Endocrinol 1999;152:147–159.
31. Wu H, Devi R, Malarkey WB. Expression and localization of prolactin messenger ribonucleic acid in the human immune system. Endocrinology 1996;137:349–353.
32. Bellone G, Astarita P, Artusio E, et al. Bone marrow stroma-derived prolactin is involved in basal and platelet- activating factor-stimulated in vitro erythropoiesis. Blood 1997;90:21–27.
33. Kooijman R, Gerlo S, Coppens A, Hooghe-Peters EL. Growth hormone and prolactin expression in the immune system. Ann N Y Acad Sci 2000;917:534–540.
34. Dogusan Z, Book ML, Verdood P, Yu-Lee LY, Hooghe-Peters EL. Prolactin activates interferon regulatory factor-1 expression in normal lympho-hemopoietic cells. Eur Cytokine Netw 2000;11:435–442.
35. Touraine P, Leite-de-Moraes MC, Dardenne M, Kelly PA. Expression of short and long forms of prolactin receptor in murine lymphoid tissues. Mol Cell Endocrinol 1994;104:183–190.
36. Dardenne M, de Moraes M, Kelly PA, Gagnerault MC. Prolactin receptor expression in human hematopoietic tissues analyzed by flow cytofluorometry. Endocrinology 1994;134:2108–2114.
37. Bellone G, Geuna M, Carbone A, et al. Regulatory action of prolactin on the in vitro growth of CD34+ve human hemopoietic progenitor cells. J Cell Physiol 1995;163:221–231.
38. Montgomery DW, LeFevre JA, Ulrich ED, Adamson CR, Zukoski CF. Identification of prolactin-like proteins synthesized by normal murine lymphocytes. Endocrinology 1990;127:2601–2603.
39. O'Neal KD, Montgomery DW, Truong TM, Yu Lee LY. Prolactin gene expression in human thymocytes. Mol Cell Endocrinol 1992;87:R19–R23.
40. Montgomery DW, Shen GK, Ulrich ED, Steiner LL, Parrish PR, Zukoski CF. Human thymocytes express a prolactin-like messenger ribonucleic acid and synthesize bioactive prolactin-like proteins. Endocrinology 1992;131:3019–3026.
41. Pellegrini I, Lebrun JJ, Ali S, Kelly PA. Expression of prolactin and its receptor in human lymphoid cells. Mol Endocrinol 1992;6:1023–1031.
42. Hooghe-Peters EL, Hooghe R. Growth hormone, prolactin and IGF-I as lymphohemopoietic cytokines. Austin/New York: R.G. Landes/Springer-Verlag, 1995.
43. Montgomery DW, Zukoski CF, Shah GN, Buckley AR, Pacholczyk T, Russell DH. Concanavalin A-stimulated murine splenocytes produce a factor with prolactin-like bioactivity and immunoreactivity. Biochem Biophys Res Commun 1987;145:692–698.
44. Kenner JR, Holaday JW, Bernton EW, Smith PF. Prolactin-like protein in murine splenocytes: morphological and biochemical evidence. Prog Neuroendocrinimmunol 1990;3:188–195.
45. Sabharwal P, Glaser R, Lafuse W, et al. Prolactin synthesized and secreted by human peripheral blood mononuclear cells: An autocrine growth factor for lymphoproliferation. Proc Natl Acad Sci USA 1992;89:7713–7716.
46. Larrea F, Martinez-Castillo A, Cabrera V, et al. A bioactive 60-kilodalton prolactin species is preferentially secreted in cultures of mitogen-stimulated and nonstimulated peripheral blood mononuclear cells from subjects with systemic lupus erythematosus. J Clin Endocrinol Metab 1997;82:3664–3669.

47. Matera L, Bellone G, Lebren JJ, et al. Role of prolactin in the in vitro development of interleukin-2- driven anti-tumoural lymphokine-activated killer cells. Immunology 1996;89:619–626.
48. Gingras MC, Margolin JF. Differential expression of multiple unexpected genes during U937 cell and macrophage differentiation detected by suppressive subtractive hybridization. Exp Hematol 2000;28:350.
49. Hooghe R, Merchav S, Gaidano G, Naessens F, Matera L. A role for growth hormone and prolactin in leukaemia and lymphoma? Cell Mol Life Sci 1998;54:1095–1101.
50. Matera L, Cutufia M, Geuna M, et al. Prolactin is an autocrine growth factor for the Jurkat human T- leukemic cell line. J Neuroimmunol 1997;79:12–21.
51. Matera L, Geuna M, Pastore C, et al. Expression of prolactin and prolactin receptors by non-Hodgkin's lymphoma cells. Int J Cancer 2000;85:124–130.
52. Kooijman R, Gerlo S, Coppens A. Myeloid leukemic cells express and secrete bioactive pituitary-sized 23 kDa prolactin. J Neuroimmunol 2000;110:252–258.
53. Hatfill SJ, Kirby R, Hanley M, Rybicki E, Bohm L. Hyperprolactinemia in acute myeloid leukemia and indication of ectopic expression of human prolactin in blast cells of a patient of subtype M4. Leuk Res 1990;14:57–62.
54. Gout PW, Beer CT, Noble RL. Prolactin-stimulated growth of cell cultures established from malignant Nb rat lymphomas. Cancer Res 1980;40:2433–2436.
55. Oberholtzer E, Contarini M, Veglia F, et al. Prolactin increases the susceptibility of primary leukemia cells to NK and LAK effectors. Adv Neuroimmunol 1996;6:233–247.
56. Hooghe R, Dogusan Z, Martens N, Velkeniers B, Hooghe-Peters EL. Signal transduction and modulation of gene expression by prolactin in human leukocytes. In: Matera L, Rapaport R, editors. Growth and Lactogenic Hormones. Neuroimmune Biology, Vol. 2. Elsevier, 2002;123–136.
57. Gutierrez MA, Molina JF, Jara LJ, et al. Prolactin and systemic lupus erythematosus: prolactin secretion by SLE lymphocytes and proliferative (autocrine) activity. Lupus 1995;4:348–352.
58. Brennan P, Hajeer A, Ong KR, et al. Allelic markers close to prolactin are associated with HLA-DRB1 susceptibility alleles among women with rheumatoid arthritis and systemic lupus erythematosus. Arthritis Rheum 1997;40:1383–1386.
59. Stevens A, Ray D, Alansari A, et al. Characterisation of a prolactin gene polymorphism and its associations with systemic lupus erythematosus. Arthritis Rheum 2001; accepted for publication.
60. Steinfeld S, Rommes S, Francois C, et al. Big prolactin 60 kDa is overexpressed in salivary glandular epithelial cells from patients with Sjogren's syndrome. Lab Invest 2000;80:239–247.
61. Steinfeld S, Maho A, Chaboteaux C, et al. Prolactin up-regulates cathepsin B and D expression in minor salivary glands of patients with Sjogren's syndrome. Lab Invest 2000;80:1711–1720.
62. Nagafuchi H, Suzuki N, Kaneko A, Asai T, Sakane T. Prolactin locally produced by synovium infiltrating T lymphocytes induces excessive synovial cell functions in patients with rheumatoid arthritis. J Rheumatol 1999;26:1890–1900.
63. Hedman M, Nilsson E, de la Torre B. Low blood and synovial fluid levels of sulpho-conjugated steroids in rheumatoid arthritis. Clin Exp Rheumatol 1992;10:25–30.
64. Clapp C, Weiner RI. A specific, high affinity, saturable binding site for the 16-kilodalton fragment of prolactin on capillary endothelial cells. Endocrinology 1992;130:1380–1386.

65. Walker AM, Montgomery DW, Saraiya S, et al. Prolactin-immunoglobulin G complexes from human serum act as costimulatory ligands causing proliferation of malignant B lymphocytes. Proc Natl Acad Sci U S A 1995;92:3278–3282.
66. Andersen JS, Mann M. Functional genomics by mass spectrometry. FEBS Lett 2000;480:25–31.

III. HEMOPIESIS AND DEVELOPMENT

Growth and Lactogenic Hormones
Edited by L. Matera and R. Rapaport

Prolactin as a Promoter of Growth and Differentiation of Hemopoietic Cells

GRAZIELLA BELLONE

Department of Clinical Physiopathology, University of Turin, Italy

ABSTRACT

Evidence implicating prolactin (PRL) in the development of blood cell lineages was originally provided by classical studies in hormone-deficient animals. The mass of supporting data collected since then, over the past 20 years, calls for re-examination of the role of PRL in the hemopoietic process in the light of recent advances in this field. Identification of functional PRL-receptors on human hemopoietic cells, together with the local production of extra pituitary PRL within the hemopoietic microenvironment, points to the existence of both endocrine and paracrine mechanisms of PRL action that, in synergy with specific hemopoietic growth factors, contribute to modulating basal and activated hemopoieisis.

Preclinical studies, demonstrating the pharmacological ability of PRL to promote hemopoietic recovery after myelotoxic treatments, suggest that the hormone may be considered a therapeutic agent in the management of immunohematologic deficiencies.

1. INTRODUCTION

Mature hemopoietic cells that populate hemopoietic organs and circulate in the blood are derived from a minute population of totipotent stem cells that are found in the bone marrow (BM) of mammals. A constant number of hemopoietic cells – most of which have a limited life span – are maintained by these stem cells through processes of cell proliferation, migration, homing, attachment, and differentiation. Blood cell development is primarily regulated by local production of hemopoietic growth factors (HGFs) within the BM microenvironment [1,2]. Four major HGFs have been characterized and cloned in humans, namely (IL)[a]-3 [3], GM-CSF[b] [4], G-CSF [5], M-CSF [6] that, together with the circulating hormone erythropoietin (EPO) [7] and thrombopoietin (TPO) [8], control the formation and function of granulocytes, macrophages, erythrocytes and platelets. In addition to these factors, which are relatively specific in their action and in

[a] Interleukin.
[b] Granulocyte-macrophage (GM) - colony stimulating factor (CSF).

the distribution of their receptors, several lymphokines [9–11], immunomodulators [12,13] and hormones [14–18] can act in concert on hemopoietic progenitors, either directly or via accessory cells, to produce synergistic, additive or antagonistic responses to specific factors.

Evidence has recently accumulated suggesting that prolactin (PRL), originally regarded as a pituitary hormone implicated in reproduction and lactation, may belong to the host soluble factors involved in the regulation of the immune and hemopoietic systems [19–29]. The relevance of PRL in this context was initially suggested by the association between defects in immunity and in hemopoiesis, and deficiencies in PRL production or secretion [30–32]. In particular, the role of PRL in blood cell development has recently been strengthened by the inclusion of PRL receptors (R) in the class 1 hemopoietin/cytokine receptor superfamily [33–35], based on structural similarities [36] and use of common or related signaling pathways [37]. Specific PRL membrane-binding sites are expressed by lymphoid and hemopoietic cells [38–42] and the interaction of the hormone with these receptors is coupled with changes in the proliferative and/or maturation state of these cells. In addition, besides pituitary PRL, PRL mRNA and protein have been identified at other sites where the hormone most likely mediates local function [43].

During the last decade, considerable literature has accumulated pointing to PRL as a critical immunoregulatory factor. This issue has been treated in numerous comprehensive reviews, to which the reader's attention is directed [19,20,22–26,28].

The present review focuses on new topics arising from the growing mass of evidence supporting the role of PRL in normal hemopoiesis.

2. IN VIVO EFFECTS OF PRL ON HEMOPOIESIS

Proof of specific involvement of PRL in the development of the hemopoietic system comes from early studies performed on rodents in which pituitary deficiency was congenitally present due to a defect in their anterior pituitary gland or was experimentally induced by drugs or surgical ablation of the gland [44–46].

DW/J dwarf mice, lacking acidophilic anterior pituitary cells, exhibit decreased peripheral blood cell counts affecting all lineages (erythrocytic, leukocytic, and platelets) [44]. Hypophysectomized rats typically display hemopoietic defects, including suppressed numbers of splenic hemopoietic progenitors cells, reduction of the red cell mass and hypoplasia of BM [46]. Implant of syngeneic pituitary gland under the kidney capsule or treatment with exogeous PRL restored BM function almost completely and erythroid hypoplasia in these rats [30]. Moreover, PRL antibodies given to normal animals produced anemia and death due to hemopoietic failure [45].

Increased erythropoiesis during pregnancy and lactation, which is manifested in elevated blood volume, has long been known. PRL treatment stimulated erythropoiesis in normal and in postpartum lactating or non-lactating mice [47,48]. PRL administration has also been reported to have a stimulatory effect on red cells and plasma volume in hyperoxic mice [49] and on ^{59}Fe uptake by the erythrocytes of hypoxia or transfusion-induced polycythaemic mice [50]. Moreover, normal resting mice receiving recombinant human (rh) PRL exhibit a significant increase of both frequency and absolute numbers of BM colony forming unit (CFU)-GM and burst forming unit (BFU)-erythroid (E) [51].

The precise action of PRL in hemopoiesis in vivo, however, remained poorly defined, as our understanding was essentially based on the administration of exogenous hormone to animals or on experimental models that had multiple hormonal deficiencies or from which it was impossible to remove potentially important local sources of extra-pituitary PRL.

To circumvent these limitations, mice with disruptions in genes for PRL [52], PRL-R [53] and PRL-stimulated Signal Transducers and Activators of Transcription (STAT)5a [54] and STAT5b [55] have recently been generated, greatly facilitating our understanding of the role of PRL in physiological and developmental processes [54–56]. In contrast to essential roles of PRL in reproduction and lactation, the presence of a normal hematocrit and normal numbers of myeloid and lymphoid cells in the BM of PRL-/- mice indicate that PRL is not required for normal steady-state hemopoiesis [52].

Nevertheless, the expression of the PRL receptor on the majority of developing blood cells provides for the possibility that PRL can have subtle effects on these populations.

It is of interest that, in humans, hypo- or hyper-prolactinemia are not clinically associated with significant hemopoietic detrimental effects. Only mild hypoproliferative anemia is frequently seen in pituitary hormone deficiencies, as the result of decreased erythropoietin stimulation related to the hypometabolic state. Treatment of pituitary dwarfs with growth hormone (GH) increased EPO levels and myeloid and erythroid progenitors [57], but it is not clear whether this is a direct action of GH or a secondary effect in response to increased oxygen consumption.

Since current evidence does not convincingly support PRL's playing an obligatory role in blood cell formation, the involvement of this hormone in hemopoiesis in vivo may be considered to be only a modulating one under normal conditions or a compensatory one in HGF deficiency.

3. IN VITRO EFFECTS OF PRL ON HEMOPOIESIS

The establishment of semi-solid culture systems for the formation of different types of blood cells in vitro [58], together with the availability of pure rHGFs [59] and the development of methodologies to purify CD34+ hemopoietic stem cells [60], has made it possible to assess the single or combined effects of soluble factors on colony formation by cultured purified target cells. It should be emphasized that cells in semi-solid or liquid cultures are imprecise models of the hemopoietic microenvironment, and that the growth factor activities observed in vitro may not accurately reflect their activities in vivo. However, these tools have made a considerable contribution to defining the direct modulatory role of PRL in hemopoietic development, based on i) identification of the corresponding receptors on primitive hemopoietic progenitor cells [61], and ii) availability of extra-pituitary PRL at the site of hemopoietic process in vivo [62]. Unless specified, data refer to the human system.

3.1. PRL receptors on hemopoietic progenitor cells

In the early 1990s, sequence comparison with newly identified membrane receptors led to the identification of a new family of receptors, including both PRL receptor (R) and GH-R. This superfamily includes receptors for IL-2 (β and γ chain), IL-3, IL-4, IL-6, IL-7, IL-9, IL-12, IL-15, GM-CSF, G-CSF, EPO, TPO, gp130, Leukemia Inhibitory Factor, Oncostatin M, Obesity Factor Leptin, and Ciliary Neurotropic Factor [33–36,63]. Although all these membrane chains are apparently genetically unrelated, they contain stretches of highly conserved amino acids, both in the extracellular and the intracellular domains. This suggests that their cognate cytokines may have common structural features and/or can mediate biological activities on different hemopoietic cells using similar signal transduction mechanisms [37,64]. Multiple isoforms of membrane-bound PRL-R, resulting from alternative splicing of the primary transcript, have been identified in several species. These different PRL-R isoforms differ in the length and composition of their cytoplasmic tail, and are referred to as short, intermediate, or long PRL-R [65]. The short and long forms of the receptor are differentially expressed or regulated in different cell compartments, which suggests the individuality of the PRL effect on a given cell system [66].

As for the cytokine/hemopoietic growth factors, the interaction of PRL with its receptor in the target cells leads to the activation of members of the cytoplasmic Janus tyrosine kinase (JAK) family, namely Tyk2, JAK1, JAK2, JAK3 [67,68], and other receptor-associated kinases of the Src family [69]. One major signaling pathway involves phosphorylation of cytoplasmic STAT proteins, which translocate to the nucleus and bind to specific promoter elements on PRL-responsive genes [70]. In addition, the Ras/Raf/MAP kinase pathway is also activated by PRL and may be involved in the proliferative effects of the hormone. Detailed description of signal transduction by the PRL-R is available in previous reviews [71,72].

The structural base for the putative immuno-hemopoietic regulatory role of PRL has been provided by the description of widespread distribution of PRL-R and corresponding gene expression in various human as well as murine lymphoid and hemopoietic tissues [38–42,73–75]. In particular, a large proportion of BM cells, which includes all hematopoietic lineage precursors, express high levels of PRL-R [76]. We recently reported the presence, on a subset of primitive hemopoietic cells, freshly isolated from human BM and peripheral blood, of a membrane structure recognized by two different mAb, raised against rat and rabbit PRL-R, but crossreacting with human cells [61]. Biochemical characterization assigned to this molecule, immunoprecipitated from metabolically labeled BM CD34+ cells, an apparent Mw of 43-Kda, under reducing conditions. The size of this molecule is very close to the short form of the PRL-R reported on classic PRL target cells [73]. These PRL-binding sites are functionally active, since the interaction with the hormone results in a synergistic response of the hemopoietic progenitor cells to multipotential and lineage restricted HG, in terms of both proliferation and maturation [61].

3.2. Effect of exogenous PRL on myelopoiesis

The effects of PRL on myeloid differentiation have been described for human BM, cord blood and circulating CD34+ cells [61]. PRL alone cannot support myeloid colony formation. However, at the supraphysiological concentration of 50 ng/ml PRL has been shown to improve both proliferation and terminal maturation of myeloid progenitors in the presence of specific HGFs, such as IL-3 and GM-CSF. Limiting dilution analysis indicated that this phenomenon is apparently direct and not mediated by secondary cytokines produced by contaminating cells [61].

Recent studies performed on normal mice show that, as in humans, the addition of recombinant human PRL to the BM hemopoietic progenitor cell culture, under defined serum poor conditions, induces a slight increase in CFU-GM formation only in the presence of murine GM-CSF and IL-3 [51]. This enhancement is not to be ascribed to the ability of PRL to protect hemopoietic cells from apoptosis, since the hormone was not observed to promote myeloid progenitor survival.

The *in vitro* lack of an individual effect of PRL on CFU-GM colony formation, found in murine and human studies, is in line with the reported normal myelopoiesis in mice with targeted disruption of the PRL gene [52], and further corroborates with the notion that the hormone is not critical for primary development of myeloid cells. However, the *in vitro* myelopoietic effect of PRL in synergism with IL-3 and GM-CSF supports the existence of a functional interplay between the hormone and lineage-restricted HGFs, providing an explanation for its ability to accelerate myelopoiesis in resting or normal animals and in those that have undergone myelosuppressive therapy [51].

3.3. Effect of exogenous PRL on erythropoiesis

Although PRL was found to increase myeloid colony formation, our in vitro studies demonstrated that the most striking effect was observed on the maturation of erythroid progenitors, in agreement with earlier literature pointing to a role for the hormone in normal erythropoiesis.

We have recently shown that exogenous PRL acts in synergism with IL-3, GM-CSF and EPO at an early stage of BFU-E colony formation, at both the divisional and the differentiating levels [61]. It was shown in earlier studies that there is an age structure amongst BFU-E and that their EPO sensitivity increases with maturation. Moreover, in intact animals, BFU-E proliferation is influenced by stimuli to marrow regeneration but not by changes in EPO concentration [77]. PRL, at a concentration slightly above the normal range, was found to potentiate the IL-3 and GM-CSF-dependent transition from EPO-insensitive BFU-E to EPO-responsive CFU-E by promoting the expression of receptors for the progression factor EPO [61]. Several types of evidence suggest functional interaction between PRL and EPO in controlling erythropoiesis both *in vitro* and *in vivo*. PRL-R and severely truncated EPO-R support differentiation of erythroid progenitors, and the PRL-R rescues EPO-R -/- erythroid progenitors and replaces EPO-R, in a synergistic interaction with c-kit [78,79]. A direct erythropoietic effect in vitro of plasma from pregnant and lactating mice and an enhanced erythropoiesis in pregnant women have been observed [48,80]. Furthermore, pituitary grafts under the kidney capsule, an in vivo model of hyperprolactinemia, favor the development of Friend murine virus-induced leukemias and switch their isotype from predominantly lymphoid to erythroid [81]; conversely, regression of erythroblastic leukemia has been observed in a significant number of rats after hypophysectomy [82].

We recently used enriched circulating hemopoietic progenitor cells from hyperprolactinemic hemodialysis patients as a model to investigate the interactive action of PRL with EPO in vivo [83]. Due to the paramount role of the kidneys as a source of EPO, anemia is very frequent in dialyzed patients, often accompanied by hyperprolactinemia [84,85]. We showed that the increase in PRL serum levels in dialyzed patients is associated with the presence of EPO-responsive erythroid precursors in the blood, which are present neither in normoprolactinemic patients nor in healthy subjects [83]. It is not clear whether this phenomenon reflects an extra-BM maturation process. The PRL-dependent enhancement of transition from BFU-E to CFU-E required, at least in vitro, the presence of IL-3 and GM-CSF [61], which are undetectable in the blood [86]. Therefore, the most likely explanation for this phenomenon is an abnormal dismission of these cells from the BM. Enlargement of the pool of EPO-responsive erythroid precursors may counteract the EPO deficiency in these patients, thus optimizing the erythroid maturation process. The latter interpretation is in line with the reported normalization of PRL levels in hyperprolactinemic dialyzed patients after EPO-treatment [85]. A formal assessment of this hypothesis would require the experimental demonstration of the effect of decreasing the PRL levels on the hematologic parameters. Tentatively one may suggest that the hyperprolactinemia in this pathological condition is a compensatory mechanism for the reduced availability of the erythroid specific factor EPO.

3.4. Effects of exogenous PRL on megakaryocytopoiesis

As far as the writer is aware, no studies have been published that examine in depth the effect of PRL on megakaryocytopoiesis in murine or human in vitro models. It has been shown that the thrombocytopenia associated with the anemia and leukopenia in hypophysectomized rats, or in mice infused with syngeneic BM cells after a myeloablative dose of total body irradiation, can be corrected by anterior pituitary transplants or administration of PRL, respectively [18,87]. When megakaryocyte (Meg)-CSF activity of PRL was assayed on hemopoietic progenitor cells in plasma clot cultures containing IL-3, GM-CSF, IL-6 and EPO, we found no enhancing effect on CFU-Meg-colony formation [61]. However, since TPO is the specific factor for Meg terminal maturation [88], the possibility of a cooperative action of PRL and the megakaryocytic progression factor, in analogy with the functional interplay with the obligatory erythroid growth factor, EPO, is not to be excluded without careful investigation of this topic.

4. BM STROMA CELLS AS EXTRA-PITUITARY SOURCE OF PRL

In the adult, hemopoiesis occurs in the BM in association with a stromal cell meshwork [89]. Evidence indicates that the stromal cells, consisting primarily of fibroblasts, endothelial cells, adipocytes, epithelial cells, macrophages and dendritic cells [90], are not passive bystanders but themselves influence hemopoietic cell development, presumably by supplying the necessary cell matrix and diffusible short-range acting HGFs [91].

Although most of the biological actions of PRL are believed to be sustained by pituitary release, recent evidence points to its paracrine and autocrine role during immune response [24]. In animal models, low levels of PRL can still be detected in the circulation after hypophysectomy. Lethal impairment of the lympho-hemopoietic system occurs only when complete depletion of PRL is achieved by treatment with anti-PRL antibody [45]. This finding raises the questions of whether BM stroma cells can produce PRL and, in that case, whether the local production participates in the hemopoietic process. We addressed these issues by developing a semisolid culture system in which irradiated BM stroma cell monolayers were used as feeders for colony formation from autologous CD34+ hemopoietic progenitors in the presence of exogenous IL-3, GM-CSF and EPO [62]. We found that, in these experimental conditions, formation of BFU-E colonies was slightly, but significantly, blocked by an anti-PRL antibody, whereas CFU-GM were unaffected, suggesting that PRL-sustained paracrine mechanisms may also operate during basal erythropoiesis. Hemopoiesis is regulated by external stimuli, which act primarily by recruiting cells of the hemopoietic microenvironment to produce HGFs [92]. We evaluated the possible role of exogenous Platelet-Activating Factor (PAF) [93] on the modulation of PRL production in BM stroma environment, based on the premise that this mediator i) is secreted during inflammation [94], ii) regulates pituitary PRL release [95], iii) stimulates DNA synthetic activity in freshly isolated BM cells [96], and iv) can interact with BM stroma cells that have been shown to express PAF-R mRNA [62,97]. PAF stimulation was found to strongly increase PRL release, PRL synthesis, and PRL gene transcription as detected by ELISPOT, Western blotting and in situ hybridization, respectively [62]. Concomitant enhancement of BFU-E colony formation was neutralized by anti-PRL antibody. Interestingly, most of the PRL produced by the unstimulated BM stroma cells was of the glycosylated (G)-form (25.5 kDa), while the non-glycosylated (NG)-type (~23kDa) predominated after PAF stimulation [62]. Mature pituitary PRL is a polypeptide of 199 amino acids with a molecular mass of 23.5 kDa.

Structural variants of PRL, deriving from post-translational modifications, including glycosylation, phosphorylation, dimerization, or proteolitic cleavage, have been identified [98]. Most of the biological activity of PRL has been associated with the NG form [99], while selective down-regulation of PRL action at individual target tissues has been ascribed to the G-variant, because of its lower biological potency and receptor binding capacity [100]. Based on these premises, it is tempting to speculate that prevalent constitutive production by BM stroma cells of G-PRL may be regarded as a mechanism by which the promoting activity of PRL on BFU-E formation is maintained at subliminal levels during basal hemopoiesis. The fact that PAF-induced activation of PRL-mediated colony promoting activity is accompanied by synthesis of the NG form strongly supports the hypothesis that the major contribution of PRL in the regulation of hemopoiesis occurs during states of increased blood cell demand.

5. CONCLUSIONS AND FUTURE THERAPEUTIC PROSPECTS

Given the great body of *in vivo* and *in vitro* evidence, PRL is now accepted as one of the physiological modulators of hemopoiesis. Clues in this direction are provided by the identification of PRL-R on primitive hemopoietic progenitors and by the observation that interaction with the endogenous extra-pituitary hormone, produced by BM stroma cells, is accompanied by a modulation of HGF-induced growth and differentiation of these cells.

PRL appears to contribute to the complex hemopoietic process by promoting the biological activity of specific HGFs, rather than, by itself, inducing, basal hemopoiesis. The up-modulation of local PRL production and activity by the inflammatory mediator PAF strongly suggests that PRL may be primarily involved in activated hemopoiesis. Moreover, the observation that in hemodialyzed patients hyperprolactinemia is associated with an increased number of EPO-positive and, thus, EPO-responsive erythropoietic presursors, suggests a compensatory role for PRL in situations of reduced availability of EPO.

The mechanisms sustaining the mimesis or the synergism between PRL and other specific HGFs are not completely clear. PRL and PRL-R share structural and functional features with hemopoietin/cytokine peptides and their cognate receptors, respectively. PRL has been shown either to interact with lymphohemopoietic growth factors or to mimic their action [101]. The existence of crosstalk between hemopoietin/cytokines superfamily receptors has also been suggested [102].

Thus, the functional interactions may occur either at the receptor or at the post-receptor signaling level. The most convincing hypothesis is that PRL affects the response of the other HGFs through modulation of receptor activity, strongly influencing their biological action. Proof of this ability is the fact that PRL up-modulates the EPO-R on CD34+ hemopoietic progenitors [61] and protects from GM-CSF-R down-regulation on monocytic dendritic cell precursors, during in vitro maturation process [103]. Another possibility is that the combined effects may be triggered by activation of similar target signaling molecules and genes.

Extension of the studies focused on pre and post-receptor events will be needed to precisely delineate the role of PRL in the hemopoietic cytokine network. Whatever the possible mechanisms of PRL action on hemopoiesis, preclinical studies support the concept that rh PRL may be of benefit as a hemopoietic factor in clinical situations accompanied with hematologic deficiencies. PRL administration to mice after a myeloablative dose of total body irradiation, followed by a syngeneic BM transplantation, promotes accelerated recovery of myeloid and erythroid progenitors in the BM as well as of circulating granulocytes, erythrocytes and platelets [29]. In an other murine model, PRL has been shown to exert hemopoietic growth-promoting effects, counteracting the myelosuppression induced by azidothymidine, a reverse transcriptase inhibitor currently used in the treatment of acquired immunodeficiency syndrome [51]. Minimal PRL-related toxicity was observed in all animal experiments [51].

In humans, the most common use of hemopoietic stem cell transplantation is in the treatment of cancer. In a transplant setting, rapid hemopoietic reconstitution after high-dose chemotherapy and/or radiation is accompanied by decreased morbidity. Moreover, immune function recovery is also crucial, not only to prevent the incidence and severity of opportunistic infections, but also for the potential use of immunotherapy in eliminating minimal residual tumor. The fact that rh PRL can exert hemopoietic as well as immunomodulatory activities makes it particularly attractive as a therapeutic agent, either alone or in combination with other cytokines and growth factors, in shortening the period of aplasia associated with this clinical situation, augmenting host immune function, and permitting the continued use of potentially beneficial myelosuppressive therapies, which would otherwise result in dose-limiting side effects. Though PRL is promising for use in the hemopoietic transplant setting, care should be taken to avoid promotion of tumor growth. PRL-R transcripts or PRL-like production have been demonstrated in several human primary tumor cells, acute myeloid leukemia blasts (M4) [104,105] and non-hemopoietic tumors, such as esophagus, lung, ovary, breast and kidney carcinomas, melanoma, osteosarcoma and glioblastoma [106], suggesting a growth-dependence on endogenous PRL for these tumors.

Moreover, PRL may also be useful in the treatment of a broad range of clinical situations associated with iatrogenic or disease-related immunohematologic suppression.

Even if mice with a disruption of the PRL gene have no apparent defects in hemopoiesis [52], the evidence to date indicates that PRL can exert significant hemopoietic effects, and initial preclinical studies on the utility of PRL in reversing myelosuppression have been extremely encouraging [87]. There is little information concerning the functional interaction between PRL with other HGFs in regulating blood cell development. The post-receptor cascade of events occurring in hemopoietic cell targets and the possible effects on transformed cell growth still hamper the rational clinical use of PRL at the present time. It is hoped that more knowledge in these areas will facilitate the future use of PRL as a therapeutic agent.

REFERENCES

1. Williams ME, Quesenberry PJ. Hematopoietic growth factors. Hematol Pathol 1992;6:105–124.
2. Dorshkind K. Regulation of hemopoiesis by bone marrow stromal cells and their products. Annu Rev Immunol 1990;8:111–137.
3. Morris CF, Young IG, Hapel AJ. Molecular and cellular biology of interleukin-3. Immunol Ser 1990;49:177–214.
4. Tkatch LS, Tweardy DJ. Human granulocyte colony-stimulating factor (G-CSF), the premier granulopoietin: biology, clinical utility, and receptor structure and function. Lymphokine Cytokine Res 1993;12:477–488.
5. Steward WP. Granulocyte and granulocyte-macrophage colony-stimulating factors. Lancet 1993;342:153–157.
6. Kawasaki ES, Ladner MB. Molecular biology of macrophage colony-stimulating factor. Immunol Ser 1990;49:155–176.
7. Lacombe C. Mayeux P. Biology of erythropoietin. Haematologica 1998;83:724–732.

8. Drachman JG. Role of thrombopoietin in hematopoietic stem cell and progenitor regulation. Curr Opin Hematol 2000;7:183–190.
9. Alexander WS. Cytokines in hematopoiesis. Int Rev Immunol 1998;16:651–682.
10. Moore MA, Muench MO, Warren DJ, Laver J. Cytokine networks involved in the regulation of haemopoietic stem cell proliferation and differentiation. Ciba Found Symp 1990;148:43–58.
11. Sachs L, Lotem J. The network of hematopoietic cytokines. Proc Soc Exp Biol Med 1994;206:170–175.
12. Broxmeyer HE, Kim CH. Regulation of hematopoiesis in a sea of chemokine family members with a plethora of redundant activities. Exp Hematol 1999;27:1113–1123.
13. Fortunel NO, Hatzfeld A, Hatzfeld JA. Transforming growth factor-beta: pleiotropic role in the regulation of hematopoiesis. Blood 2000;96:2022–2036.
14. Golde DW, Bersch N, Chopra IJ, Cline MJ. Thyroid hormones stimulate erythropoiesis in vitro. Br J Haematol 1977;37:173–177.
15. Golde DW, Bersch N, Li CH. Growth hormone: species-specific stimulation of erythropoiesis in vitro. Science 1977;196:1112–1113.
16. Merchav S, Tatarsky I, Hochberg Z. Enhancement of erythropoiesis in vitro by human growth hormone is mediated by insulin-like growth factor I. Br J Haematol 1988;70:267–271.
17. Merchav S, Tatarsky I, Hochberg Z. Enhancement of human granulopoiesis in vitro by biosynthetic insulin-like growth factor I/somatomedin C and human growth hormone. J Clin Invest 1988;81:791–797.
18. Berczi I, Nagy E. Placental lactogen is a haemopoietic hormone. Br J Haematol 1991;79:355–358.
19. Gala RR. Prolactin and growth hormone in the regulation of the immune system. Proc Soc Exp Biol Med 1991;198:513–527.
20. Reber PM. Prolactin and immunomodulation. Am J Med 1993;95:637–644.
21. Hooghe R, Delhase M, Vergani P, Malur A, Hooghe-Peters EL. Growth hormone and prolactin are paracrine growth and differentiation factors in the haemopoietic system. Immunol Today 1993;14:212–214.
22. Ferrag F, Lebrun JJ, Touraine P, Nagano M, Dardenne M, Kelly PA. Prolactin and the immune system. Immunomethods 1994;5:21–30.
23. Murphy WJ, Rui H, Longo DL. Effects of growth hormone and prolactin immune development and function. Life Sci 1995;57:1–14.
24. Matera L. Endocrine, paracrine and autocrine actions of prolactin on immune cells. Life Sci 1996;59:599–614.
25. Berczi I. Pituitary hormones and immune function. Acta Paediatr Suppl 1997;423:70–75.
26. Yu-Lee LY. Molecular actions of prolactin in the immune system. Proc Soc Exp Biol Med 1997;215:35–52.
27. Merchav S. The haematopoietic effects of growth hormone and insulin-like growth factor-I. J Pediatr Endocrinol Metab 1998;11:677–685.
28. Velkeniers B, Dogusan Z, Naessens F, Hooghe R, Hooghe-Peters EL. Prolactin, growth hormone and the immune system in humans. Cell Mol Life Sci 1998;54:1102–1108.
29. Welniak LA, Tian ZG, Sun R, et al. Effects of growth hormone and prolactin on hematopoiesis. Leuk Lymphoma 2000;38:435–445.
30. Nagy E, Berczi I. Pituitary dependence of bone marrow function. Br J Haematol 1989;71:457–462.

31. Berczi I, Nagy E, Asa SL, Kovacs K. Pituitary hormones and contact sensitivity in rats. Allergy 1983;38:325–330.
32. Nagy E, Berczi I, Friesen HG. Regulation of immunity in rats by lactogenic and growth hormones. Acta Endocrinol 1983;102:351–357.
33. Cosman D. The hematopoietin receptor superfamily. Cytokine 1993;5:95–106.
34. Bazan JF. Haemopoietic receptors and helical cytokines. Immunol Today 1990;11:350–354.
35. Bazan JF. Structural design and molecular evolution of a cytokine receptor superfamily. Proc Natl Acad Sci 1990;87:6934–6938.
36. Bazan JF. A novel family of growth factor receptors: a common binding domain in the growth hormone, prolactin, erythropoietin and IL-6 receptors, and the p75 IL-2 receptor beta-chain. Biochem Biophys Res Commun 1989;164:788–795.
37. Leonard WJ, Lin JX. Cytokine receptor signaling pathways. J Allergy Clin Immunol 2000;105:877–888.
38. Gagnerault MC, Touraine P, Savino W, Kelly PA, Dardenne M. Expression of prolactin receptors in murine lymphoid cells in normal and autoimmune situations. J Immunol 1993;150:5673–5681.
39. Russell DH, Kibler R, Matrisian L, Larson DF, Poulos B, Magun BE. Prolactin receptors on human T and B lymphocytes: antagonism of prolactin binding by cyclosporine. J Immunol 1985;134:3027–3031.
40. Matera L, Muccioli G, Cesano A, Bellussi G, Genazzani E. Prolactin receptors on large granular lymphocytes: dual regulation by cyclosporin A. Brain Behav Immun 1988;2:1–10.
41. Bellussi G, Muccioli G, Ghe C, Di Carlo R. Prolactin binding sites in human erythrocytes and lymphocytes. Life Sci 1987;41:951–959.
42. Dardenne M, de Moraes Md, Kelly PA, Gagnerault MC. Prolactin receptor expression in human hematopoietic tissues analyzed by flow cytofluorometry. Endocrinology 1994;134:2108–2114.
43. Ben-Jonathan N, Mershon JL, Allen DL, Steinmetz RW. Extrapituitary prolactin: distribution, regulation, functions, and clinical aspects. Endocr Rev 1996;17:639–669.
44. Murphy WJ, Durum SK, Anver MR, Longo DL. Immunologic and hematologic effects of neuroendocrine hormones. Studies on DW/J dwarf mice. J Immunol 1992;148:3799–3805.
45. Nagy E, Berczi I. Hypophysectomized rats depend on residual prolactin for survival. Endocrinology 1991;128:2776–2784.
46. Duquesnoy RJ, Pedersen GM. Immunologic and hematologic deficiencies of the hypopituitary dwarf mouse. In: Gershwin ME, Merchant B, editors. Immunologic defects in laboratory animals. New York: Plenum, 1981:309–312.
47. Jepson JH, Lowenstein L. Erythropoiesis during pregnancy and lactation in the mouse. Proc Soc Exp Biol Med 1966;121:457–460.
48. Jepson JH, Lowenstein L. Erythropoiesis during pregnancy and lactation. I. Effect of various hormones on erythropoiesis during lactation. Proc Soc Exp Biol Med 1965;120:500–504.
49. Lindemann R, Trygstad O, Halvorsen S. Pituitary control of erythropoiesis. Scand J Haematol 1969;6:77–86.
50. Alippi RM, Barcelo AC, Bozzini CE. Enhanced erythropoiesis induced by hypoxia in hypertransfused, post-hypoxic mice. Exp Hematol 1983;11:878–883.

51. Woody MA, Welniak LA, Sun R, et al. Prolactin exerts hematopoietic growth-promoting effects *in vivo* and partially counteracts myelosuppression by azidothymidine. Exp Hematol 1999;27:811–816.
52. Horseman ND, Zhao W, Montecino-Rodriguez E, et al. Defective mammopoiesis, but normal hematopoiesis, in mice with a targeted disruption of the prolactin gene. EMBO J 1997;16 :6926–6935.
53. Ormandy CJ, Binart N, Kelly PA. Mammary gland development in prolactin receptor knockout mice. J Mammary Gland Biol Neoplasia 1997;2:355–364.
54. Liu X, Robinson GW, Wagner KU, Garrett L, Wynshaw-Boris A, Hennighausen L. Stat5a is mandatory for adult mammary gland development and lactogenesis. Genes Dev 1997;11:179–186.
55. Udy GB, Towers RP, Snell RG, et al. Requirement of STAT5b for sexual dimorphism of body growth rates and liver gene expression. Proc Natl Acad Sci 1997;94:7239–7244.
56. Ormandy CJ, Camus A, Barra J, et al. Null mutation of the prolactin receptor gene produces multiple reproductive defects in the mouse. Genes Dev 1997;11:167–178.
57. Kotzmann H, Riedl M, Clodi M, et al. The influence of growth hormone substitution therapy on erythroid and myeloid progenitor cells and on peripheral blood cells in adult patients with growth hormone deficiency. Eur J Clin Invest 1996;26:1175–1181.
58. Bradley TR, Metcalf D. The growth of mouse bone marrow cells in vitro. Aust J Exp Biol Med Sci 1966;44:287–299.
59. Platzer E. Human hemopoietic growth factors. Eur J Haematol 1989;42:1–15.
60. Gabbianelli M, Sargiacomo M, Pelosi E, Testa U, Isacchi G, Peschle C. "Pure" human hematopoietic progenitors: permissive action of basic fibroblast growth factor. Science 1990;249:561–564.
61. Bellone G, Geuna M, Carbone A, et al. Regulatory action of prolactin on the in vitro growth of CD34+ve human hemopoietic progenitor cells. J Cell Physiol 1995;163:221–231.
62. Bellone G, Astarita P, Artusio E, et al. Bone marrow stroma-derived prolactin is involved in basal and platelet-activating factor-stimulated in vitro erythropoiesis. Blood 1997;90:21–27.
63. Drachman JG, Kaushansky K. Structure and function of the cytokine receptor superfamily. Curr Opin Hematol 1995;2:22–28.
64. Ihle JN, Kerr IM. Jaks and Stats in signaling by the cytokine receptor superfamily. Trends Genet 1995;11:69–74.
65. Goffin V, Binart N, Clement-Lacroix P, et al. From the molecular biology of prolactin and its receptor to the lessons learned from knockout mice models. Genet Anal 1999;15:189–201.
66. Davis JA, Linzer DI. Expression of multiple forms of the prolactin receptor in mouse liver. Mol Endocrinol 1989;3:674–680.
67. Dusanter-Fourt I, Muller O, Ziemiecki A, et al. Identification of JAK protein tyrosine kinases as signaling molecules for prolactin. Functional analysis of prolactin receptor and prolactin-erythropoietin receptor chimera expressed in lymphoid cells. EMBO J 1994;13:2583–2591.
68. Lebrun JJ, Ali S, Goffin V, Ullrich A, Kelly PA. A single phosphotyrosine residue of the prolactin receptor is responsible for activation of gene transcription. Proc Natl Acad 1995;92:4031–4035.
69. Clevenger CV, Medaglia MV. The protein tyrosine kinase P59fyn is associated with prolactin (PRL) receptor and is activated by PRL stimulation of T-lymphocytes. Mol

Endocrinol 1994;8:674–681.

70. DaSilva L, Rui H, Erwin RA, et al. Prolactin recruits STAT1, STAT3 and STAT5 independent of conserved receptor tyrosines TYR402, TYR479, TYR515 and TYR580. Mol Cell Endocrinol 1996;117:131–140.
71. Bole-Feysot C, Goffin V, Edery M, Binart N, Kelly PA. Prolactin (PRL) and its receptor: actions, signal transduction pathways and phenotypes observed in PRL receptor knockout mice. Endocr Rev 1998;19:225–268.
72. Clevenger CV, Freier DO, Kline JB. Prolactin receptor signal transduction in cells of the immune system. J Endocrinol 1998;157:187–197.
73. Dardenne M, Kelly PA, Bach JF, Savino W. Identification and functional activity of prolactin receptors in thymic epithelial cells. Proc Natl Acad Sci 1991;88:9700–9704.
74. Gala RR, Shevach EM. Identification by analytical flow cytometry of prolactin receptors on immunocompetent cell populations in the mouse. Endocrinology 1993;133:1617–1623.
75. Touraine P, Leite de Moraes MC, Dardenne M, Kelly PA. Expression of short and long forms of prolactin receptor in murine lymphoid tissues. Mol Cell Endocrinol 1994;104:183–190.
76. Dardenne M, de Moraes Md, Kelly PA, Gagnerault MC. Prolactin receptor expression in human hematopoietic tissues analyzed by flow cytofluorometry. Endocrinology 1994;134:2108–2114.
77. Iscove NN. The role of erythropoietin in regulation of population size and cell cycling of early and late erythroid precursors in mouse bone marrow. Cell Tissue Kinet 1977;10:323–334.
78. Socolovsky M, Dusanter-Fourt I, Lodish HF. The prolactin receptor and severely truncated erythropoietin receptors support differentiation of erythroid progenitors. J Biol Chem 1997;272:14009–14012.
79. Socolovsky M, Fallon AE, Lodish HF. The prolactin receptor rescues EpoR-/- erythroid progenitors and replaces EpoR in a synergistic interaction with c-kit. Blood 1998;92:1491–1496.
80. Widness JA, Clemons GK, Garcia JF, Schwartz R. Plasma immunoreactive erythropoietin in normal women studied sequentially during and after pregnancy. Am J Obstet Gynecol 1984;149:646–650.
81. Fontanini G, Basolo F, Garzelli C, Squartini F, Toniolo A. Fr-MLV infection induces erythroleukaemia instead of lymphoid leukaemia in mice given pituitary grafts. Br J Cancer 1990;61:841–845.
82. Huggins C, Oka H. Regression of stem-cell erythroblastic leukemia after hypophysectomy. Cancer Res 1972;32:239–242.
83. Bellone G, Rollino C, Borsa S, Ferrero I, Martina G, Carbone A, et al. Association between elevated prolactin levels and circulating erythroid precursors in dialyzed patients. Proc Soc Exp Biol Med 2000;223:367–371.
84. Winearls CG, Oliver DO, Pippard MJ, Reid C, Downing MR, Cotes PM. Effect of human erythropoietin derived from recombinant DNA on the anaemia of patients maintained by chronic haemodialysis. Lancet 1986;2:1175–1178.
85. Schaefer RM, Kokot F, Kurner B, Zech M, Heidland A. Normalization of elevated prolactin levels in hemodialysis patients on erythropoietin. Nephron 1988;50:400–401.
86. Sato H, Ohkubo M, Nagaoka T. Levels of serum colony-stimulating factors (CSFs) in patients on long-term haemodialysis. Cytokine 1994;6:187–194.
87. Richards SM, Murphy WJ. Use of human prolactin as a therapeutic protein to potentiate

immunohematopoietic function. J Neuroimmunol 2000;109:56–62.
88. Kaushansky K, Lok S, Holly RD, Broudy VC, Lin N, Bailey MC, et al. Promotion of megakaryocyte progenitor expansion and differentiation by the c-Mpl ligand thrombopoietin. Nature 1994;369:568–571.
89. Wilson FD. Studies of the hematopoietic microenvironment. Hematopoietic stroma and microenvironment. Bibl Haematol 1984;48:210–292.
90. Clark BR, Keating A. Biology of bone marrow stroma. Ann NY Acad Sci 1995;770:70–78.
91. Mayani H, Guilbert LJ, Janowska-Wieczorek A. Biology of the hemopoietic microenvironment. Eur J Haematol 1992;49:225–233.
92. Muller-Sieburg CE, Deryugina E. The stromal cells' guide to the stem cell universe. Stem Cells 1995;13:477–486.
93. O'Flaherty JT, Wykle RL. Biology and biochemistry of platelet-activating factor. Clin Rev Allergy 1983;1:353–367.
94. Camussi G, Tetta C, Baglioni C. The role of platelet-activating factor in inflammation. Clin Immunol Immunopathol 1990;57:331–338.
95. Camoratto AM, Grandison L. Platelet-activating factor stimulates prolactin release from dispersed rat anterior pituitary cells in vitro. Endocrinology 1989;124:1502–1506.
96. Rougier F, Dupuis F, Cornu E, Dulery C, Praloran V, et al. Platelet-activating factor and antagonists modulate DNA synthesis in human bone marrow stromal cell cultures. J Lipid Mediat Cell Signal 1997;16:147–153.
97. Desplat V, Besse A, Faucher JL, Praloran V, Denizot Y. Expression of platelet-activating factor receptor transcript-1 but not transcript-2 by human bone marrow cells. Stem Cells 1999;17:121–124.
98. Sinha YN. Structural variants of prolactin: occurrence and physiological significance. Endocr Rev 1995;16:354–369.
99. Price AE, Logvinenko KB, Higgins EA, Cole ES, Richards SM. Studies on the microheterogeneity and in vitro activity of glycosylated and nonglycosylated recombinant human prolactin separated using a novel purification process. Endocrinology 1995;136:4827–4833.
100. Hoffmann T, Penel C, Ronin C. Glycosylation of human prolactin regulates hormone bioactivity and metabolic clearance. J Endocrinol Invest 1993;16:807–816.
101. Clevenger CV, Altmann SW, Prystowsky MB. Requirement of nuclear prolactin for interleukin-2-stimulated proliferation of T lymphocytes. Science 1991;253:77–79.
102. Ihle JN. Cytokine receptor signalling. Nature 1995;377:591–594.
103. Matera L, Galetto A, Geuna M, Vekemans K, Ricotti E, Contarini M, et al. Individual and combined effect of granulocyte-macrophage colony-stimulating factor and prolactin on maturation of dendritic cells from blood monocytes under serum-free conditions. Immunology 2000;100:29–36.
104. Hatfill SJ, Kirby R, Hanley M, Rybicki E, Bohm L. Hyperprolactinemia in acute myeloid leukemia and indication of ectopic expression of human prolactin in blast cells of a patient of subtype M4. Leuk Res 1990;14:57–62.
105. Oberholtzer E, Contarini M, Veglia F, et al. Prolactin increases the susceptibility of primary leukemia cells to NK and LAK effectors. Adv Neuroimmunol 1996;6:233–247.
106. Rosen SW, Weintraub BD, Aaronson SA. Nonrandom ectopic protein production by malignant cells: direct evidence in vitro. J Clin Endocrinol Metab 1980;50:834–841.

Growth and Lactogenic Hormones
Edited by L. Matera and R. Rapaport

Growth Hormone/Insulin-like Growth Factors and Hematopoiesis

ROBERT MOGHADDAS and ROBERT RAPAPORT

Diabetes Center, Mount Sinai Hospital, New York, NY, USA

ABSTRACT

The growth hormone/insulin-like growth factor (GH/IGF) system plays an essential role in cell growth and tissue differentiation during development. However, the role of GH/IGF-I in the regulation of production of red blood cells, is not completely understood. Erythropoiesis is regulated mainly by erythropoietin. The combination of erythropoietin and GH/IGF-I results in growth of red blood cell colonies that are less than the sum of each factor alone. GH/IGF-I membrane receptors have been detected on rat and human erythrocytes. GH/IGF-I stimulate erythropoiesis under various *in vitro* and *in vivo* conditions. There seems to be a relationship between hemoglobin/hematocrit and GH/IGF-I as established by *in vitro* and in animal experiments.

We begin with a brief introduction to the function and regulation of GH/IGF-I and of hematopoiesis. Thereafter, we discuss the result of research done on this topic starting from *in vitro* results, followed by the presentation of classical studies in animals and finally we discuss the result of studies done on children and adults. We focus mainly on hemoglobin/hematocrit and how it is affected in children and adults with or without growth hormone deficiency. A summary of our current understanding of the relationship between GH/IGF-I and hematopoiesis is provided.

1. INTRODUCTION

1.1. The GH/IGF axis

The role of Growth hormone (GH), insulin-like growth factors (IGFs) in the regulation of human erythropoiesis is not completely understood. Growth hormone, a 191 amino acid polypeptide, stimulates the synthesis of peptide hormones with insulin-like chemical structure and biological properties called insulin-like growth factors (IGF) in the liver and in other tissues. IGF-I, whose gene is located on chromosome 12q22-q24.1, is a 70-amino acid peptide and the IGF-II gene located on Chromosome11p15, and codes for a 67-amino acid peptide.

Both IGF-I and IGF-II show 50% amino acid identity with insulin [1] suggesting a common evolutionary precursor. The liver was traditionally considered to be the only producer of IGF-I and IGF-II. It has recently been shown that many tissues at various stages of development

Table I Growth hormone / IGF-I and hematopoiesis in animals.

Author (year)	animal	remarks	Ref.
Guler et al. (1988)	rats	decreased IGF-I/Hb	[21]
Kurtz et al. (1988)	rats	IGF-I stimulates Hb in hypophysectomized rats	[22]
Philips et al. (1988)	rats	IGF-I stimulates CFU-E	[31]
Berczi et al. (1989)	rats	pituitary graft increase Hb in hypophysectomized rats	[25]
Kurtz et al. (1990)	rats	elevated IGF/Hb during growth spurt	[23]
Murphy et al. (1992)	mice	GH reverses AZT induced anemia	[32]
Murphy et al. (1994)	mice	human IGF-I increase CFU-E in mice	[27]
Fang et al. (2001)	mice	GHRH increase CFU-E in transgenic mice	[56]
In vitro (animals)			
Thomopoulos et al. (1981)	rats	IGF-I receptor on rat erythrocytes	[28]

express IGF-I or IGF-II receptors [1].

The current view supports both an endocrine and a local autocrine/paracrine effect for GH,IGF-I and IGF-II system, in part by affecting cellular differentiation and proliferation, in normal postnatal growth [1]. In fetal and neonatal life, the synthesis of IGF-I or IGF-II appears not to be under control of GH. IGF-I, is the mediator of GH activity in peripheral tissues and IGF-II, which is more insulin like in its mode of action, is not considered to have any major role in IGF signal transduction [1]. Insulin like growth factor binding proteins (IGFBP), a group of at least six regulatory proteins that bind IGF-I and/or IGF-II, facilitate IGF endocrine action by increasing the half-life of the IGFs in the circulation, and have also been shown to have some stimulatory effect on erythroid progenitor cell growth in vitro [2]. Plasma IGFBP-I level seem to be at least four-fold higher in patients with polycythemia vera than in the normal population [3]. The biological actions of IGF-I and IGF-II are mediated by specific receptors. The type I IGF receptor gene is located on human chromosome 15q26 (1), a tyrosine kinase receptor, highly homologous to the insulin receptor, which binds IGF-I and IGF-II with high affinity. The IGF-II receptor, on the other hand, is a single highly glycosylated transmembrane peptide with no tyrosine kinase activity. GH and IGF-I act either independently or synergistically, through paracrine or endocrine IGF-I pathways, with hematopoietic cytokines on erythroid cell lines [4]. This effect is mediated via a direct mechanism involving the type I IGF receptor [5]. Their role in somatic growth has been mainly established in mice by ablation of the genes for IGF-I and IGF-II (null mice).

IGF-I knockout mice exhibit extreme embryonic and postnatal growth retardation, neurologic defects and prenatal mortality [6]. Both IGF-I and IGF-II null mice are growth impaired. However the IGF-II null mice show growth retardation only in utero and continue to grow postnatally as opposed to IGF-I null mice [6]. However, little is known about their immune or hematopoietic systems.

Table II Growth hormone / IGF-I and hematopoiesis in vitro (human).

Author (year)	remarks	Ref.
Golde et al. (1997)	GH species specific stimulation of erythropoiesis	[11]
Golde et al. (1980)	GH no change in BFU-E in Laron patients	[46]
Polychranos et al. (1983)	IGF-I receptor on human erythrocytes	[29]
Catanese et al. (1986)	mechanism of action of IGF-I receptor	[30]
Claustres et al. (1987)	supra physiological dose of GH, physiological dose of IGF-I enhance erythropoiesis	[14]
Merchav et al. (1988)	enhancement of erythropoiesis by GH mediated by IGF-I	[4]
Correa et al. (1991)	IGF-I results in increased erythropoiesis	[17]
Merchav et al. (1992)	both IGF-I and IGF-II stimulate erythropoiesis	[13]
Merchav et al. (1993)	GH/IGF-I no change in BFU-E of thalassemia patients	[44]

2. HEMATOPOIESIS

The process of hematopoiesis, occuring mostly in bone marrow, involves proliferation and differentiation of hematopoietic stem cells into erythroid progenitor cells. Hematopoietic cell growth and differentiation is regulated by local production of various cytokines and growth factors within the bone marrow [7].

Erythropoiesis is regulated by erythropoietin (EPO), a hormone responsible for red blood cell differentiation and proliferation of their progenitor cells. There are two types of erythroid progenitor cells. Mature or late-stage erythroid progenitor cells are referred to as erythroid colony-forming units (CFU-E). IGF-I and/or insulin as well as EPO are required by human CFU-E colonies for erythroid development [8]. Early erythroid progenitor cells known as erythroid burst-forming units (BFU-E) require EPO for proliferation as well as burst-promoting factors (BPA) such as granulocyte-macrophage colony-stimulating factor (GM-CSF), interleukin-3 and interleukin-9 [9].

CFU-E is a rapidly dividing cell and gives rise to erythroblast colonies of 8–49 cells in 7 days in humans. BFU-E is a much more immature cell that develops into grouped clusters of erythroblasts or larger colonies of >200 erythroblasts in 15 days in humans. The fact that EPO alone fails to stimulate BFU-E in serum-free conditions, as compared with serum containing cultures indicates that serum contains some growth factors that are necessary for the BFU-E development [9].

3. GH/IGF AND HEMATOPOIESIS

The relationship between the growth hormone (GH) and the hematopoietic system was described long ago [10]. The effect of GH on hematopoiesis of erythroid precursor cells is species specific since human erythroid cells respond to human but not bovine GH [11]. Many of the erythropoietic effects of GH are mediated through IGF-I [12]. Growth hormone, IGF-I and IGF-II stimulate erythropoiesis under various in vitro conditions [10,12–18]. GH and IGF-I enhance erythropoiesis *in vitro* using cultures of human bone marrow cells; the effect of GH is blocked

Table III Growth hormone / IGF-I and hematopoiesis in children.

Author (year)	N	age (years)	Dx*	duration GH tx/ observation (years)	results	Ref.
Barak et al. (1992)	11	6.5–14.5	GHD	2	increase in BFU-E	[50]
Antilla et al. (1994)	60	11.1–13.3	NL**	0.75	Hb increased over 9 months observation positive correlation Hb & IGF-I/IGFBP-3	[47]
Vihervuori et al. (1996)	36	1.6–13.3	short Stature	1	increase in Hb	[48]
Tenore et al. (1997)	19	2.08–15	GHD	0.5	increase in Hb	[52]
Koc et al. (1997)	23	newborns	NL	0.16	positive correlation between Hb, IGF-I/IGFBP3 in 2 months	[43]
Rapaport et al. (2000)	31	3–18	GHD	2.6	no relationship between Hgb & IGF-I	[54]

* Dx = diagnosis
** NL = normal

by antibodies against the IGF-I receptor [12]. Both IGF-I and IGF-II stimulate erythropoiesis of human bone marrow cultures *in vitro*, an effect mediated through the IGF-I receptor [13].

In a study on human bone marrow cells obtained from 16 pre-pubertal children, both GH and IGF-I stimulated erythropoiesis; only supra-physiological concentration of GH, and physiological concentration of IGF-I, were enough to stimulate erythroid progenitor cells [14]. IGF-I is an essential component of proliferation and maturation of late stage erythroid progenitor cells with EPO and IGF-I being equally important factors in erythropoiesis [15]. In the presence of EPO, IGF-I markedly enhance hemoglobin synthesis [15]. The morphology of the cells cultured with EPO alone is markedly defective, compared with that of normal erythroblasts in cultures with EPO and IGF-I [15]. IGF-I stimulates primarily BFU-E or late mature stages of bone marrow erythroid progenitor cells. IGF-1R receptor has been detected on cells isolated from day 6 BFU-E [16]. IGF-I and IGF-II in physiologic doses directly stimulate neonatal cord blood erythroid progenitor cells resulting in up to four-fold enhancement in erythroid colony formation [16]. IGF-I has EPO-like activity that targets circulating early erythroid progenitor cells, provide strong evidence for the existence of an EPO-independent pathway for normal human erythropoiesis [17]. Erythroid progenitor cells from adult human peripheral blood form erythroid colonies in the presence of IGF-I, even in the absence of EPO.

The combination of EPO and IGF-I results in growth of cell colonies that are less than the sum of colonies grown with each factor alone, indicating a possible overlap between the two systems [17]. Polycythemia vera is a chronic myeloproliferative disorder characterized by hyperplasia of the three myeloid cell lineages, particularly the erythroid lineage. The erythroid progenitor cells are twice as sensitive to IGF-I than are normal cells in terms of colony burst formation, an effect mediated through the IGF-I receptor [18].

IGF-I acts synergistically with other cytokines such as granulocyte-macrophage colony stimulating factor(GM-CSF) and EPO, which are required for the initial and terminal stages of erythroid progenitor cell growth and maturation [3].

Table IV Growth hormone / IGF-I and hematopoiesis in adults (human).

Author (year)	N	age (years)	Dx*	GH treatment duration (months)	effect	Ref.
Kotzman et al. (1996)	11	27–60	GHD	24	no change in Hb but increase in BFU-E	[40]
Christ et al. (1997)	13	24–66	GHD	3	increase in Hb	[41]
Ten Have et al. (1997)	17	30–59	GHD	35	increase in Hb & Red Cell Mass	[42]

* Dx = diagnosis

IGF-II also seems to play a role in erythropoiesis. The erythropoietic activity of IGF-II is synergistic with the actions of insulin and erythropoietin. In the absence of erythropoietin, the stimulation of erythroid colony formation from umbilical cord blood is greater with IGF-II than with IGF-I. In contrast, the stimulation of adult marrow cells is greater with IGF-I than with IGF-II [19]. This suggests that IGF-II might be the prominent regulator of neonatal erythropoietin-independent erythropoiesis, while IGF-I is the primary stimulant of adult erythroid cell proliferation [19].

Both IGF-I and IGF-II induced a similar dose-dependent magnitude of erythroid colony enhancement in serum-free cultures of human marrow CFU-E and BFU-E. Effect of GH/IGF is completely aborgated by monoclonal antibody (alpha IR-3), directed against the type I IGF receptor, indicating that the erythroid-promoting effects of these peptides are mediated by the type I IGF receptor [12].

Although the *in vitro* erythroid potentiating effects of GH, IGF-I and IGF-II are clearly evident, *in vivo* effects have been shown mostly in anemic animals. In classical studies in rodents the removal of the pituitary gland affected the blood cell formation [20]. Growth arrest in hypophysectomized rats is not only accompanied by low IGF-I serum levels but also by reduction of erythropoiesis [21]. The administration of GH and IGF-I to hypophysectomized rats caused the resumption of growth and a significant enhancement of erythropoiesis as monitored by increased iron incorporation into red blood cells and by an increase in the number and percentage of reticulocytes, without any change in hematocrit levels [22]. During accelerated growth in rats, there is a correlation between IGF-I , erythropoiesis and erythropoietin [23].

The low hematocrit of hypophysectomized animals is reversed by administration of human growth hormone or pituitary graft transplantation [24–26]. In mice, human IGF-I increases hematopoietic activity in bone marrow cells [27]. IGF-I and IGF-II membrane receptors have been detected on rat and human erythrocytes [28–30]. IGF-I and IGF-II enhance the *in vitro* growth of mature and primitive human erythroid progenitors from the bone marrow and peripheral blood [19]. IGF-I, in the absence of GH, stimulates erythropoiesis *in vivo* in an endocrine fashion. It is IGF-I that mediates the effect of GH on erythropoiesis and EPO synthesis [22]. IGF-I stimulates erythropoiesis, with elevation of erythropoietin, in anemic hypophysectomized rats [31]. Although GH does not affect the peripheral blood counts of normal adult mice, it is capable of reversing the anemia induced by azidothymidine (AZT) [32]. Intra-peritoneal recombinant GH cause a significant increase in bone marrow erythroid progenitor cells. GH reverses the significant myelotoxic effect of AZT [31].

IGF-I increases bone marrow erythropoietic precursors in neonatal rats, resulting in the acceleration of red cell production, without any significant changes in reticulocyte counts or packed cell volumes [31]. IGF-I, increases the body weight as well as red blood cell production, in hypophysectomized rats. However, no increase in packed cell volume or hemoglobin concentration is seen. IGF-I and IGF-I receptor (IGF-I-R) knockout mice are born with multiple abnormalities, including hypoplastic muscles, delayed bone development and thin epidermis [33]. The appearance of ossification centers is delayed by an average of 2 embryonic days in cranial and facial bones, and between 1 and 2 days in long bones and trunk [33]. To the best of our knowledge no data is available regarding red blood cells or erythroid progenitor cells in IGF-I knockout mice.

In humans there are no known homozygous IGF-I receptor mutations. However a severely affected patient with a partial IGF-I gene deletion is reported who has a significant prenatal and postnatal growth failure, sensorineural deafness, developmental delay and severe microcephaly [34]. No data is available regarding erythroid cells in that patient. GH and GH deficiency has been reported to have a wide variety of effects on the immune system and the hematopoietic system [35–39]. Children as well as adults with GH deficiency are not severely anemic suggesting that locally produced GH, IGF-I and/or prolactin maybe sufficient for hematopoiesis.

No relationship between GH treatment and hemoglobin exists, in a study of 11 adults with GH deficiency after 24 months of GH treatment [40]. GH stimulated erythropoiesis, increased red cell mass and plasma volume, in a 3-month, double blind, placebo-controlled trial of GH treatment of 13 adult patients with GH deficiency [41]. Increase in Hemoglobin concentration was noted in a study of 17 adults with GH deficiency treated with GH for 2.7 years without any changes in hemoglobin noted until 9 months of treatment. Thereafter hemoglobin concentration increased over the baseline, with peak at 2.5 years [42]. In a study of 23 healthy term infants, increase in hemoglobin correlates positively with IGF-I and IGFBP-3 at birth but not at 2 months of age when physiologic anemia of infancy occurs [43]. In children with beta-thalassemia, the in vitro effect of GH and IGF-I on erythroid colonies, BFU-E, is normal [44]. GH treatment of children with thalassemia does not improve the anemia of these patients [45]. In children with Laron syndrome, who have short stature with high concentration of GH, low IGF-I level, peripheral BFU-E are found not to be responsive to *in vitro* treatment with human GH [46]. In one study of 60 normal pre-pubertal boys, Hb concentration positively correlated with serum IGF-1 and IGFBP-3 [47]. This correlation is more significant for IGFBP-3 ($r = 0.53$, $p < 0.001$) and less significant for IGF-I ($r = 0.36$, $p = 0.008$) [47].

Increase in Hb concentration in response to GH treatment, for a year, is reported in 36 children with short stature without GH deficiency [48]. GH treatment raises RBC and decreases all red cell indexes (MCV, MCH, MCHC) and serum ferritin concentrations [48]. These results are compatible with the interpretation that IGF induces increase in the erythrocyte production rate, leading to increased demand for iron [49]. Administration of either GH or IGF-I results in a significant increase in BFU-E colonies, of erythroid progenitor cells of 11 children with GH deficiency, but not in short children, without GH deficiency [50]. IGF-I has also been shown, in the presence of EPO, to induce the proliferation of erythroid colonies from cells in bone marrow in vitro [51]. After three and six months of GH treatment, 19 GH deficient children, with either isolated GH deficiency or multiple pituitary hormone deficiency, showed increase in hemoglobin, and hematocrit [52]. Human bone marrow cells expressing IGF-I receptor were cultured in vitro in another study [53]. These cells did not give rise to BFU-E or CFU-E cells in mediums containing recombinant human IGF-I [53]. Therefore, it was concluded to be unlikely that IGF-I plays a role in normal hematopoiesis [53]. No relationship between IGF-I and hemoglobin level was shown in a study of 31 children with GH deficiency, during treatment with human GH for 2–10 years [54].

It is apparent therefore that the correlation between IGF-I and Hb is incremental [49–54]. The role of IGF-II in hematopoiesis is also being more clearly established. IGF-II is expressed in bone marrow of children and adults in a bi-allelic manner, whereas it is expressed and imprinted in a mono-allelic manner in peripheral blood cells [55]. Growth hormone releasing hormone (GHRH) also has a role in hematopoiesis [56]. GHRH transgenic mice, which express a GHRH precursor called GHRH-related peptide, have increased BFU-E colonies and increased hematopoietic erythroid progenitor cells [56].

4. SUMMARY AND CONCLUSIONS

In most animal models, severe GH/IGF-I deficiency does lead to anemia and/or decreased hematopoiesis, which is reversible by GH and/or IGF-I administration. Human studies have shown no significant alteration in hematopoiesis in children and adults with GH/IGF-I deficiency or excess. GH/IGF-I treatment leads to either no effect or an increase in markers of hematopoiesis. Long term GH treatment however, does not result in a sustained correlation between IGF-I and hematpoiesis. Optimum erythropoiesis appears to be dependent on several factors including EPO, stem cell factor, IGF-I, IGF-II, interleukins, kit-ligand and multiple other cytokines. There seems to be a relationship between Hemoglobin/Hematocrit and growth hormone/IGF-I in vitro and in animals. GH and IGF-I have stimulatory effect on erythroid precursor progenitor cells but only a marginal effect on peripheral red blood cells. This relationship and its possible significance in humans is currently under intense scrutiny.

REFERENCES

1. Leroith D, Bondy C, Yakar S, Liu JL, Butler A. Somatomedin hypothesis 2001. Endo Review 2001;22(1):53–74.
2. Hwa V, Oh Y, Rosenfeld RG. The insulin-like growth factor-binding protein (IGFBP) superfamily. Endocrine Reviews 1999;20(6):761–787.
3. Mirza AM, Ezzat S, Axelrad AA. IGFBP-I is elevated in patients with polycythemia vera and stimulates erythroid burst foramtion in vitro. Blood1997;89:1862–1866.
4. Merchav S. The hematopoietic effects of growth hormone and IGF-I. J Ped Endocrinol Metab 1998;11(6):677–685.
5. LeRoith D, Werner H, Beitner Johnson D, Roberts CT. Molecular and cellular aspects of the insulin-like growth factor I receptor. Endocrine Reviews 1995;16:143–163.
6. Lamothe B, Baudry A, Christofferson CT, De Meyts P, Jami J, Bucchini D, Joshi RL. Insulin receptor-deficient cells as a new tool for dissecting complex interplay in insulin and insulin-like growth factors. Federation of European biochemical societies letters1998;426:381–385.
7. Wu H, Klingmuller U, Besmer P, Lodish HF. Interaction of the erythropoietin and stem-cell factor receptors. Nature 1995;377:242–246.
8. Sawada K, Krantz SB, Dessypris EN, Koury ST, Sawyer ST. Human colony forming units erythroid don't require accessory cells, but do require direct interaction with insulin like growth factor I and/or insulin for erythroid development. Journal of Clinical Investigation 1989;83:1701–1709.

9. Krantz SB. Erythropoietin. Blood 1991;77:419–434.
10. Gordon AS. Endocrine influences upon the formed elements of blood and blood forming organs. Recent Progress in Hormone Research 1954;10:339–394.
11. Golde DW, Bersch N, Hao LI. GH species specific stimulation of erythropoiesis in vitro. Science 1977;196:1112–1113.
12. Merchav S, Tatarsky I, Hochberg Z. Enhancement of erythropoiesis in vitro by human GH is mediated by IGF-1. Br J Haematology 1988;70:267–271.
13. Merchav S, Silvian-Drachsler I, Tatarsky I, Lake M, Skottner A. Comparative studies of the erythroid-potentiating effects of biosynthetic human IGF-I and II. J of Clin Endocrinol Metab 1992;74:447–452.
14. Claustres M, Chatelain P, Sultan C. IGF-I stimulates human erythroid colony formation in vitro. J Clin Endocrinol Metab 1987;65:78–82.
15. Muta K, Krantz SB, Bondurant MC, Wickrema A. Distinct roles of erythropoietin, IGF-I and stem cell factor in the development of erythroid progenitor cells. Journal of Clinical Investigation 1994;94:34–43.
16. Ratajczak J, Zhang Q, Pertusini E, Wojczyck BS, Wasik MA, Ratajczak MZ. The role of insulin and IGF-I in regulating erythropoiesis. Studies *in vitro* under serum-free conditions – comparison to other cytokines and growth factors. Leukemia 1998;12:371- 381.
17. Correa PN, Axelrad AA. Production of erythropoietic bursts by progenitor cells from adult human peripheral blood in an improved serum-free medium: role of IGF-I. Blood 1991;78:2823–2833.
18. Correa PN, Eskinazi D, Axelrad AA. Circulating erythroid progenitors in polycythemia vera are hypersensitive to IGF-I *in vitro*. Studies in an improved serum-free medium. Blood 1994;83:99–112.
19. Sanders M, Sorba S, Dainiak N. Insulin-like growth factors stimulate erythropoiesis in serum substituted umbilical cord blood cultures. Exp Hematol 1993;21:25–30.
20. Crafts RC, Meineke HA. The anemia of hypophysectomized animals. Ann NY Acad Sci 1959;77:501–517.
21. Guler HP, Zapf J, Scheiwiller E, Froesch ER. Recombinant human IGF-I stimulates growth and has distinct effects on organ size in hypophysectomized rats. Proc Natl Acad Sci 1988:85:4889–4893.
22. Kurtz A, Zapf J, Eckhard K, Clemons G, Froesch R, Bauer C. IGF-I stimulates erythropoiesis in hypophysectomized rats. Proc Natl Acad Sci USA 1988;85:7825–7829.
23. Kurtz A, Matter R, Eckhardt KU, Zapf J. Erythropoiesis, serum erythropoietin and serum IGF-I in rats during accelerated growth. Acta Endocrinologica 1990;122:323–328.
24. Berczi I, Nagy E. The effect of prolactin and GH on hemolymphopoietic tissue and immune function. In: Berczi I, Kovacs K, editors. Hormones and Immunity. Lancaster: MTP press, 1987;144–171.
25. Nagy E, Berczi I. Pituitary dependence of bone marrow function. British J Hematology 1989;71:457–462.
26. Jepson JH, McGarry EE. Hemopoiesis in pituitary dwarfs treated with human GH and testosterone. Blood 1972;39:238–248.
27. Tsarfaty G, Longo DL, Murphy WJ. Human IGF-I exerts hematopoietic growth-promoting effect after *in vivo* administration. Experimental Hematology 1994;2:1273–1277.
28. Thomopoulos P, Postal-Vinay MC, Testa U, Guyda HJ, Posner BI. Receptors for IGF peptides in rat reticulocytes and erythrocytes. Endocrinology 1981;108:1087–1090.
29. Polychronakos C, Guyda HJ, Posner BI. Receptors for IGF on human erythrocytes. J Clin

Endocrinol Metab 1983;57:436–438.
30. Catanese VM, Grigorescu F, King GL, Kahn CR. The human erythrocyte IGF-I receptor: characterization and demonstration of ligand-stimulated autophosphorylation. J Clin Endocrinol Metab 1986;62:692–699.
31. Philipps AF, Persson B, Hall K, Lake M, Skottner A, Sanengen T, Sara VR. The effects of biosynthetic IGF-I supplementation on somatic growth, maturation and erythropoiesis on the neonatal rat. Pediatric Research 1988;23:298–305.
32. Murphy WJ, Tsarfaty G, Longo DL. GH exerts hematopoietic growth-promoting effects in vivo and partially counteracts the myelosuppressive effects of azidothymidine. Blood 1992;80:1443–1447.
33. Liu JP, Baker J, Perkins AS, Robertson EJ, Efstratiadis A. Mice carrying null mutations of the genes encoding IGF-I and type I IGF receptor. Cell 1993;75:59–72.
34. Woods KA, Camachin-Huebuer C, Savage MO, Clark AJ. Intra-uterine growth retardation, postnatal growth failure, association with deletion of IGF-I gene. N E J Med 1996;335:1363–1367.
35. Rapaport R, Oleske J, Ahdieh H, Soloman S, Delfaus C, Denny T. Suppression of immune function in growth hormone deficient children during treatment with human growth hormone. J Pediatrics 1986;109:434–439.
36. Rapaport R, Oleske J, Ahdieh H, Skuza K, Holland BK, Passannte MR, Denny T. Effect of human GH on immune function: in vitro studies on cells of normal and growth hormone deficient children. Life Science 1987;41:2319–2324.
37. Rapaport R, Oleske J. Effect of GH therapy on immune function. J Pediatrics 1987;110:663–664.
38. Rapaport R, Petersen B, Skuza KA, Heim M, Goldstein S. Immune function during treatment of GH deficient children with biosynthetic human GH. Clin Pediatrics 1991;30:22–27.
39. Rapaport R, Moghaddas R. GH therapy and immune function. In: Berczi I, Gorczynski R, editors. Neuroimmune biology Vol 1: New foundation of biology. Amsterdam: Elsevier Science, 2001;403–408.
40. Kotzman H, Riedl M, Clodi M, Barnas U, Kaider A, Hocker P, Luger A. The influence of GH substitution therapy on erythroid and myeloid progenitor cells and on peripheral blood cells in adult patients with GH deficiency. 1996;26:1175–1181.
41. Christ ER, Cummings MH, Westwood NB, et al. The importance of GH in the regulation of erythropoiesis, red cell mass, and plasma volume in adults with GH deficiency. J Clin Endocrinol Metab 1997;82:2985–2990.
42. Ten Have SMTH, Van Der lely AJ, Lamberts SWJ. Increase in Hb concentration in GH-deficient adults during human recombinant GH replacement therapy. Clinical Endocrinology 1997;47:565–570.
43. Koc E, Bideci A, Cinaz P, Unlu M, Atalay Y, Turkyilmaz C. IGF-I and IGF-BP3 in relation to Hemoglobin concentration in healthy term infants. Journal of Pediatric Endocrinol Metab 1997;10:609–613.
44. Merchav S, Graif Z, Skottner A. *In vitro* response of erythroid progenitors from children with thalassemia major to recombinat IGF-I and human GH. Clin Endocrinol 1993;39:207–211.
45. Low LC, Kwan EY, Lim YJ, Lee AC, Tam CF, Lam KS. GH treatment of short chinese children with beta-thalassemia major without GH deficiency. Clin Endocrinol 1995;42:359–363.

46. Golde DW, Bersch N, Kaplan SA, Rimoin DL, Li CH. Peripheral unresponsiveness to human GH in Laron syndrome. N E J Med 1980;303:1156–1159.
47. Anttila R, Koistinen R, Seppala M, Koistinen H, Siimes MA. IGF-I and IGF-BP3 as determinants of blood Hb concentration in healthy subjects. Pediatric Research 1994;36:745–748.
48. Vihervuori E, Virtanen M, Koistinen H, Koistinen R, Seppala M, Siimes MA. Hb level is linked to GH-dependent proteins in short children. Blood 1996;87:2075–2081.
49. Shimon I, Shpilberg O. The insulin like growth factor system in regulation of normal and malignant hematopoiesis. Leukemia Research 1995;19:233–240.
50. Barak Y, Zadik Z, Haron Y, Hahn T. Enhanced response to human circulating erythroid progenitor cells to human GH and IGF-I in children with insufficient GH secretion. Pediatric Research 1992;32(3):282–284.
51. Miyagawa S, Kobayashi M, Konishi N, Sato T, Ueda K. Insulin and IGF-I support the proliferation of erythroid progenitor cells in bone marrow through sharing of receptor. Br J Hematology 2000;109:555–562.
52. Valerio G, Di Maio S, Salerno M, Argensiano A, Badolato R, Tenore A. Association of RBC indices in GH treated children. Hormone Research 1997;47:62–66.
53. Ratajczak MZ, Kuczynski WI, Onodera K, Moore J, Ratajczak J, Kregenow DA, DeRiel K, Gewirtz AM. A reappraisal of the role of IGF-I in the regulation of Human hematopoiesis. J Clin Invest 1994;94:320–327.
54. Rapaport R, Davidow AL, Eisenstein D. IGF-I and Hemoglobin in children with GH deficiency during treatment with Human GH. 2000 The Endocrine society, 82nd annual meeting, abstract 1983.
55. Morrison I, Eccles MR, Reeve AE. Imprinting of IGF-II is modulated during hematopoiesis. Blood 2000;96(9):3023–3027.
56. Fang S, Steinmetz R, Walker King D, Zeng P, Vogelmeid C, Cooper S, Hangcoc G, Bronx meyer HE, Pescovitz OH. Development of transgenic mouse that overexpress a novel product of GH related hormone gene. Endocrinology 2000;141(4):1377–1383.

Growth and Lactogenic Hormones
Edited by L. Matera and R. Rapaport

Uteroplacental Prolactin Family: Immunological Regulators of Viviparity[a]

RUPASRI AIN[1,b], HEINER MÜLLER[2], NAMITA SAHGAL[1,3], GUOLI DAI[1,c] and MICHAEL J. SOARES[1]

Departments of Molecular & Integrative Physiology[1] and Pediatrics[3], University of Kansas Medical Center, Kansas City, KS 66160, USA
[2]Department of Obstetrics and Gynecology, University of Rostock, Rostock, Germany

ABSTRACT

Rodents possess an expanded prolactin (PRL) gene family. These genes encode for proteins that are structurally – but not necessarily functionally – related to PRL. They are hormones/cytokines of pregnancy and they are abundantly expressed in the uteroplacental compartment. In this short review, we focus on a subset of two PRL family members involved in regulating the maternal immune system, decidual/trophoblast PRL-related protein (d/tPRP) and PRL-like protein-A (PLP-A). D/tPRP is dually expressed in uterine decidual tissue and in trophoblast cells. This uteroplacental cytokine associates with the extracellular matrix via heparin containing molecules and targets eosinophils. PLP-A is expressed by trophoblast cells of the chorioallantoic placenta and specifically interacts and regulates uterine natural killer cell functions. Collectively, these two members of the rodent PRL family contribute to immunological adjustments ensuring viviparity.

1. INTRODUCTION

Viviparity is a mode of reproduction that occurs within the female reproductive tract. It has necessitated the acquisition of specialized maternal and extraembryonic tissues. In primates and rodents the maternal/fetal interface includes the uterine decidua of the mother and the placenta of extraembryonic origin. Hemochorial placentation, as occurs in both primates and rodents,

[a] Supported by HD37123 and HD38430 from the National Institutes of Child Health and Human Development (MJS), the J.B. Reynolds Foundation (MJS), Phillip Astrowe Foundation (NS), and the Deutsche Forschungsgemeinschaft, DFG, Mu 1183/3-1 (HM).
[b] Supported by a postdoctoral fellowship from the American Heart Association.
[c] Supported by a Junior Faculty Development Award from the Andrew Mellon Foundation.

results in the establishment of a close connection between maternal and embryonic/fetal tissues [1]. This close connection provides optimal gas exchange, supply of nutrients and disposal of wastes but also inherently results in maternal immunologic challenges to the genetically disparate extraembryonic and embryonic tissues. Maternal adjustments to the demands of pregnancy are paramount and include immunologic, endocrine, metabolic, and cardiovascular adaptations. Cells situated at the maternal/fetal interface orchestrate requisite changes in maternal physiology. Specifically, decidual and placental cells produce a variety of hormones, cytokines, and growth factors targeted to key maternal tissues. Among these various regulatory signals secreted by the decidua and placenta are a prominent family of proteins related to pituitary prolactin (PRL) and known as the uteroplacental PRL family [2,3]. The ancestral PRL structure has proven to be a malleable template for the generation of a diverse group of regulatory factors. Most interestingly, the size and composition of the PRL family is species specific. In some animals such as the rat, mouse, and cow the PRL family has expanded considerably, while in other species such as the pig, the PRL family contains but a sole member, PRL [3,4]. Primates offer an interesting variation, in that the related growth hormone (GH), which in primates dually activates both GH and PRL signaling pathways, has served as a template for an expanded family of pregnancy-dependent hormones [5,6]. Nomenclature for members of the PRL family reflects biological activities (placental lactogens), structural relationships with PRL (PRL-like proteins, PRL-related proteins), or associations with proliferation (proliferin).

The biology of the PRL family is intriguing. Some members of the PRL family are effectively mimics of PRL action [2,3]. These hormones have been referred to as *classical* members of the PRL family [7,8]. They represent functional PRL analogues possessing unique patterns of expression, circulatory profiles, and access to maternal and/or fetal compartments. Ligands for the PRL receptor are present throughout gestation. Most members of the PRL family do not activate the PRL receptor [2,3]. These hormones are referred to as *nonclassical* members of the PRL family [7,8]. Although, their biology is only beginning to be understood, it is already evident that nonclassical members of the PRL family are key contributors to the regulation of the maternal environment. Among their targets is the maternal reproductive tract, organs controlling metabolism, the vasculature, hematopoiesis, and the immune system [3].

In this short review, we discuss two nonclassical members of the rodent PRL family, decidual/trophoblast PRL-related protein (d/tPRP) and PRL-like protein-A (PLP-A) emphasizing their involvement as modulators of the maternal immune system during pregnancy. D/tPRP is a major secretory product of uterine decidual cells and the chorioallantoic placenta produces PLP-A. Both ligands target cellular constituents of the maternal immune system.

2. DECIDUA, D/TPRP, AND UTERINE EOSINOPHIL BIOLOGY

2.1. Decidual cells and the establishment of pregnancy

Successful embryonic development within the female reproductive tract requires the generation of a specialized maternal structure, the decidua. Decidual cells are modified uterine endometrial stromal cells. During gestation, decidual cells are located at the interface separating invading trophoblast cells from the maternal environment. A number of important functions have been attributed to decidua [9]: i) a protective role in controlling trophoblast cell invasion, ii) a nutritive role for the developing embryo, iii) a role in preventing immunological rejection of genetically disparate embryonic/fetal tissues, and iv) an endocrine/paracrine role in controlling maternal adaptations required for the establishment and maintenance of pregnancy. Pregnancy is dependent upon decidual cell acquisition of each of these specialized functions. The ovaries and blastocyst provide signals responsible for initiating changes in the uterus. Differentiation of decidual cells is among the earliest uterine adaptations to pregnancy [10,11] and is exquisitely sensitive to the regulatory actions of progesterone [12–15]. Decidual cells have profound effects on the local uterine environment [9,11]. The uterus shows dramatic changes in its vascularization and the distribution and function of its immune and inflammatory cell constituents following decidualization [11,16–18].

2.2. Decidual PRL family signaling

Decidual cell signaling is mediated, at least in part, through the production of cytokines related to PRL [13,19–21]. Gibori and Rothchild [22,23] first advanced the concept of a decidual PRL-like protein. These researchers determined that luteal progesterone production could be maintained in pseudopregnant rats treated with inhibitors of pituitary PRL secretion, only if their uterine stroma was decidualized. These early observations were the impetus to search for a decidual PRL in the human [24,25]. Human decidual PRL was found to be identical to human pituitary PRL [26–28]. Human decidual PRL is a heparin-binding cytokine [29] and its targets are likely intrauterine [21]; however, its role in the physiology of pregnancy is yet to be fully resolved. Gibori and coworkers further characterized the actions of the rat decidual PRL-like protein on the ovary and uterus [30] and most recently have demonstrated that its structural characteristics are identical with pituitary PRL [31,32]. In the rat and/or mouse, four members of the decidual PRL subfamily have been cloned and at least partially characterized. They include d/tPRP, PRL-like protein-B (PLP-B; [33–36]), PRL-like protein-J (PLP-J; [37,38]), and PRL [31]; and should collectively be viewed as downstream mediators of progesterone action. PLP-B and PLP-J are considered 'orphan' ligands because their biological targets have not been resolved.

2.3. D/tPRP discovery & expression

Our laboratory discovered d/tPRP during a search for a decidual luteotropin in the rat [39]. Mouse d/tPRP was subsequently identified and characterized [40,41]. D/tPRP expression patterns during pregnancy parallel the differentiation of antimesometrial decidual cells [41–45]. Following the formation of the chorioallantoic placenta, the antimesometrial decidua degenerates and d/tPRP expression shifts to trophoblast cells and continues until term [41,44]. Patterns of d/tPRP expression are similar in the mouse and rat.

2.4. D/tPRP action

D/tPRP avidly binds to heparin-containing molecules, including the decidual extracellular matrix, and does not circulate at appreciable levels in maternal blood [46,47]. Human decidual PRL also binds heparin [29]. In contrast to PRL, d/tPRP does not utilize the PRL receptor [46,47]. In a transplantation model, CHO cells expressing d/tPRP more readily form tumors in athymic mice than do control CHO cells [46]. The *in vivo* growth differences in the two CHO cell populations cannot be accounted for by *in vitro* growth rates and are independent of the effects of d/tPRP on the host vasculature. These observations suggested that d/tPRP participates in the regulation of host immune cell responses. Our newest insights regarding d/tPRP have come from the use of an alkaline phosphatase-d/tPRP (AP-d/tPRP) fusion protein. Using this

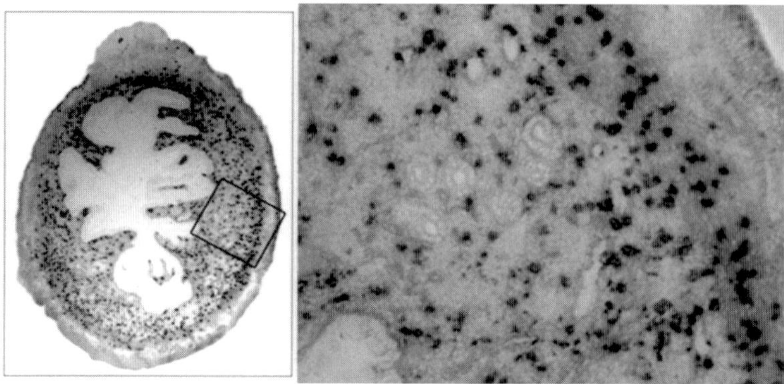

Figure 1. D/tPRP binding to uterine eosinophils. D/tPRP binding was detected with an alkaline phosphatase (AP)-d/tPRP fusion protein in rat uterine sections from Day 0.5 of pregnancy. Resolution of AP-d/tPRP binding to heparin-containing molecules from other potential interactions was achieved by post-AP-d/tPRP incubation of tissue sections with heparin (250 μg/ml). AP-d/tPRP binding was detected by AP histochemistry. *Left panel*, low magnification photomicrograph of a transverse section of the uterus. *Right panel*, high magnification of the region in the left panel outlined in the box. Please note the extensive binding (dark staining cells) present throughout the uterine stroma and myometrium.

approach, we have determined that d/tPRP interacts with eosinophils (Ref. [47], Figure 1). Thus, d/tPRP has two biologically-relevant features: i) interactions with heparin and ii) interactions with eosinophils.

2.5. D/tPRP-heparin interactions

Heparin and heparin-related structures are abundant regulatory signals widely distributed and involved in an array of physiological processes. These molecules participate in cell adhesion, migration, growth regulation, basement membrane properties, differentiation, etc. [48]. During early pregnancy, the uterus is a site of extensive remodeling that undoubtedly involves heparin/heparin-related structures. D/tPRP's high affinity for heparin places it in a key position to modulate heparin-dependent events during the establishment of pregnancy. Furthermore these heparin-dependent actions of d/tPRP represent a conserved function with the heparin-binding human decidual PRL [29].

2.6. D/tPRP-eosinophil signaling

Eosinophils have been the focus of research in the rodent uterus and more recently the human uterus. In rodents, there is a striking entry of eosinophils into the uterus at proestrous and estrous that is controlled by estrogens; and there is a striking exit/death/functional restraint during the luteal phase and pregnancy, which is mediated by progesterone (Figure 2, [49–56]). In the human, eosinophils accumulate in the uterus during the late secretory phase of the menstrual cycle as circulating progesterone levels decline, and participate in menstruation [57–59]. Steroid hormones are likely acting indirectly by influencing the expression of locally acting cytokines and chemokines [58,60,61].

Eosinophils are primary responders to certain types of infections (especially parasitic) and the presence of foreign cells/tissues [62–65]. Eosinophils also are actively involved in tissue

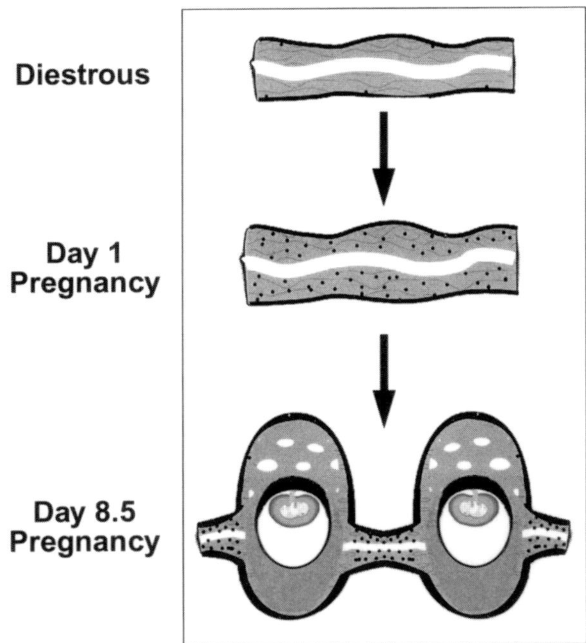

Figure 2. Eosinophil trafficking within the uterus. Schematic representation of eosinophil distributions in a uterus from diestrous (*top panel*), a uterus from day 1 of pregnancy (*middle panel*), and uteroplacental units from day 8.5 of pregnancy (*bottom panel*). Please note the increased numbers of eosinophils within the uterus at the beginning of pregnancy and their subsequent exclusion from the decidua, region surrounding the developing embryo, following implantation (day 8.5 of gestation). Eosinophils are depicted as black spots on the schematic tissue sections.

transplant and solid tumor rejection [66–69]. They produce an array of cytokines and cytotoxic molecules that can effectively eradicate infectious and foreign agents and genetically disparate cells. However, as a byproduct they also can cause considerable damage to normal cells. It would appear to be advantageous to have eosinophils present in the uterus around the time of mating when there is a possibility for the introduction of potentially infectious and foreign agents. Likewise, it would also appear efficacious to remove or restrain eosinophils once pregnancy has been initiated in order to prevent their attack of the semi-allogeneic embryo. Interestingly, it has been proposed that pregnancy resembles aspects of a host-parasite interaction [70]. The mechanisms controlling uterine eosinophils are not well understood.

Uterine eosinophils are critically involved in uterine remodeling events during the estrous and menstrual cycles [54,57,71] and during pathological processes such as endometriosis [72,73]. Eosinophils contribute to tissue remodeling via their production of cytokines and matrix metalloproteinases.

Eosinophils are targets of d/tPRP [47]. Based on our transplantation studies and the reciprocal intrauterine patterns of d/tPRP and eosinophil distributions, we predict that eosinophil exposure to d/tPRP leads to their exit, death, and/or functional restraint. Therefore d/tPRP mediates, at least in part, progesterone's anti-inflammatory actions during pregnancy. This effect is amplified because of the progesterone-dependent formation of decidual cells, which abundantly express d/tPRP.

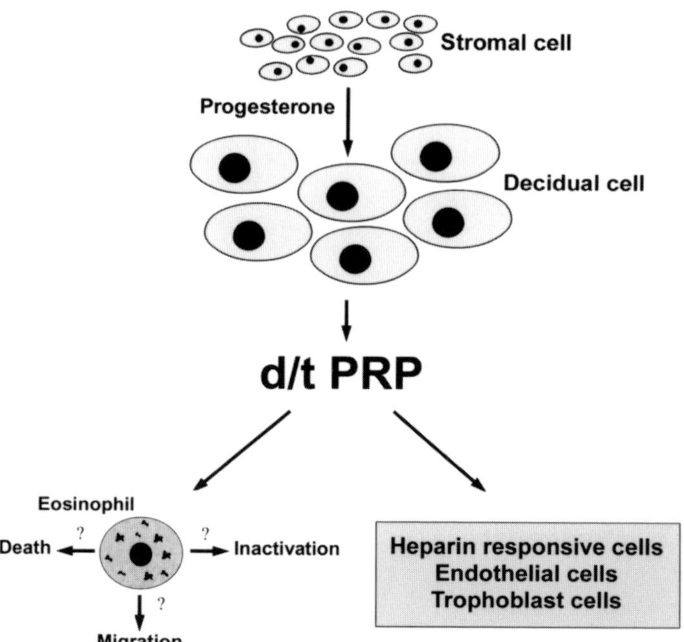

Figure 3. Progesterone-decidual cell-d/tPRP-heparin-eosinophil model. Progesterone is responsible for the differentiation of decidual cells, including their capacity to produce d/tPRP. D/tPRP is a downstream mediator of progesterone's anti-inflammatory actions on the uterus. Eosinophils and heparin containing substrates are targets for the action of d/tPRP.

In humans like rodents, decidual cells develop under the control of progesterone and represent the proximal maternal barrier to the developing embryo [10,11,14]. Progesterone possesses key anti-inflammatory actions within the uterus [74–76], including influences on the intrauterine cytokine/chemokine milieu and affects on the uterine distribution of leukocytes, including eosinophils [58]. Both human and rodent decidual cells synthesize and secrete members of the PRL cytokine family [19–21]. These members of the PRL family are heparin-binding cytokines and mediators of progesterone's actions on the uterine environment (Figure 3, [29,46,47]).

3. PLACENTA, PLP-A, AND UTERINE NATURAL KILLER CELLS

3.1. Placental organization

The placenta is a specialized structure developing in concert with the embryo. The placenta allows the embryo to access maternal resources without being harmed. Trophoblast cells are the parenchymal cells of the placenta. They are specialized, exhibit distinct phenotypes, and arise via a multilineage differentiation process [7,77,78]. Trophoblast lineages go on to contribute to the formation of two placental structures in the rodent [79,80]: i) the choriovitelline placenta and ii) the chorioallantoic placenta. These structures are responsible for controlling fetal and maternal environments during pregnancy. The choriovitelline placenta is a relatively simple structure consisting of trophoblast cells adhered to parietal endoderm. It forms shortly after implantation and degenerates shortly after midgestation. In contrast, the chorioallantoic placenta is a more complex structure of the latter half of pregnancy. It is organized into an invasive/endocrine component located at the maternal interface referred to as the junctional zone and a region responsible for maternal-fetal bidirectional transport and limited endocrine activity located at the fetal interface referred to as the labyrinth zone. Two trophoblast cell types express members of the PRL family: i) trophoblast giant cell and ii) spongiotrophoblast cell.

3.2. Placental PRL family signaling

At least 19 different members of the PRL family are expressed in the rodent placenta [2,3]. Their expression patterns are cell and temporal specific and their actions, where known, are vital for the establishment and maintenance of pregnancy [2,3]. PLP-A was discovered by Duckworth and Friesen [81] and represents one of the first nonclassical members of the PRL family to be characterized. The remainder of the discussion focuses on the biology of PLP-A.

3.3. PLP-A discovery and expression

PLP-A was originally identified during a search for a placental lactogen cDNA [81]. Expression of PLP-A protein and mRNA are restricted to spongiotrophoblast cells and trophoblast giant cells of the junctional zone of the rat and mouse chorioallantoic placenta [36,40,81–87]. The ontogeny of PLP-A production coincides with the development of the chorioallantoic placenta. In vitro differentiating spongiotrophoblast and trophoblast giant cells secrete abundant amounts of PLP-A [88,89]. PLP-A circulates in fetal and maternal compartments as a high molecular weight complex [84].

3.4. PLP-A actions

Even though PLP-A has significant structural homologies to PRL it is not capable of binding to PRL receptors or activating the PRL receptor signaling pathway [85]. The generation of alkaline phosphatase-PLP-A (AP-PLP-A) fusion proteins led to the identification of specific PLP-A binding to natural killer cells within the mesometrial compartment of the uterus from pregnant rats and mice (Figure 4, [90]). These observations were supported by the co-distribution of PLP-A targets with cells expressing the rat natural killer cell surface marker, gp42, the absence of PLP-A binding in conceptuses from natural killer cell deficient tgε26 mice, and the specific interaction of PLP-A with rat natural killer cells. Based on these observations we surmised that PLP-A likely regulates uterine natural killer cells during pregnancy.

3.5. Uterine natural killer cells and PLP-A-signaling during the establishment of pregnancy

Natural killer cells are components of our natural/innate immunity and participate in immune surveillance. They effectively target virus-infected cells, tumor cells, and potentially other cells for destruction without prior sensitization [91]. Natural killer cells can be identified based on their morphological appearance, their expression of cell surface markers, and their ability to lyse targets such as YAC-1 cells [92,93]. Natural killer cells are prominent residents of the uterus of rodents and humans [17,92–94].

The phenotype of uterine natural killer cells changes during the establishment of pregnancy

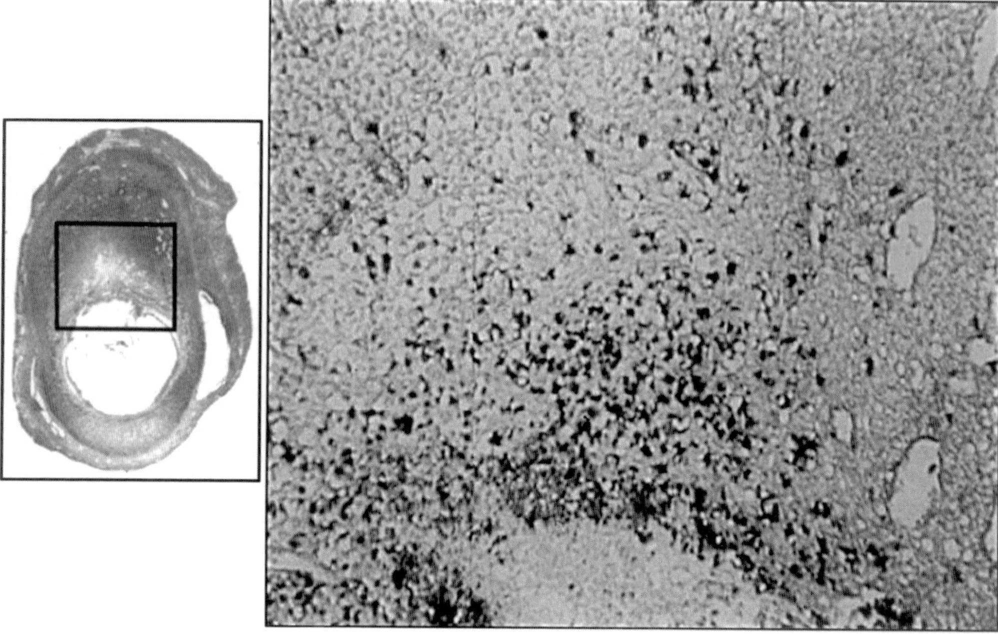

Figure 4. PLP-A binding to uterine natural killer cells. PLP-A binding was detected with an alkaline phosphatase (AP)-PLP-A fusion protein in a tissue section from the day 9.5 rat conceptus. AP-PLP-A binding was detected by AP histochemistry. *Left panel*, low magnification photomicrograph of a transverse section from a control day 9.5 rat conceptus. *Right panel*, AP-PLP-A binding to a region of the day 9.5 rat conceptus similar to that outlined in the box shown within the left panel. Please note the extensive binding (dark staining cells) present throughout the mesometrial decidua overlying the developing chorioallantoic placenta.

[92,93,95]. Natural killer cells increase in number and redistribute to specific locations within the uterus (Figure 5). The initial natural killer cell expansion occurs in close proximity to the developing chorioallantoic placenta. These natural killer cells undergo considerable morphological and functional changes creating, in effect, a natural killer cell with a unique phenotype. Uterine natural killer cells of pregnancy are conspicuous in their relative absence of cytolytic activities and their enhanced production of specific bioeffector molecules [92,93,96,97] . Natural killer cells are also redirected away from the developing placenta and form a new highly vascular structure, the metrial gland, which is embedded in the mesometrial myometrium [92,93,95].

We propose that trophoblast cells communicate with natural killer cells situated within the uteroplacental compartment via secretion of PLP-A. PLP-A possesses three features, which support it having a modulatory role on the uterine natural killer cell phenotype [90]: i) PLP-A production is spatially and temporally coincident with the appearance of natural killer cells in the mesometrial decidua, ii) PLP-A specifically binds to natural killer cells, and iii) PLP-A specifically inhibits natural killer cell cytolytic activities. Natural killer cell surface molecules interacting with PLP-A are yet to be identified. PLP-A interactions with natural killer cells result in a rapid mobilization of intracellular calcium [98]. Interleukin-15, a known natural killer cell modulatory cytokine, has similar effects on intracellular calcium mobilization [98]. The actions of PLP-A on natural killer cells may lead to natural killer cell differentiation towards a pregnancy-

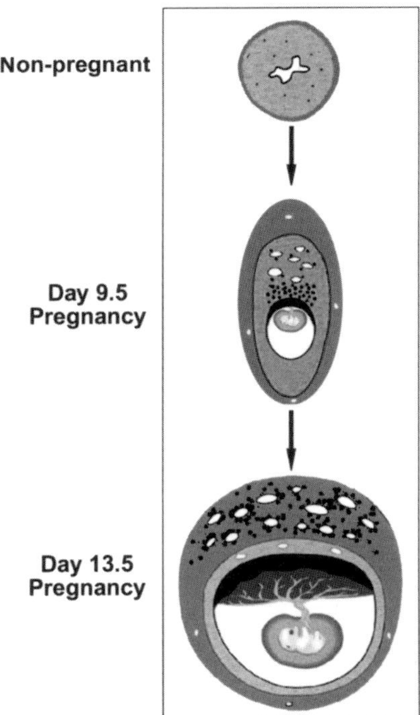

Figure 5. Natural killer cell trafficking within the uterus. Schematic representation of natural killer cell distributions in a uterus from a nonpregnant animal (*top panel*), a conceptus from day 9.5 of pregnancy (*middle panel*), and a conceptus from day 13.5 of pregnancy (*bottom panel*). Natural killer cells increase in numbers within the uterus following implantation. Please note the natural killer cell expansion in the mesometrial decidual compartment immediately overlying the developing chorioallantoic placenta. As gestation advances, natural killer cells move away from the chorioallantoic placenta and colonize a richly vascular structure in the mesometrial compartment referred to as the metrial gland. Natural killer cells are depicted as black spots on the schematic tissue sections.

associated phenotype (Figure 6). Suppression of natural killer cell killing activity facilitates survival of genetically disparate extraembryonic and embryonic tissues. Natural killer cells release nitric oxide, which directly influences uterine vasculature, facilitating the delivery of nutrients to the developing placenta, while their elaboration of CSF-1 and LIF promote the growth and maturation of the chorioallantoic placenta [96,97]. Systemic effects of PLP-A on extrauterine natural killer cells are likely obviated by the association of PLP-A with circulating PLP-A binding proteins.

4. CONCLUSIONS

Rodents and primates share similarities in the organization of their uteroplacental interface. Immune cells, including eosinophils and natural killer cells, are conspicuous residents of the uterine stromal compartment in both groups of mammals. These cells have an important protective role within the uterus as well as other tissues. They possess keen abilities to recognize the intrusion of foreign cells and to facilitate the eradication of these potential threats. Pregnancy

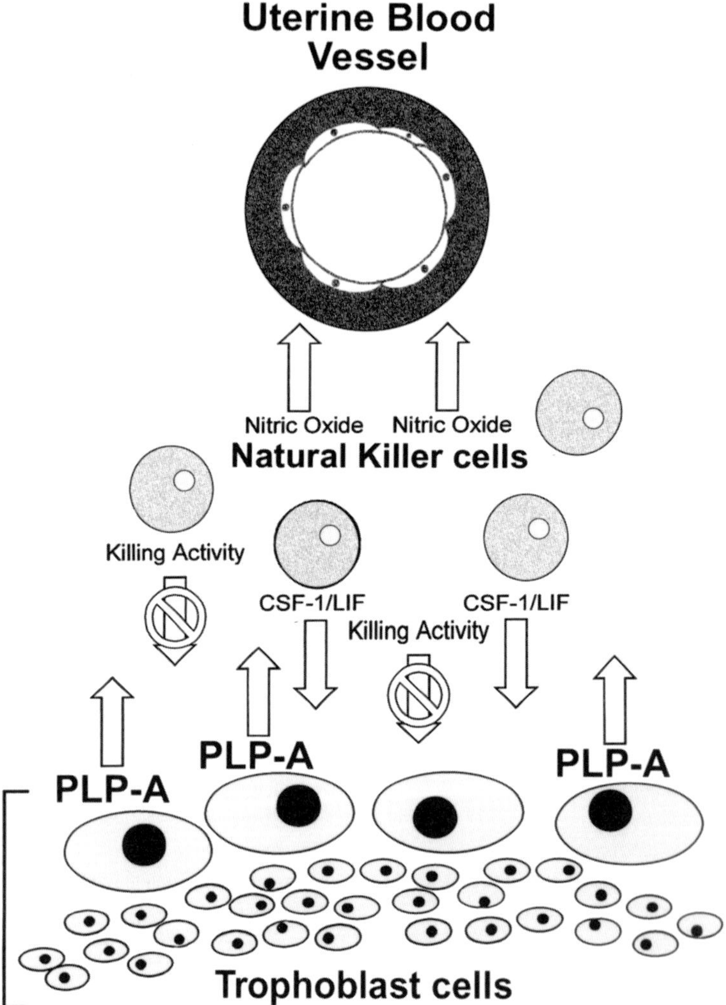

Figure 6. A model of trophoblast-natural killer cell signaling via PLP-A. Trophoblast cells communicate with natural killer cells situated within the uteroplacental compartment via the secretion of PLP-A. We propose that PLP-A impacts the phenotype of uterine natural killer cells. The movement of natural killer cells away from the developing chorioallantoic placenta, suppression of killing activity, enhanced nitric oxide synthase activity, and a specific cytokine expression profile characterize this pregnancy-associated phenotype. Collectively, these changes in natural killer cell function ensure the success of pregnancy.

requires enhanced discriminatory processes by the maternal immune system. Specialized tissues of pregnancy, including the uterine decidua and the placenta, modulate the functions of immune cells. In some cases, the immune cells are banished from the uteroplacental compartment (e.g. eosinophils), while in other situations their efforts are redirected toward activities that benefit placental/embryonic development (e.g. natural killer cells). Members of the rodent PRL family contribute to the pregnancy-specific regulation of the intrauterine immune environment. D/tPRP and PLP-A represent examples of two potential effectors of intrauterine immune cells.

We have much to learn about the biological roles of d/tPRP and PLP-A as immune cell modulators during pregnancy. Gain-of-function and loss-of-function in vivo analyses need to be performed. These experiments will permit an evaluation of the role of these cytokines in the physiology of pregnancy. We also need to identify the immune cell receptors and dissect the signal transduction pathways utilized by these PRL family cytokines. Some functional redundancy within the PRL family and among other cytokines and chemokines seems likely. Immune cell effectors may be also included among the dozen or more current orphan PRL family ligands [3].

Understanding the physiology of the PRL family in the mouse and rat provides access to important regulatory processes in the human. In some instances, cross species similarities may prevail, while in other cases the differences may be most compelling. Nonetheless, our appreciation for the biology of pregnancy increases. We hypothesize that members of the uteroplacental PRL family evolved to subserve important biological roles during pregnancy. Functional homologies among species will likely exist and may include the ligands, their receptors, or components of their signal transduction pathways.

ACKNOWLEDGEMENTS

We would like to thank past and present members of our laboratory and our collaborators.

REFERENCES

1. Enders AC, Welsh AO. Structural interactions of trophoblast and uterus during hemochorial placenta formation. J Exp Zool 1993;266:578–587.
2. Soares MJ, Dai G, Orwig KE, Peters TJ, Müller H. The uteroplacental prolactin family and pregnancy. Biol Reprod 1998;58:273–284.
3. Soares MJ, Linzer DIH. Rodent prolactin family and pregnancy. In: Horseman N, editor. Prolactin. Boston: Kluwer Academic, 2001;139–167.
4. Talamantes F, Ogren L, Markoff E, Woodard S, Madrid J. Phylogenetic distribution, regulation of secretion, and prolactin-like effects of placental lactogens. Fed Proc 1980;39:2582–2587.
5. Cooke NE, Liebhaber SA. Molecular biology of the growth hormone-prolactin gene system. Vitamins Horm 1995;50:385–459.
6. Handwerger S, Brar A. Human uteroplacental lactogens: physiology and molecular biology. In: Horseman N, editor. Prolactin. Boston: Kluwer Academic, 2001;169–187.
7. Soares MJ, Chapman BM, Rasmussen CA, Dai G, Kamei T, Orwig KE. Differentiation of trophoblast endocrine cells. Placenta 1996;17:277–289.
8. Soares MJ, Dai G, Cohick CB, Müller H, Orwig KE. The rodent placental prolactin family and pregnancy. In: Bazer FW, editor. The Endocrinology of Pregnancy. Totowa, NJ: Humana Press, 1998;145–176.
9. Bell SC. Decidualization: regional differentiation and associated function. Oxford Rev Reprod Biol 1983;5:220–271.
10. DeFeo VJ. Decidualization. In: Wynn RM, editor. Cellular Biology of the Uterus. New York: Appleton-Century-Crofts, 1967;191–290.

11. Parr MB, Parr EL. The implantation reaction. In: Wynn RM, Jollie WP, editors. Biology of the Uterus. New York: Plenum Press, 1989;233–278.
12. Glasser SR, Clark JH. A determinant role for progesterone in the development of uterine sensitivity to decidualization and ovo-implantation. Symp Soc Dev Biol 1975;33:311–345.
13. Tang B, Guller S, Gurpide E. Mechanism of human endometrial stromal cell decidualization. Ann NY Acad Sci 1994;734:19–25.
14. Lydon JP, DeMayo FJ, Funk CR, Mani SK, Hughes AR, Montgomery CA, Shyamala G, Conneely OM, O'Malley BW. Mice lacking progesterone receptor exhibit pleiotropic reproductive abnormalities. Genes Dev 1995;9:2266–2278.
15. Brar AK, Frank GR, Kessler CA, Cedars MI, Handwerger S. Progesterone-dependent decidualization of the human endometrium is mediated by cAMP. Endocrine 1997;6:301–307.
16. Hunt JS. Immunologically relevant cells in the uterus. Biol Reprod 1994;50:461–466.
17. Croy BA, Whitelaw PF, Engelhardt H. The influences of immune cells on the success of pregnancy. In: Bazer FW, editor. The Endocrinology of Pregnancy. Totowa, NJ: Humana Press, 1998;229–289.
18. Hunt JS, Petroff MG, Burnett TG. Uterine leukocytes: key players in pregnancy. Sem Cell Dev Biol 2000;11:127–137.
19. Orwig KE, Rasmussen CA, Soares MJ. Decidual signals in the establishment of pregnancy: the prolactin family. Troph Res 1997;10:329–343.
20. Telgmann R, Gellersen B. Marker genes of decidualization: activation of the decidual prolactin gene. Hum Reprod Update 1998;4:472–479.
21. Jabbour HN, Critchley HO. Potential roles of decidual prolactin in early pregnancy. Reproduction 2001;121:197–205.
22. Gibori G, Rothchild I, Pepe GI, Morishige WK, Lam P. Luteotrophic action of decidual tissue in the rat. Endocrinology 1974;95:1113–1118.
23. Rothchild I, Gibori G. The luteotrophic effect of decidual tissue: the stimulating effect of decidualization on serum progesterone level of pseudopregnant rats. Endocrinology 1975;97:838–842.
24. Golander A, Hurley T, Barrett J, Hize A, Handwerger S. Prolactin synthesis by human chorion-decidual tissue: a possible source of amniotic fluid prolactin. Science 1978;202:311–312.
25. Riddick D, Luciano A, Kusmik W, Maslar I. De novo synthesis of prolactin by human decidua. Life Sci 1978;23:1913–1929.
26. Golander A, Hurley T, Barrett J, Handwerger S. Synthesis of prolactin by human decidua in vitro. J Endocrinol 1979;82:263–267.
27. Gellersen B, DiMattia GE, Friesen HG, Bohnet HG. Prolactin (PRL) mRNA from human decidua differs from pituitary PRL mRNA but resembles the IM-9-P3 lymphoblast PRL transcript. Mol Cell Endocrinol 1989;64:127–130.
28. DiMattia GE, Gellersen B, Duckworth ML, Friesen HG. Human prolactin gene expression. The use of an alternative noncoding exon in decidua and the IM-9-P3 lymphoblast cell line. J Biol Chem 1990;265:16412–16421.
29. Khurana S, Kuns R, Ben-Jonathan N. Heparin-binding property of human prolactin: a novel aspect of prolactin biology. Endocrinology 1999;140:1026–1029.
30. Gibori G, Jayatilak PG, Khan I, Rigby B, Puryear T, Nelson S, Herz Z. Decidual luteotropin secretion and action: its role in pregnancy maintenance in the rat. In: Mahesh

VB, Dhindsa DS, Anderson E, Kalra SP, editors. Regulation of Ovarian and Testicular Function. New York: Plenum Press, 1987;379–397.

31. Prigent-Tessier A, Tessier C, Hirosawa-Takamori M, Boyer C, Ferguson-Gottschall S, Gibori G. Rat decidual prolactin. Identification, molecular cloning, and characterization. J Biol Chem 1999;274:37982–37982.
32. Prigent-Tessier A, Barkai U, Tessier C, Cohen H, Gibori G. Characterization of a rat uterine cell line, UIII cells: prolactin (PRL) expression and endogenous regulation of PRL-dependent genes;estrogen receptor β, α2-macroglobulin, and decidual PRL involving the Jak-Stat5 pathway. Endocrinology 2001;142:1242–1250.
33. Duckworth ML, Peden LM, Friesen HG. A third prolactin-like protein expressed by the developing rat placenta. Mol Endocrinol 1988;2:912–920.
34. Croze F, Kennedy TG, Schroedter IC, Friesen HG. Expression of rat prolactin-like protein-B in deciduoma of pseudopregnant rat and in decidua during early pregnancy. Endocrinology 1990;127:2665–2672.
35. Cohick CB, Xu L, Soares MJ. Placental prolactin-like protein-B: heterologous expression and characterization of placental and decidual species. J Endocrinol 1997;152:291–302.
36. Müller H, Ishimura R, Orwig KE, Liu B, Soares MJ. Homologues for prolactin-like protein-A and B are present in mouse. Biol Reprod 1998;58:45–51.
37. Toft DJ, Linzer DIH. Prolactin (PRL)-like protein J, a novel member of the PRL/growth hormone family, is exclusively expressed in maternal decidua. Endocrinology 1999;140:5095–5101.
38. Dai G, Wang D, Liu B, Kasik JW, Müller H, White RA, Hummel GS, Soares MJ. Three novel paralogs of the rodent prolactin gene family. J Endocrinol 2000;166:63–75.
39. Roby KF, Deb S, Gibori G, Szpirer C, Levan G, Kwok SCM, Soares MJ. Decidual prolactin-related protein: identification, molecular cloning, and characterization. J Biol Chem 1993;268:3136–3142.
40. Lin J, Poole J, Linzer DIH. Three new members of the mouse prolactin/growth hormone family are homologous to proteins expressed in the rat. Endocrinology 1997;138:5541-5549.
41. Orwig KE, Ishimura R, Müller H, Liu B, Soares MJ. Identification and characterization of a mouse homolog for decidual/trophoblast prolactin-related protein. Endocrinology 1997;139:5511–5517.
42. Gu Y, Soares MJ, Srivastava RK, Gibori G. Expression of decidual prolactin-related protein in the rat decidua. Endocrinology 1994;135:1422–1427.
43. Orwig KE, Dai G, Rasmussen CA, Soares MJ. Decidual/trophoblast prolactin related protein: characterization of gene structure and cell-specific expression. Endocrinology 1997;138:2491–2500.
44. Rasmussen CA, Orwig KE, Vellucci S, Soares MJ. Dual expression of prolactin-related protein in decidua and trophoblast tissues during pregnancy. Biol Reprod 1997;55:647–654.
45. Spencer F, Chi L, Zhu MX. Hydroxyurea inhibition of cellular and developmental activities in the decidualized and pregnant uteri of rats. J Appl Toxicol 2000;20:407–412.
46. Rasmussen CA, Hashizume K, Orwig KE, Xu L, Soares MJ. Decidual prolactin-related protein: heterologous expression and characterization. Endocrinology 1996;137:5558–5566.
47. Wang D, Ishimura R, Walia DS, Müller H, Dai G, Hunt JS, Lee NA, Lee JJ, Soares MJ. Eosinophils are cellular targets of the novel uteroplacental heparin-binding cytokine

decidual/trophoblast prolactin-related protein. J Endocrinol 2000;167:15–28.
48. Jackson RL, Busch SJ, Cardin AD. Glycosaminoglycans: molecular properties, protein interactions, and role in physiological processes. Physiol Rev 1991;71:481–539.
49. Rytomaa T. Organ distribution and histochemical properties of eosinophil granulocytes in rat. Acta Path Microbiol Scand Suppl 1960;140:11–118.
50. Bassett EG. Infiltration of eosinophils into the modified connective tissue of oestrous and pregnant animals. Nature 1962;1259–1261.
51. Bjersing L, Borglin NE. Effect of hormones on incidence of uterine eosinophilia in rats. Acta Path Microbiol Scand 1964;60:353–364.
52. Ross R, Klebanoff SJ. The eosinophilic leukocyte. Fine structure studies of changes in the uterus during the estrous cycle. J Exp Med 1966;124:653–659.
53. King WJ, Allen TC, DeSombre ER. Localization of uterine peroxidase activity in estrogen-treated rats. Biol Reprod 1981;25:859–870.
54. Tchernitchin AT. Eosinophil-mediated non-genomic parameters of estrogen stimulation – a separate group of responses mediated by an independent mechanism. J Steroid Biochem 1983;19:95–100.
55. Lee YH, Howe RS, Sha S-J, Teuscher C, Sheehan DM, Lyttle CR. Estrogen regulation of an eosinophil chemotactic factor in the immature rat uterus. Endocrinology 1989;125:3022–3028.
56. Howe RS, Lee YH, Fischkoff SA, Teuscher C, Lyttle CR. Glucocorticoid and progestin regulation of eosinophil chemotactic factor and complement C3 in the estrogen-treated rat uterus. Endocrinology 1990;126:3193–3199.
57. Jeziorska M, Salamonsen LA, Woolley DE. Mast cell and eosinophil distribution and activation in human endometrium throughout the menstrual cycle. Biol Reprod 1995;53:312–320.
58. Salamonsen LA, Lathbury LJ. Endometrial leukocytes and menstruation. Hum Reprod Update 2000;140:111–118.
59. Zhang J, Lathbury LJ, Salamonsen LA. Expression of the chemokine eotaxin and its receptor, CCR3, in human endometrium. Biol Reprod 2000;62:404–411.
60. Pollard JW, Lin EY, Zhu L. Complexity in uterine macrophage responses to cytokines. Biol Reprod 1998;58:1466–1475.
61. Robertson SA, Allanson M, Mau VJ. Molecular regulation of uterine leukocyte recruitment during early pregnancy in the mouse. Troph Res 1998;11:101–119.
62. Weller PF. Lipid, peptide, and cytokine mediators elaborated by eosinophils. In: Smith H, Cook RM, editors. Immunopharmacology of Eosinophils. New York: Academic Press, 1993;25–42.
63. Venge P. Human eosinophil granule proteins: structure, function and release. In: Smith H, Cook RM, editors. Immunopharmacology of Eosinophils. New York: Academic Press, 1993;43–55.
64. Hirai K, Miyamasu M, Takaishi T, Morita Y. Regulation of the function of eosinophils and basophils. Crit Rev Immunol 1997;17:325–352.
65. Rothenberg M. Eosinophilia. New England J Med 1998;338:1592–1600.
66. Tepper RI, Coffman RL, Leder P. An eosinophil-dependent mechanism for the antitumor effect of interleukin-4. Science 1992;257:548–551.
67. Tepper RI. The eosinophil-mediated antitumor activity of interleukin-4. J Allergy Clin Immunol 1994;94:1225–1231.
68. LeMoine A, Flamand V, Demoor F-X, Noel J-C, Surquin M, Kiss R, Nahori M-A,

Pretolani M, Goldman M, Abramowicz D. Critical roles for IL-4, IL-5, and eosinophils in chronic skin allograft rejection. J Clin Invest 1999;103:1659–1667.
69. LeMoine A, Surquin M, Demoor F-X, Noel JC, Nahori M-A, Pretolani M, Flamand V, Braun MY, Goldman M, Abramowicz D. IL-5 mediates eosinophilic rejection of MHC class II-disparate skin allografts in mice. J Immunol 1999;163:3778–3784.
70. Clark DA, Arck PC, Chaouat G. Why did you mother reject you? Immunogenetic determinants of the response to environmental selective pressure expressed at the uterine level. Am J Reprod Immunol 1999;41:5–22.
71. Tchernitchin AT, Mena MA, Soto J, Unda C. The role of eosinophils in the actions of estrogens and other hormones. Med Sci Res 1989;17:5–10.
72. Blumenthal RD, Samoszuk M, Taylor AP, Brown G, Alisauskas R, Goldenberg DM. Degranulating eosinophils in human endometriosis. Am J Pathol 2000;156:1581–1588.
73. Hornung D, Dohrn K, Sotlar K, Greb RR, Wallwiener D, Kiesel L, Taylor RN. Localization in tissues and secretion of eotaxin by cells from normal endometrium and endometriosis. J Clin Endocrinol Metab 2000;85:2604–2608.
74. Siiteri P, Stites DP. Immunologic and endocrine interrelationships in pregnancy. Biol Reprod 1982;26:1–14.
75. Finn CA. Implantation, menstruation, and inflammation. Biol Rev 1986;61:313–328.
76. Tibbitts TA, Conneely Om, O'Malley BW. Progesterone via its receptor antagonizes the pro-inflammatory activity of estrogen in the mouse uterus. Biol Reprod 1999;60:1158–1165.
77. Gardner RL, Beddington RSP. Multi-lineage 'stem' cells in the mammalian embryo. J Cell Sci 1988;10 (Suppl):11–27.
78. Rossant J. Development of the extraembryonic lineages. Sem Dev 1995;6:237–247.
79. Davies J, Glasser SR. Histological and fine structural observations on the placenta of the rat. Acta Anat 1968;69:542–608.
80. Enders AC. A comparative study of the fine structure of the trophoblast in several hemochorial placentas. Am J Anat 1965;116:29–67.
81. Duckworth ML, Peden LM, Friesen HG. Isolation of a novel prolactin-like cDNA clone from developing placenta. J Biol Chem 1986;261:10879–10884.
82. Deb S, Youngblood T, Rawitch A, Soares MJ. Placental prolactin-like protein-A: identification and characterization of two major glycoprotein species with antipeptide antibodies. J Biol Chem 1989;264:14348–14353.
83. Campbell WJ, Deb S, Kwok SCM, Joslin JA, Soares MJ. Differential expression of placental lactogen-II and prolactin-like protein-A in the rat chorioallantoic placenta. Endocrinology 1989;125:1565–1574.
84. Deb S, Soares MJ. Characterization of placental prolactin-like protein-A in intracellular and extracellular compartments. Mol Cell Endocrinol 1990;74:163–174.
85. Deb S, Hamlin GP, Roby KF, Kwok SC, Soares MJ. Heterologous expression and characterization of prolactin-like protein-A. J Biol Chem 1993;268:3298–3305.
86. Duckworth ML, Schroedter IC, Friesen HG. Cellular localization of rat placental lactogen-II and rat prolactin-like proteins A and B by in situ hybridization. Placenta 1990;11:143–155.
87. Manzella SM, Dharmesh SM, Cohick CB, Soares MJ, Baenziger JU. Developmental regulation of a pregnancy-specific oligosaccharide structure, NeuAcα2,6GalNAcβ1,4GlcNAc on select members of the rat placental prolactin family. J Biol Chem 1997;272:4775–4782.
88. Hamlin GP, Lu X-J, Roby KF, Soares MJ. Recapitulation of the pathway for trophoblast

giant cell differentiation in vitro: stage-specific expression of members of the prolactin gene family. Endocrinology 1994;134:2390–2396.

89. Lu X-J, Deb S, Soares MJ. Spontaneous differentiation of trophoblast cells along the spongiotrophoblast pathway: expression of the placental prolactin gene family and modulation by retinoic acid. Dev Biol 1994;163:86–97.
90. Müller H, Dai G, Croy BA, Head JR, Liu B, Soares MJ. Uterine natural killer cells are regulated by a trophoblast-specific cytokine, prolactin-like protein-A. Endocrinology 1999;140:2711–2720.
91. Yokoyama WM. Natural killer cells and the NK gene complex. In: Weir's Handbook of Experimental Immunology, 5th Ed., Vol. II. Cambridge, MA: Blackwell Sciences, 1996;65.1–65.21.
92. Liu C-C, Parr EL, Young JD-E. Granulated lymphoid cells of the pregnant uterus: morphological and functional features. Int Rev Cytol 1994;153:105–136.
93. Croy BA, Luross JA, Guimond M-J, Hunt JS. Uterine natural killer cells: insights into lineage relationships and functions from studies of pregnancies in mutant and transgenic mice. Nat Immun 1996;15:22–33.
94. King A, Burrows T, Loke YW. Human uterine natural killer cells. Nat Immun 1996;15:41–52.
95. Head JR. Uterine natural killer cells during pregnancy in rodents. Nat Immun 1996;15:7–21.
96. Hunt JS, Miller L, Vassmer D, Croy BA. Expression of the inducible nitric oxide synthase gene in mouse uterine leukocytes and potential relationships with uterine function during pregnancy. Biol Reprod 1997;57:827–836.
97. Croy BA, Guilbert LJ, Browne MA, Gough NM, Stinchcomb DT, Reed N, Wegmann TG. Characterization of cytokine production by the metrial gland and granulated metrial gland cells. J Reprod Immunol 1991;19:149–166.
98. Ain R, Tash J, Soares MJ. Trophoblast-derived cytokine prolactin-like protein-A: a regulator of uterine natural killer cells during the establishment of pregnancy. FASEB 2001, Orlando, FL.

IV. GLH AND THE IMMUNE RESPONSE

Growth and Lactogenic Hormones
Edited by L. Matera and R. Rapaport

Effect of Prolactin on Natural Killer and MHC-restricted Cytotoxic Cells

LINA MATERA[1], STEFANO BUTTIGLIERI[1], FRANCESCO MORO[2] and MASSIMO GEUNA[3]

[1]*Lab. of Tumor Immunology, Dept. of Internal Medicine University of Turin, Italy*
[2]*U.O.A.D.U. of Oncological Surgery, Dept. of Oncology University of Turin, Italy*
[3]*Lab. Oncological Immunology, Institute for Cancer Research and Treatment (IRCC) Candiolo Turin, Italy*

ABSTRACT

A remarkable immune property of the hormone/cytokine prolactin (PRL) is its ability to modulate the cytotoxic activity of natural killer (NK) cells. This effect is observed with purified CD56+ cell populations stimulated by PRL or with peripheral blood mononuclear cells (PBMC) stimulated by PRL plus suboptimal concentrations of IL-2. The effect of PRL on NK cells can be referred to: i) increased proliferation of the γ/δ TCR positive NK subpopulation (CD16+CD3+), ii) increased release of IFN-γ, which acts as a potent autocrine maturation factor for lymphokine activated killers (LAK) and iii) activation of IRF-1, which regulates transcription of perforin, one of the effectors molecules of cell-mediated cytotoxicity. Modulation of MHC-restricted cytotoxic T lymphocytes (CTL) by PRL has recently been described. Acquisition of antigen presenting capabilities by monocytes is increased by physiological PRL in conjunction with GM-CSF, which results in increased IFN-γ response of T lymphocytes to alloantigen. IFN-γ release and cytotoxicity of CTL sensitized *in vitro* against the tumor associated carcinoembrionic antigen (CEA) are also potentiated by the same concentrations of PRL. As a general trend, both NK cell mediated and MHC-restricted cytotoxicities are inhibited by high concentrations of PRL, so is Th1 cytokines production. CEA positive tumors produce PRL, which does not seem to be necessary for autocrine growth stimulation. It is tempting to speculate that by producing excess PRL the tumor hampers the local host tumor defense, thus creating a microenvironment favourable to its undisturbed growth.

1. INTRODUCTION

Evidence for a role of prolactin (PRL) in the regulation of immune functions has been growing at a fast rate in the past ten years, with molecular data confirming earlier functional observations. Defective development of lymphoid tissues was reversed by exogenous PRL, while lowering the levels of PRL either by hypophysectomy or by treatment with the dopamine agonist bromocriptine (BCR) affected B and T lymphocyte functions and protected animals from transplant rejection and autoimmune diseases (reviewed in Refs. [1–9]). Hemopoietic cells and lymphocytes express PRL-receptors [10–16], which are linked to signaling pathways commmon to citokines and hemopoietic factors [17–20]. Although accessory factors and populations may mediate the effect of PRL *in vivo*, *in vitro* studies with purified cell populations have firmly demonstrated that PRL acts as an autocrine/paracrine factor for immune cells and is therefore very likely to play a role at the site of the immune response. The action of PRL on lympho-hemopoiesis is best seen in conjunction with other cytokines [21,22] or hemopoietic factors [23–25]. Mimicry or cooperation as opposed to absolute requirement may underly the immunocompetence of PRL/PRL-R knockout models [26–27]. Defense of the immune system against virus-transformed and tumor cells involves both native [monocytes and NK cells] and acquired (cytotoxic T lymphocytes, CTL) cellular mechanisms. Although the correlation between tumor progression and patient's blood/peritumoral PRL has not been addressed yet, increased PRL has been described in mice after injection of tumor cells [28].

The effect of PRL on tumoricidal activity of monocytes is treated in a separate chapter [29]. Here we will review the functional interactions of PRL with NK and T lymphocytes, the effectors of MHC-unrestricted and MHC-restricted antitumor killing, respectively.

2. THE PLAYERS OF THE ANTITUMOR IMMUNE RESPONSE

2.1. Monocytes

Monocytes are at the edge between native and acquired immunity, in that they can mature to a tumoricidal state, or can phagocytose tumor cells and present the tumor antigen in a processed form to T lymphocytes [30]. PRL-R have been described on monocytes [31]. Cytofluorimetric analysis from our laboratory has shown that PRL-R, recognized by the mAb, B6.2, raised against the long form of the PRL-R [32], are expressed on 45% of $CD14^+$ cells. The frequency of PRL-R positive cells is much higher among $CD14^+$ compared to CD14-cell populations (1.5–5%), so is the density of the receptors (unpublished data).

Dendritic cells (DC), the most potent antigen presenting cells, are generated from $CD14^+$ blood monocytes under the influence of GM-CSF and IL-4. Near-physiological concentrations of PRL acted synergistically with low doses GM-CSF, thus substituting for IL-4 during the maturation of monocytes to DC [22]. This effect appeared to be secondary to the PRL-mediated regulation of the GM-CSF-R. The reverse was not true, since maturation towards the granulocyte pathway, induced by adherence followed by culturing in GM-CSF medium, or towards the DC pathway, induced by GM-SF plus IL-4, reduced to 50% the number of the CD14+ cells expressing the PRL-R (our unpublished data). PRL alone induced a DC phenotype (CD14-CD80+CD86+MHC II+) in 6-day cultured CD14+ monocytes with a dose-dependent effect peaking at 200 ng/ml [22], an unusual pattern for PRL action on immune cells.

2.2. Natural killer cells

Extensive investigations on animal or tumor/host systems strongly suggest that NK cells are early participants in the immune response and are particularly effective at eliminating blood-borne metastases. NK cells lack classical antigen receptors, but are still capable of target recognition by alternate receptors. Other receptors recognize MHC molecules on their targets. The presence of self-MHC inhibits the distruction of the target cell [33–34]. NK cells constitutively express the beta-chain of the IL-R and after culturing with IL-2 develop so called lymphokine activated killer (LAK) activity, directed against NK resistant target cells [35–37].

Killing by NK cells is accomplished by necrosis of the target cells through the release of specialized granules which contain the cytotoxic molecules perforins and granzymes. NK resistant, LAK susceptibile targets are killed through both the secretory pathway and the non-secretory Fas (CD95) apoptosis pathway. LAK cells but not NK cells express Fas-L on their membrane, which represents the agonist ligand molecule for FAS-mediated death [38–40].

Structural evidence for a PRL/NK interplay was provided by the identification of PRL-binding sites. Scathcard analysis of ^{125}I-PRL labeling showed 660 receptors/cells on NK lymphocytes with a Kd of 10^{-10} to 10^{-9} M [11], which is close to the circulating levels of the hormone. Cytofluorimetric analysis, which allows an estimate of the frequency of PRL-R bearing cells, revealed PRL-R positivity on 35–46% NK cells and on 2–5% T lymphocytes, confirming our earlier data of lower PRL-R expression on T lymphocytes [11]. The specific activation of NK and T cells by IL-2 or anti-CD3 induced a moderate augmentation of PRL-R on NK cells and a strong increase (39–60%) on T cells (unpublished data). The different expression of PRL-R on these sub-populations is in keeping with the higher responsiveness of NK cells to PRL.

The cytotoxicity of NK cells purified (>90%) from peripheral blood is significantly increased after a 16 hr incubation with PRL and this pattern of activation is bell-shaped with a peak at the upper physiological concentrations of 12–25 ng/ml [41]. This effect is accompanied by increased release of the pore-forming molecule perforin (unpublished data), the mediator of NK cytoxicity. IFN-γ is also induced, as documented by increased transcriptional activity of IFN-γ mRNA, and by the release of the protein in the medium. Anti-IFN-γ Ab blocked the enhancing effect of PRL on NK activity [42].

As mentioned, both PRL-R and IL-2-R are highly expressed on resting NK cells, which may explain functional mimicry and cooperation between their ligands. The same concentrations of PRL that enhanced NK activity (12–25 ng/ml) also induced LAK activity, which was independent of IFN-γ release and was observed only with purified NK cells [42]. In a separate study we have addressed the individual and combined effects of PRL, IL-2 and IL-12. These cytokines are potent stimulators of NK cells [43], and of IFN-γ release. IL-2 induced synthesis of IFN-γ was increased by physiological concentrations of PRL but was not affected or inhibited by high concentrations. By contrast, optimal PRL enhancement of IL-12-induced IFN-γ release was observed with T, but not NK, cells. Unexpectedly, an interaction between PRL and IL-12 only occurred at high concentrations of PRL [44].

A synergistic effect of PRL with low, ineffective doses of IL-2 was also observed on the generation of LAK cells from unfractionated peripheral blood mononuclear cells (PBMC). LAK activity was demonstrated against primary myeloid leukemia cells [45], leukemia cell lines [46] and non-Hodgkin's Burkitt lymphomas [47]. Pharmacological concentrations of PRL (200 ng/ml) were inhibitory on both NK and LAK activities. In other studies, the induction of LAK activity from peripheral blood lymphocytes (PBL) of healthy individuals and oral cancer patients was observed after 5-day incubation with 50 ng/ml PRL [48]. In our study, not only was

LAK activity against fresh myeloid cells stimulated by 12–25 ng/ml of PRL, but the susceptibility of leukemia cells to untreated LAK effectors was also increased by these concentrations [45]. The basal resistance of fresh myeloid cells to NK activity was not changed by PRL treatment. Since the FAS-L is involved in LAK-mediated but not NK-mediated killing [38–40], increased target susceptibility is likely to be secondary to PRL-induced up-regulation of the FAS receptor. Studies are ongoing to address this point.

Endogenous PRL is required during IL-2-induced LAK maturation of NK cells [21]. Functional mimicry or cooperation of PRL with IL-2 may be explained by the use of common signalling pathways, which may lead to activation of genes, like perforin, which is under the control of the transcription factor IRF-1, a common target for IL-2, IFNγ, IL-12 and PRL [49–52]. PRL is required during the IL-2 triggered proliferation of T lymphocytes [53,54] and is a growth factor for a population of purified NK cells [41,55]. We have attempted to characterize the NK subpopulations triggered by the proliferative and differentiating signals conveyed by PRL and IL-2. CD3+CD56+ NK cells belong to the subset of T lymphocytes which express the γ/δ TCR and display "NK-like" activity. They express common markers with NK cells such as CD16, CD56 and different inhibitory NK cell receptors (NKR) for HLA class I Ags. These lymphocytes respond to bacterial toxins (superantingens) and to transformed antigens, which they recognize in the context of the class I like molecule CD1a [56–60]. We studied the distribution of CD56+CD3+ and CD56+CD3- among a population of CD56+ NK cells, and the development of LAK activity against the Daudi cell line, at different times after the addition of IL-2 and/or PRL [61]. As expected, PRL alone increased the LAK activity in a bell-shaped manner with a peak at 12–25 ng/ml, IL-2 had a dose-dependent effect and low dose IL-2 acted synergistically with PRL. The percentage of CD56+CD3- phenotypes were increased by IL-2 in a dose dependent manner, while CD56+CD3+ cells were decreased. An opposite effect was exerted by PRL, with strong augmentation of the CD56+CD3+ cells at 12–25 ng/ml concentrations and decrease of the basal or IL-2 induced CD56+CD3- cell fraction at the highest concentrations teted (200 ng/ml). The increase of the CD56+CD3- cells in the IL-2 stimulated cell cultures was associated with a dose-dependent enhancement of cytotoxicity. By contrast, PRL-induced cytotoxicity was independent of CD56+CD3+/CD56+CD3- changes. However, the synergistic action of the PRL/IL-2 combinations was always accompained by an inversion of the CD56+CD3+/CD56+CD3- ratio compared to PRL alone [67]. It may thus be concluded that, i) most of functional LAK precursors reside in the CD56+CD3- subpopulation and their IL-2 induced, but not PRL induced, maturation is preceded by a mitotic phase, ii) PRL seems to act as a differentiating factor for CD56+CD3- cells and as a growth factor for CD56+CD3+ cells and iii) high concentrations of PRL inhibit both the proliferation and the functional activation of the IL-2 triggered CD56+CD3- effector cells.

2.3. PRL in the polyclonal or cytokine-mediated stimulation of T lymphocytes

The adaptive immune response is the result of coordinate signals coupled to clonally distributed receptors. After interaction with the specific antigen, T and B lymphocytes become competent to respond to further signals conveyed by growth/differentiating factors. The extent of the response is regulated by these locally acting mediators, collectively referred to as cytokines. In this context a paramount role is played by IL-2. Resting lymphocytes express the intermediate affinity (β and γ chains) IL-2-R, whereas the high affinity (α, β, γ) functional IL-2-R is only expressed after T cell activation [62]. The polyclonal activator, phytohemagglutin (PHA), is directed against the nonpolymorphic portion of the T cell receptor, the lectin Concanavalin A, and PMA, an activator of protein kinase C (PKC), all mimick antigen-mediated clonal activation of T lymphocytes, and induce IL-2 and IL-2-R expression, thereby activating autocrine T cell proliferation. *In vitro* experiments with such polyclonal lymphocyte activators showed the importance of PRL in the cytokine network. Neutralization experiments with blocking antibodies [63,64] and direct identification of endogenously produced PRL [64–66] have demonstrated that the proliferative response of peripheral blood T lymphocytes to the mitogens PHA and ConA is dependent on an autocrine stimulation by PRL. PRL has been shown to modulate the T cell antigen receptor expression and to phosphorylate the kinases activated by TCR ligation [67–68]. Activation of genes involved in T cell clonal amplification (IL-2 and IFN-γ) has also been demonstrated [12,69,70]. Receptors have been described on blood cells [10–16], which are made up of about 70% T lymphocytes. Binding experiments with labeled PRL have shown a low number (360/cell) of PRL-R on resting T cells [11], which is below the threshold of flowcytometry analysis. As mentioned above, both the number and the percentage of positive cells are increased after CD3 triggering with a specific antibody. Resting T lymphocytes are almost unresponsive to PRL alone, but proliferate in medium containing suboptimal concentrations of the cell lineage-specific stimulators [55].

3. PRL IN THE T LYMPHOCYTE CLONAL RESPONSE

T cells are the main effector cells responsible for specific long-lasting immunity. Antigen recognition by a T cell occurs via the MHC-restricted engagement of the TCR with the antigenic peptide on the surface of antigen presenting cells (APC). Specifically, MHC class II molecules bind peptides consisting of approximately 12 residues, which are derived by APC by enzymatic cleavage of exogeneous proteins and present them to CD4+ T lymphocytes. CD4+ T cells are specific for a particular MHC class II-peptide complex, whereas MHC class I molecules bind 8–10 residue peptides derived from the cleavage of endogeneous proteins and present them to the TCR on CD8+ lymphocytes [71].

Antitumor CTLs recognise nonapeptide antigen epitopes presented by MHC class I molecules on the surface of a tumor or antigen presenting cells. The identification of tumor associated antigens (TAA) [72–76] that can be specifically recognized by autologous cytotoxic T lymphocytes (CTL) has renewed the interest in immune surveillance and has provided the theoretical basis for the generation of protective anti-tumor responses in patients. So far, nearly all of the defined TAA that stimulate CTL responses consist of a short immunodominant peptide(s), which is associated non-covalently with an MHC class I molecule. These complexes, which are displayed on the surface of the tumor cells, are the ligands for specific, clonally distributed, T-cell receptors (TCR) on the surface of CTLs [72–77]. For TCR ligation to produce effective tumor killing, CTL must be primed by professional antigen presenting cells (APC). The most powerful APC are dendritic cells (DC) [78–79]. They cross-present exogenous antigens in the MHC class I and class II pathways for recognition by CD4+ and CD8+ T lymphocytes [80]. HLA-A2-restricted peptides derived from the prostate antigen PSA has been successfully employed to treat patients with hormone-refractory prostate cancer [81] and vaccination with autologous DCs pulsed with peptides derived from HER2/neu protooncogene and the epithelial mucin MUCI have been shown to induce a response in patients with breast and ovarian cancer [82,83]. Dendritic cells pulsed with a peptide specific for the carcinoembrionic (CEA) antigen have been shown to induce CEA specific CTLs [84,85] and are now on clinical trials [86].

4. EFFECT OF PROLACTIN ON CEA-SPECIFIC CD8+ T LYMPHOCYTE RESPONSE TRIGGERED BY PEPTIDE-LOADED DENDRITIC CELLS

In our previous work, the T cell response to alloantigen induced by IFN-γ was increased by additional treatment of DC with PRL [22], indicating a role for the hormone in the expression of molecules involved in the antigen-MHC/TCR interaction. Tumor progression is characterized by a shift from the Th1- to the Th2-type T cell response [87]. Factors which promote the production of cytokines that stimulate cellular immunity mediated by Th1 T cells, give protection against tumors. Although no proof exists for a direct influence of PRL on tumor progression, PRL has been shown to induce the release of IFN-γ (a Th1 cytokine) from human NK cells [42], of IL-2 from murine T cells [70]. Furthermore, PRL reversed the abnormal Th2 response of patients with atopic dermatitis to the Th1 cytokine profile [7]. That PRL may have a place in the host response to tumor is also suggested by increased PRL in mice injected with tumor cells [88].

In a recent study, we have addressed the role of PRL on the activity of long-term cultured antigen-specific CTL. Lymphocytes were sensitized against the CEA antigen, which is best expressed on gastrointestinal tumors. The nonapeptide epitopes with high binding affinity to HLA-A*0201 (HLA-A2) (Cap-1, aa:571–579, YLSGANLNL) was presented by DC which were generated *in vitro* from blood monocytes [22,89–91]. Autologous T lymphocytes from blood received three rounds of stimulation with CEA-pulsed DC. Cytotoxiciy testing followed against CEA pulsed T2 cells in a cold target inibition ^{51}Cr release assay, where the standard NK target, K562, acted as the competitor. Killer wells were then stimulated with irradiated PBMCs and IL-2 in the presence of 20 ng/ml anti-CD3 monoclonal antibody (mAb). On day 14, when the cell count/culture was approximately 1–3 × 10^6, CTLs were incubated for 24 hr in serum-free medium containing grading concentrations of PRL, before being tested for cytotoxicity in the ^{51}Cr release assay [92] and for IFN-γ production in an elispot assay [92–93].

Incubation with PRL increased the cytotoxicity against CEA-loaded/naturally expressing (KATO) target cells. The effect of PRL was maximal at 12–25 ng/ml with a second, lower peak at 100 ng/ml and a marked, consistent inhibition at 200 ng. PRL-induced CEA-CTL did not recognize unloaded/CEA negative target cells and their activity was MHC-restricted, since it was 50% inhibited in the presence of the anti-MHC class I mAb W6.32.

5. MECHANISMS OF PRL-INDUCED INHIBITION

As a general pattern, supraphysiological concentrations of PRL stimulated constitutive or induced (LAK) activity of NK cells, whereas concentrations ten-fold the physiological levels consistently inhibited both the development of LAK activity [46,55] and the proliferation of NK cells after IL-2 activation [55]. As described above, opposite effects are also exerted by supra-physiological and high concentrations of PRL on the activity of CTLs. Possible explanations for inhibition of NK mediated and MHC–restricted cytotoxicities (described here) are: i) the previously reported homologous down-regulation of receptors [94], and ii) PRL-R signaling paralysis due to formation of monomeric PRL-PRL-R complexes [95]. Indeed, effective PRL-R triggering appears to be a key event during IL-2 activation, as indicated by the increased trafficking of these receptors in IL-2 T cell clones [53] and primary NK cells [21] and inhibition of T-cell proliferation by anti-PRL Abs [63,64]. This double faceted action of PRL is likely to serve a homeostatic role in the maintenance of mature lymphocyte function under physiological conditions and also to suppress excessive lymphocyte activation during situations associated with transient hyperprolactinemia, such as pregnancy and acute stress.

6. PRL IN THE TUMOR MICROENVIRONMENT

As suggested by the *in vitro* studies described above, the concentrations of PRL in the tumor microenvironment must be kept within a narrow window (12–25 ng/ml) for antitumor host defense to be maximally effective. The results with NK cells prompted the suggestion that the *in vivo* activity of PRL can be compartmentalized, with a moderate increase above physiological levels acting on NK cells at the site of infection/tumor growth through a paracrine mechanism. High pituitary PRL release acts systemically to inhibit native or IL-2 activated NK cell function. This hypothesis found support in the observation that LAK activity in 6-day PBMC cultures stimulated with IL-2 is profoundly reduced by a neutralizing anti-PRL antiserum [21]. Although *de novo* synthesis of PRL was demonstrated in these cultures, no PRL mRNA was found in NK cells before or after IL-2 stimulation, indicating that PRL promotes LAK maturation through a paracrine mechanism.

In tumor bearing hosts excess PRL with immunosuppressive potential may also derive from the PRL synthetic activity of the tumor itself. The production of PRL by non pituitary tumor cells has been described and may provide a model to understand the role of PRL in the biology of cancer. PRL has been shown to act as an autocrine T cell growth factor, independently of IL-2, for the JURKAT human T-leukemic cell line [96]. Furthermore, PRL is produced by the lymhoblastoid cell line IM9-P3 [97], which was originally established form a patient with multiple myeloma. PRL mRNA was also detected by RT-PCR in ten of twelve non Hodgkin's lymphoma (NHL) cell lines. PRL was present in immunoprecipitates of ^{35}S-methionine-labeled cell lysates and in the supernatants, which paralleled mRNA expression revealed by Western blotting analysis. Both pituitary/lymphocyte non-glycosylated (23.5 kDa) and glycosylated (25 kDa) isoforms of PRL were present in Hodgkin's lymphoma cells. PRL-R, as defined by multiple antibodies, was detected in 3 of these cell lines. However, exogenous PRL, when tested at a wide range of concentrations (0.75–50 ng/ml), was not mitogenic on these lymphomas. Moreover, experiments with blocking antibodies showed that the NHL cell lines expressing both PRL and PRL-R were not dependent on endogenosus PRL [98,99]. Since normal T lymphocytes produce PRL when activated and PRL may be necessary for them to perform some functions, its expression by these cell lines may be part of the normal cytokine repertoire of a cell frozen at a given stage along the B/T cell pathway, rather than result from *in vitro* acquisition of aberrant capabilities. Production of immunoreactive PRL has been shown by the blasts from acute myeloid leukemia (AML) patients [100–101]. Immunochemistry showed positivity for PRL on colon carcinoma cells [102–103]. In a recent study, we demonstrated PRL mRNA expression and PRL production by one colon cancer cell line (HT-29) out of five gastrointestinal cell lines tested. When compared with early reports on the IM9-P3 cell line [97], the estimated amount of PRL produced by HT-29 cells ranged between 50 and 70 ng/ml. No expression of PRL-R was found, thus ruling out the existence of an autocrine loop. Thus, except for the lymphocyte cell line JURKAT, PRL does not seem to act as a tumor growth factor and its role in the tumor microenvironment remains to be elucidated.

CTL sensitized to widely distributed tumor associated antigens infiltrate the tumor *in vivo*. NK and LAK cells also infiltrate the tumor and their activity is likely to be influenced by the tumor microenvironment. If high concentrations of PRL are inhibitory to cytotoxic lymphocytes at the site of tumor growth as they are *in vitro*, establishing local hyperprolactinemia may be a way by which the tumor can neutralize host defense. Monitoring the influence of PRL-producing gastrointestinal tumor cells on the development of CEA specific CTL may help to elucidate this point. The simultaneous availability of a PRL-producing, CEA positive tumor cell line (HT-29) and CEA-specific, MHC-I matched CTL may allow a close estimate of the *in vivo* role of PRL in host defense against tumor.

7. CONCLUSIONS AND PERSPECTIVES

The double faceted action of PRL is likely to serve a control role in the maintenance of mature immune function under physiological conditions and suppression of immune function during situations associated with hyperprolactinemia, such as pregnancy and acute stress. As reported elsewhere in this volume, PRL enhances cell function, accelerates lymphoid and myeloid reconstitution and promotes hemopoiesis, which makes it an attractive candidate for therapeutic management of patients with various conditions. However, the clear inhibitory role of high concentrations of PRL on the effectors of antitumor cytotoxicity (NK, LAK and CTL) can not be neglected, especially now that a recombinant form of the hormone is being produced and evaluated in preclinical and clinical models of immunodeficiency and myelosuppression [104,105]. Further studies are needed to clarify the targets and to elucidate the role of PRL in various host defense mechanisms.

REFERENCES

1. Berczi I, Bertok I, Chow DA. Natural immunity and meuroimmune host defense. Ann NY Acad Sci 2000;917:248–257.
2. Berczi I, Chalmers IM, Nagy E, Warrington RJ. The immune effects of neuropeptides. Baillière's Clinical Rheumatol 1996;10:227–257.
3. Gala RR. Prolactin and growth hormone in the regulation of the immune system. Proc Soc Exp Biol Med 1991;198:513–527.
4. Kooijman R, Hooghe-Peters EL, Hooghe R. Prolactin, growyh hormon and insulin-like growth factor-1 in the immune system. Adv Immunol 1996;63:377–454.
5. Matera L. Endocrine, paracrine and autocrine actions of prolactin on immune cells. Life Sci. 1996;59:599–614.
6. Weigent DA. Immunoregulatory properties of growth hormone and prolactin. Pharmacol Ther 1996;69:237–257.
7. Matera L, Mori M, Geuna M, Buttiglieri S, Palestro G. Prolactin in autoimmunity and anitumor defence. J Neuroimmunol 2000;109:47–55.
8. Hooghe-Peters EL, Dogusan Z, Hooghe R Horseman ND. *In vitro* effects of prolactin on the lympho-hemopoietic system. Boston: Kluwer Academic, 2001;317–339.
9. Dorshkind K, Horseman ND. Anterior pituitary hormones, stress, and immune system homeostasis. Bioessays 2001;23:288–294.
10. Russell DH, Kibler R, Matrisian L, Larson DF, Poulos B, Magun BE. Prolactin receptors on human T and B lymphocytes: antagonism of prolactin binding by cyclosporin. J Immunol 1985;134:3027–3031.
11. Matera L, Muccioli G, Cesano A, Bellussi G, Genazzani E. Prolactin receptors on large granular lymphocytes. dual regulation by cyclosporin A. Brain Behav Immun

1988;2:1–10.
12. Pellegrini I, Lebrun JJ, Ali S, Kelly PA. Expression of prolactin and its receptor in human lymphoid cells. Mol Endocrinol 1992;6:1023–1031.
13. Gagnerault MC, Touraine P, Savino W, Kelly PA, Dardenne M. Expression of prolactin receptors in murine lymphoid cells in normal and autoimmune situations. J Immunol 1993;150:5673–5681.
14. Dardenne M, Leite de Moras MC, Kelly PA, Gagnerault M-C. Prolactin receptor expression in human hematopoietic tissues analyzed by flow cytofluorimetry. Endocrinology 1994;134:2108–2114.
15. Bresson JL, Jeay S, Gagnerault MC, et al. Growth hormone (GH) and prolactin receptors in human peripheral blood mononuclear cells: relation with age and GH-binding protein. Endocrinology 1999;140:3203–3209.
16. Bellone G, Geuna M, Carbone A, et al. Regulatory action of prolactin on the *in vitro* growth of CD34+ve human hemopoietic progenitor cells. J Cell Physiol 1995;163:221–231.
17. Yu-Lee L-y. Molecular actions of prolactin in the immune system. Proc Soc Exp Biol Med 1997;215:35–52.
18. Hooghe R, Devos S, Dogusan Z, Hooghe-Peters L. Signal transduction and modulation of gene expression by prolactin in human leukocytes. This issue.
19. Yu-Lee L. Signal transduction by prolactin receptors. This issue.
20. Clevenger CV, Rycyzyn MA, Syed F, Kline JB. Prolactin receptor signal transduction. In: Horseman ND, editor. Prolactin. Boston: Kluwer Academic, 2001;355–379.
21. Matera L, Bellone G, Lebrun J, et al. Role of prolactin in the *in vitro* development of interleukin-2-driven anti-tumoral lymphokine-activated killer cells. Immunology 1996;89:619.
22. Matera L, Galetto A, Geuna M, et al. Individual and combined effect of GM-CSF and prolactin on maturation of dendritic cells from blood monocytes under serum-free conditions. Immunology 2000;100:29–36.
23. Bellone G, Rollino C, Borsa S, et al. Association between elevated prolactin levels and circulating erythroid precursors in dyalized patients. Proc Soc Exp Biol Med 2000;223:367–371.
24. Bellone G, Astarita P, Artusio A, et al. Bone marrow stroma-derived prolactin is involved in basal and plateled activating factor-stimulated *in vitro* erythropoiesis. Blood 1997;90:21–27.
25. Bellone G. Prolactin as a promoter of growth and differentiation of hemopoietic cells. This issue.
26. Goffin V, Binart N, Clement-Lacroix P, et al. From the molecular biology of prolactin and its receptor to the lessons learned from knockout mice models. Genet Anal 1999;15:189–201.
27. Dorshkind K, Horseman ND. The roles of prolactin, growth hormone, insulin-like growth factor-I, and thyroid hormones in lymphocyte development and function: insights from genetic models of hormone and hormone receptor deficiency. Endocr Rev 2000;21:292–312.
28. Besedovsky HO, Sorkin E, Keller M, Muller J. Changes in blood hormone levels during the immune response. Proc Soc Exp Biol Med 1985;150:466–471.
29. Catthopadhyay U. Prolactin regulates macrophage and NK cell mediated inflammation and cytotoxic response against tumor. This issue.
30. Szabolcs P, Avigan D, Gezelter S, et al. Dendritic cells and macrophages can mature

independently from a human bone marrow-derived post-CFU intermediate. Blood 1996;87:4520–4530.

31. Bernton EW, Beach JE, Holaday JW, Smallridge RC, Fein HG. Release of multiple hormones by a direct action of interleukin-1 in pituitary cells. Science 1987;238:519–521.
32. Banerjee R, Ginsburg E, Vonderhaar BK. Characterization of a monoclonal antibody against human prolactin receptors. Int J Cancer 1993;55:712–721.
33. Whiteside TL, Herberman RB. The role of natural killer cells in immune surveillance of cancer. Curr Opin Immunol 1995;7:704–710.
34. Moretta A. Molecular mechanisms in cell-mediated cytotoxicity. Cell 90:13–8.
35. Rosemberg SA. Immunoyherapy and gene therapy of cancer. Cancer Res. 1991;51:5074s.
36. Vujanovic NL, Yasumura S, Hirabayashi H, et al. Antitumor activities of subsets of human IL-2-activated Natural Killer Cells in solid tissue. J Immunol 1995;154:281–287.
37. Vanhaesebroeck B, Grooten J, Fiers W. An ongoing *in vivo* immune response affects the abundancy and differentiation of lymphokine-activated killer cell precursors, but does not influence their broad spectrum target reactivity. J Immunol 1989;143:1396–1407.
38. Bradley M, Zeytun A, Rafi-Janajreh A, Nagarkatti PS. Role of spontaneous and interleukin-2-induced natural killer cell activity in the cytotoxicity and rejection of Fas+ and Fas– tumor cells. Blood 1998;92:4248–4255.
39. Medvedev AE, Johnsen AC, Haux J, et al. Regulation of Fas and Fas-ligand expression in NK cells by cytokines and the involvement of Fas-ligand in NK/LAK cell-mediated cytotoxicity. Cytokine 1997;9:394–404.
40. Frost P, Ng CP, Belldegrun A, Bonavida B. Immunosensitization of prostate carcinoma cell lines for lymphocytes (CTL, TIL, LAK)-mediated apoptosis via the Fas-Fas-ligand pathway of cytotoxicity. Cell Immunol 1997;180:70–83.
41. Cooperation of Pituitary Hormone Prolactin with IL-2 and IL-12 on Production of Interferon-γ by Natural Killer and T cells. NY Acad Sci, Vol. 917.
42. Matera L, Contarini M, Bellone G, Forno B, Biglino A. Up-modulation of interferon-gamma mediates the enhancement of spontanous cytotoxicity in prolactin-activated natural killer cells. Immunology 1999;98:386–392.
43. Chehimi J, Valiante NM, D'Andrea A, et al. Enhancing effect of natural killer cell stimulatory factor (NKSF/interleukin-12) on cell-mediated cytotoxicity against tumor-derived and virus infected cells. Eur J Immunol 1993;23(8):1826–30.
44. Matera L, Mori M. Cooperation of pituitary hormone prolactin with interleukin-2 and interleukin 12 on production of interferon- by Natural Killer and T cells. Ann NY Acad Sci 2001;917:505–513.
45. Oberholtzer E, Contarini M, Veglia F, et al. Prolactin increases the susceptibility of primary leukemia cells to NK and LAK effectors. Adv Neuroimmunol 1996;6:233–247.
46. Cesano A, Oberholtzer E, Contarini M, Geuna M, Bellone G, Matera L. Independent and synergistic effect of interleukin-2 and prolactin on development of T- and NK-derived LAK effectors. Immunopharmacol 1994;28:67–75.
47. Gaidano G, Contarini M, Pastore C, Saglio G, Matera L. AIDS-related Burkitt's-type lymphomas are a target for lymphokine-activated killers induced by interleukin-2 and prolactin. PSEBM 1996;213:196–205.
48. Chakraborty A, Chakraborty NG, Chattopadhyay U. Prolactin response of NK cells, but not of LAK cells, is deficient in patients with carcinoma of oral cavity and during aging. Int J Cancer 1996;66:65–69.
49. Duncan GS, Mattrucker HW, Kagi D, Matsuyama T, Mak TW. The transcription factor

interferon regulatory factor-1 is essential for natural killer cell function *in vivo*. J Exp Med 1996;184:2043–2048.

50. Ogasawara K, Hida S, Azimi N, Tagaya Y, Sato T, Yokochi-Fukuda T, Waldmann TA, Taniguchi T, Taki S. Requirement for IRF-1 in the microenvironment supporting development of natural killer cells. Nature 1998;391:700–703.
51. Dogusan Z, Hooghe-Peters EL, Berus D, Velkeniers B, Hooghe R. Expression of SOCS genes in normal and leukemic human leukocytes stimulated by prolactin, growth hormone and cytokines. J Neuroimmunol 2000;109:34–39.
52. Galon J, Sudarshan C, Ito S, Finbloom D, O'Shea JJ. IL-12 induces IFN regulating factor-1 (IRF-1) gene expression in human NK and T cells. J Immunol 1999;162:7256–7262.
53. Clevenger CV, Altmann SW, Prystowsky MB. Requirement of nuclear prolactin for interleukin-2-stimulated proliferation of T lymphocytes. Science 1991;253:77–79.
54. Prystowsky MB, Clevenger CV. Prolactin as a second messenger for interleukin 2. Immunomethods 1994;5:49–55.
55. Matera L, Cesano A, Muccioli G, Veglia F. Modulatory effect of prolactin on the DNA synthesis rate and NK activity of large granular lymphocytes. Int J Neurosci 1990;51:265–267.
56. Triebel F, Hercend. Subpopulations of human peripheral γδ T lymphocytes. Immunol Today 1989;10:186.
57. Hagmann M. A trigger of natural (and other) killers. Science 1999;285:645–647.
58. Norris S, Doherty DG, Collins C, et al. Natural T cells in the human liver: cytotoxic lymphocytes with dual T cell and natural killer cell phenotype and function are phenotypically heterogenous and include Valpha24-JalphaQ and gammadelta T cell receptor bearing cells. Hum Immunol 1999;60:20–31.
59. Katoh A, Ohshima K, Kanda M, et al. Gastrointestinal T cell lymphoma: predominant cytotoxic phenotypes, including alpha/beta, gamma/delta T cell and natural killer cells. Leuk Lymphoma 2000;39:97–111.
60. Borsellino G, Poccia F, Placido R, Tramonti D, Mancino G, Luchetti S, Galgani S, Bonetti B, Bach S, Cipriani B, Brosnan CF, Battistini L. Phenotypic and functional properties of gamma delta T cells from patients with Guillain Barre syndrome. J Neuroimmunol 2000;102:199–207.
61. Moro F, Geuna M, Triolo R, Matera L. Distinct effects of Prolactin and IL-2 on proliferation and functional maturation of CD56+CD3- and CD56+CD3+ NK subpopulations. Seventh Summer Symposiun in Molecular Biology, Pennstate, 1998;Abs. p.72.
62. Taniguchi T, Minami Y. The IL-2/IL-2 receptor system: a current overview. Cell 1993;73:5–8.
63. Hartmann DP, Holaday JW, Bernton EW. Inhibition of lymphocyte proliferation by antibodies to prolactin. FASEB J 1989;3:2194–2202.
64. Sabharwal P, Glaser R, Lafuse W, Varma S, Lui Q, Arkins S, Kooijman R, Kutz L, Kelly KW, Malakey WB. prolactin synthesized and secreted by human peripheral blood mononuclear cell: and autocrine growth factor for lymphoproliferation. Proc Natl Acad Sci 1992;89:7713–7716.
65. Montgomery DW, Shen GK, Ulrich ED, Steiner LL, Parrish PR, Zukoski CF. Human thymocytes express a prolactin-like messenger ribonucleic acid and synthesize bioactive prolactin-like proteins. Endocrinology 1992;131:3019–3026.
66. Kooijman R. Prolactin expression in the immune system. This issue.
67. Montgomery DW, Krumenacker SJ, Buckley AR. Prolactin stimulates phosphorylation

of the human T-cell antigen receptor complex and ZAP-70 tyrosine kinase: a potential mechanism for its immunomodulation. Endocrinology 1998;139:811–814.
68. Krumenacker JS, Montgomery DW, Buckley DJ, Gout PW, Buckley AR. Prolactin receptor signaling: shared components with the T-cell antigen receptor in Nb2 lymphoma cells. Endocrine 1998;9:313–320.
69. Cesario TC, Yousefi S, Carandang G, Sadati N, Le J, Varizi N. Enhanced yields of gamma interferon in prolactin treated human peripheral blood monocytes. Proc Soc Exp Biol Med 1994;205:89–95.
70. Viselli SM, Stanek EM, Mukherjee P, Hymer WC, Mastro AM. Prolactin-induced mitogenesis of lymphocytes from ovariectomized rats. Endocrinology 1991;129:983–990.
71. Matis LA. The molecular basis of T-cell specificity. Annu Rev Immunol 1990;8:65–82.
72. Finn OJ. Tumor-rejection antigens recognized by lymphocytes. Curr Opin Immunol 1993;5:701–708.
73. Boon T, Old LJ. Cancer tumor antigens. Curr Opin Immunol 1997;9:681–690.
74. Rosemberg SA. A new era for cancer immunotherapy based on the genes that encode cancer antigens. Immunity 1999;10:281.
75. Coulie P. Human tumor antigens recognized by T cells: new perspectives for anticancer vaccines. Mol Med Today 1997;3:261–268.
76. Van den Eynde BJ, Van der Bruggen. T cell defined tumor antigens. Curr Opin Immunol 1997;9:684–693.
77. Germain RN, Margulies DH. The biochemistry and cell biology of antigen processing and presentation. Annu Rev Immunol 1993;11:403–409.
78. Steinman RM. The dendritic cell system and its role in immunogenicity. Annu Rev Immunol 1991;9:271–296.
79. Banchereau J, Steinman R. Dendritic cells and the control of immunity. Nature 1998;392:245.
80. Galon J, Sudarshan C, Ito S, Finbloom D, O'Shea JJ, Ridge, JP, Di Rosa F, Matzinger PA. Conditioned dendritic cell can be a temporal bridge between a CD4+ T-helper and a T-killer cell. Nature 1998;393:474.
81. Murphy GP, Tjoa BA, Simmons SJ, Jarisch J, Bowes VA, Rodge H, Rogers M, Elgamal A, Kenny GM, Cobb OE, Ireton RC, Troychak MJ, Salgaller ML, Boynton AL. Infusion of dendritic cells pulsed with HLA-A2 specific prostate membrane antigen peptides: a phase II prostatic cancer vaccine trial involving patients with hormone-refractory metastatic disease. The Prostate 1999;38:73–8.
82. Bremers AJ, Andreola S, Leo E, Gallino F, et al. T cell responses in colorectal cancer patients: evidence for class II HLA-restricted recognition of shared tumor-associated antigens. Int J Cancer 2000;88:956–961.
83. Brossart P, Wirths S, Stuhler G, Reichardt VL, Kanz L, Brugger W. Induction of cytotoxic T-lymphocyte responses *in vivo* after vaccinations with peptide-pulsed dendritic cells. Blood 2000;96:3102–3108.
84. Alters SE, Gadea JR, Sorich M, et al. Dendritic cells pulsed with CEA peptide induce CEA specific CTL with restricted TCR repertoire. J Immunother 1998;21:17–26.
85. Tsang KY, Zaremba S, Nieroda CA, Zhu MZ, Hamilton JM, Schlom J. Generation of human cytotoxic T cells specific for human carcinoembryonic antigen epitopes from patients immunized with recombinant vaccinia-CEA vaccine. J Natl Cancer Inst 1995;87:982.

86. Morse MA, Deng Y, Coleman D, Hull S, Kitrell-Fisher E, Nair S, Schlom J, Ryback ME, Lyerly HK. A phase I study of active immunotherapy with carcinoembryonic antigen peptide (CAP-1)-pulsed, autologous human cultured dendritic cells in patients with metastatic malignancies expressing carcinoembryonic antigen. Clin Cancer Res 1999;5:1331–1338.
87. Chouaib S, Asselin-Paturel, Mami-Chouab F, Caignar A, Blay JY. The host-tumor immune conflict: from immunosuppression to resistance and desctruction. Immunol Today 1997;18:493–497.
88. Besedovsky HO, Sorkin E, Keller M, Muller J. Changes in blood hormone levels during the immune response. Proc Soc Exp Biol Med 1985;150:466–471.
89. Sallusto F, Lanzavecchia A. Efficient presentation of soluble antigen by cultured human dendritic cells is maintained by granulocute-macrophage colony stimulating factor plus interleukin 4 and downregulated by tumor necrosis factor alpha. J Exp Med 1994;179:1109–1994.
90. Zhou LJ, Tedder TF. CD14+ blood monocytes can differentiate into functionally mature CD83+ dendritic cells. Proc Natl Acad Sci 1996;93:2588.
91. Romani N, Reider D, Heuer M, Ener S, Eibl B, Schuler G. Generation of mature dendritic cells from human blood: an improved method with special regard to clinical applicability. J Immunol Methods 1996;196:137.
92. Romero P, Cerottini JC, Waanders GA. Novel methods to monitor antigen-specific cytotoxic T-cell responses in cancer immunotherapy. Mol Med Today 1998;4:305.
93. Herr W, Linn B, Leister N, Wandel E, Meyer Zum Buschenfelde KH, Wolfel T. The use of computer-assisted video image analysis for the quantification of CD8+ T lymphocytes producing tumor necrosis factor alpha spots in response to peptide antigens. J Immunol Methods 1997;203:141.
94. Matera L, Ciccarelli E, Muccioli G, Cesano A, Grottoli S, Oberholtzer E, Camanni F. Normal development of lymphokine activated killing (LAK) in peripheral blood lymphocytes from hyperprolactinemic patients. Int J Immunopharmacol 1992;14(7):1235–40.
95. Wells JA, de Vos AM. Structure and function of human growth hormone: implication for the hematopoietin. Ann Rev Biophys Miomol Structures 1993;22:329–351.
96. Matera L, Cutufia M, Geuna M, Contarini M, Buttiglieri S, Galin S, et al. Prolactin is an autocrine growth factor for the Jurkat human T-leukemic cell line. J Neuroimmunol 1997;79:12–21.
97. Di Mattia GE, Gellersen B, Bohnet HG, Friesen HG. A human B-lymphoblastoid cell line produces prolactin. Endocrinology 1988;122:2508–2518.
98. Hooghe R, Merchav S, Gaidano G, Naessens F, Matera L. A role for growth hormone and prolactin in leukaemia and lymphoma? Cell Mol Life Sci 1998;54:1095–1101.
99. Matera L, Geuna M, Pastore C, Buttiglieri S, et al. Expression of prolactin and prolactin receptors by non-hodgkin's lymphoma cells. Int J Cancer 2000;85:124–130.
100. Hatfill SJ, Kirby R, Hanley M, Rybicki E, Bohm L. Hyperprolactinemia in acute myeloid leukemia and indication of ectopic expression of human prolactin in blast cells of a patient of subtype M4. Leuk Res 1990;14:57–62.
101. Kooijman R, Gerlo S, Coppens A. Myeloid leukemic cells express and secrete bioactive pituitary-sized 23 kDa prolactin. J Neuroimmunol 2000;110:252–258.
102. Bhatavdekar J, Vora H, Balar D, Patel D, et al. Interrelationship of prolactin and its receptor in carcinoma of colon and rectum:a preliminary report. J Surg Oncol 1994;55:246–249.
103. Ilan Y, Sibersky O, Livni N, Gofrit O, Barack V, Goldin E. Plasma and tumor prolactin in

colorectal cancer patients. Dig Dis Sci 1995;40:2010–2015.

104. Richards SM. Human prolactin as an immunohematopoietic factor: implications for the clinic. This issue.

105. Woody MA, Welniak LA, Sun R, Tian ZG, Henry M, Richards S, et al. Prolactin exerts hematopoietic growth-promoting effects *in vivo* and partially counteracts myelosuppression by azidothymidine. Exp Hematol 1999;27:811–816.

Growth and Lactogenic Hormones
Edited by L. Matera and R. Rapaport

In Vivo Changes of PRL Levels During the T-cell Dependent Immune Response

CAROLINA PEREZ CASTRO[1], MARCELO PÁEZ PEREDA[2], JOHANNES M.H.M. REUL[2], GÜNTER K. STALLA[2], FLORIAN HOLSBOER[2] and EDUARDO ARZT[1,3]

[1]*Laboratorio de Fisiología y Biología Molecular. Departamento de Biología, FCEN, Universidad de Buenos Aires, Argentina*
[2]*Max-Planck Institute of Psychiatry, Clinical Institute, Kraepelinstr. 2–10, 80804 Munich, Germany*
[3]*Member of the Argentine National Research Council (CONICET)*

ABSTRACT

The functional interaction between the immune and neuroendocrine system is mediated by humoral mediators, cytokines, neurotransmitters, and hormones. Indeed, the immune response is accompanied by homeostatic changes in the neuroendocrine system. The neuroendocrine changes produced by T-cell independent antigens are very different from those produced by T-cell dependent antigens. Several lines of evidence indicate that prolactin (PRL) and thyrotrophin releasing hormone (TRH) play important roles in T-cell dependent immune responses.

This review focuses on the immunoregulatory functions of PRL, which is required for the induction of T-cell dependent immune response.

1. INTRODUCTION

The immune response is accompanied by homeostatic changes in the neuroendocrine system. The neuroendocrine changes occurring during the course of a T-cell dependent immune response are profoundly different from those occurring during the T-cell independent and inflammatory responses. The role of prolactin (PRL) and thyrotrophin-releasing hormone (TRH) in T-cell dependent immune response will be discussed in detail in this review.

There are many reports indicating the influence of PRL and TRH on the immune response. TRH, a tripeptide, is the major hypothalamic releasing factor for thyroid stimulating hormone (TSH) and it is the best known PRL-stimulatory factor. TRH-secreting neurons are located in the medial portions of the paraventricular nuclei of the hypothalamus and their axons terminate in the medial portion of the external layer of the median eminence [1] and stimulate vagal efferent fibers [2]. Human peripheral blood mononuclear cells and rat splenocytes express TRH-receptor (TRH-R) mRNA [3]. It is also known that TRH enhances the *in vitro* plaque-forming cell response [4]. In healthy humans an increase of IL-2 serum levels was reported after TRH injection during a standard TRH test [5]. In addition, TRH injected into rats stimulated thymocyte [6] and splenocyte [3] proliferation. Similarly, PRL plays an important role as an immunomodulator as discussed in detail in other chapters of this issue. Here we present how PRL levels change during the T-cell dependent immune response and its relevance to the induction of the T cell-immune responses (e.g. during the T-cell clonal expansion phase).

2. THE T-CELL IMMUNE RESPONSE MODEL

Many of the mechanisms involved in the immune-neuroendocrine interactions have been characterized by the administration of the bacterial agent lipopolysaccharide (LPS). LPS is a very particular mitogen and immunogen, which causes a T cell-independent response mediated by macrophage and B lymphocyte activation. In contrast to T-cell dependent antigens, LPS does not induce clonal expansion of T cells.

There are few neuroendocrine reports arising from studies on T-cell dependent responses. One of the best known T-cell dependent antigens utilized in these studies is sheep red blood cells (SRBC). This antigen does not produce sickness or anomalies that would interfere with the interpretation of the results.

When SRBC is injected into rats, a high level of anti-SRBC antibody production is observed at 7 days post-immunization. At 6 h post-immunization there was an increase in IL-2 levels, whereas at 7 days IL-2 levels in SRBC-injected animals did not differ from control animals.

In a first description using SRBC to test neuroendocrine changes, a decrease of noradrenaline turnover in the hypothalamus of rats was observed at the peak of the immune response to SRBC [7]. Changes in serotonin levels were reported, following SRBC immunization [8].

In contrast to the early hypothalamic-pituitary-adrenal axis (HPA) activation observed with LPS, SRBC led to an increase of both antibody and corticosterone levels within 5 to 7 days of the injection of antigen [9]. Accordingly, we confirmed that plasma corticosterone levels were significantly elevated at seven days after SRBC immunization, as compared to saline-injected control animals [10].

3. PRL AND TRH CHANGES DURING A T-CELL RESPONSE

In SRBC-immunized animals, the level of plasma PRL was significantly increased in comparison to control animals at 2, 6, and 24 h post-immunization. However, at 4 and 7 days post immunization, plasma PRL levels were not different in SRBC-injected animals and controls. TSH and GH plasma levels showed no significant changes throughout the entire SRBC experiment [10].

Increased levels of TRH mRNA at 4, 6, 18 and 24 h post-immunization was observed in SRBC-injected animals [10]. No changes in TRH mRNA levels were detected at 2 h or >24 h post-immunization. The SRBC-induced effects observed on TRH mRNA levels were opposite to those induced by LPS, which induces a decrease in hypothalamic TRH mRNA [11]. In addition, pituitary TRH-R mRNA levels at 6 h post-SRBC-immunization were markedly elevated, which decreased gradually by 24 h and 4 days, without changes at 7 days post-immunization [10].

The conclusion of these experiments is that the effect of LPS on TRH and PRL levels is markedly different from that of SRBC. Particularly, the TRH and PRL early peak observed during the T-cell dependent antigen response is absent in the response to LPS. In accordance with this, it was demonstrated that IL-1 (induced by LPS) inhibits both TRH mRNA [11] and PRL [12] accumulation.

Several cytokines (among them IL-2) could be a candidate for this difference. IL-2 is induced during the T-cell proliferation or locally in the CNS and may be inducing the TRH and/or PRL response. Supporting this, the presence of receptors for IL-2 in lacto/somatotrophic cells in the pituitary has been demonstrated [13–16]. Particularly, PRL-secreting cells have been shown to express the highest levels of IL-2 receptors amongst anterior pituitary cells [14]. *In vitro*, IL-2 induced PRL synthesis [17], which is inhibited by dopamine [18]. A combination of dopamine inhibition and TRH stimulation could be acting to stimulate PRL release.

Many mechanisms may be responsible for TRH and PRL induction during the immune response. It is most probable that the signals that induce the first peak at two hours, and also activate the hypothalamic TRH gene, are coming from the nervous system. It is known that SRBC is mainly processed in the spleen, which is richly innervated by the autonomous nervous system through afferent and efferent fibers, particularly sympathetic and peptidergic [19]. Furthermore, fibers from the paraventricular nucleus of the hypothalamus project to the periphery through neurons in the spinal cord [20]. It is probable that CNS-induced cytokines and other factors [21], mediate the TRH gene induction and PRL release through this pathway.

4. IN VIVO EVIDENCE INDICATING THAT INCREASED PRL AND TRH LEVELS ARE INSTRUMENTAL FOR THE T-CELL RESPONSE

The absence of a TRH and PRL peak during the LPS-IL-1 inflammatory response can be attributed to the fact that the inflammatory response does not require these hormones for its activation, as is the case for the IL-2-T-cell dependent response. This fact strongly supports the specificity and importance of TRH and PRL for T-cell dependent immune responses.

In rats the anti-SRBC antibody production was inhibited by the i.c.v. injection of antisense oligonucleotide complementary to rat TRH mRNA. The primary antibody response (IgM, IgG2a, IgG1) was severely suppressed in TRH-antisense-treated animals. A more pronounced inhibition of IgG1 was observed in comparison to IgG2a, that may reflect a greater influence of antisense treatment on the TH 2 response. The PRL levels of the TRH-antisense-treated animals, following SRBC immunization, did not increase as did in the SRBC-treated control rats [10]. These findings demonstrate the fundamental role of the early activation of the TRH-PRL axis in the primary anti-SRBC response.

As another validation of the instrumental role of PRL in the T-cell response, it has been demonstrated that treatment of rats with bromocriptine, that inhibits PRL secretion, decreased the contact sensitivity skin reaction and antibody formation to SRBC [22] and that following transplantation of chemically-induced tumors in syngenic rats an early increase in PRL is induced [23]. Hypophysectomized (hypox) rats and mice have been shown to exhibit decreased antibody response, a prolongation of graft survival [24], a decrease in lymphocyte proliferation [25] and a reduction in spleen natural killer cell activity [26]. Anterior pituitary transplantation to the kidney capsule (that results in large increase in PRL) restores the production of IgG and IgM antibodies in SRBC-injected hypox rats [27]. Similarly, *in vivo* administration of the inhibitor of PRL secretion, bromocriptine, inhibits T-cell proliferation [22]. PRL is necessary, but not sufficient, for lymphocyte proliferation; in interleukin-2- (IL-2), IL-4 or Concanavalin A (Con A) driven lymphocytes anti-PRL antisera inhibits *in vitro* proliferation [28–30] and nuclear translocation of PRL was demonstrated [31–33]. Furthermore, and as detailed in other chapters, the celular basis for PRL action are well established. The PRL receptor is expressed ubiquitously in all immune cells [34–36], and PRL may be produced by T lymphocytes [35,36]. PRL induces up-regulation of IL-2 receptors [37] and the molecular mediators of PRL action on T-lymphocyte activation in vitro were also described [38–46].

5. A MODEL FOR NEUROENDOCRINE CHANGES AND ROLE DURING THE T-CELL IMMUNE RESPONSE

The neuroendocrine changes, and their instrumental and differential role during T-cell dependent immune responses is summarized in the model proposed in Figure 1. In the first 24 h the T-cell dependent response leads to a peak in hypothalamic TRH mRNA and plasma PRL, whereas this circuit is inhibited during the LPS-IL-1 dependent response. This peak is critical for mounting an adequate response, which is inhibited by its blockade. In coordination with this, during the T-cell dependent response the immunosuppressive HPA-corticosterone axis is only activated during the late phase (e.g. at 5–7 days). It has been shown for another T-cell dependent antigen, phosphocholine-keyhole limpet hemocyanin (PC-KLH), that the response of the HPA axis depends on the dose: at low doses it does not induce any change, while at high doses it induces an activation of the HPA axis at day 5, which correlates with the peak of the anti-PC-KLH antibody response [47]. Thus, in contrast to the acute effect of the LPS-IL-1-inflammatory response, the immunosupressive effect of glucocorticoids appears, in the T-cell dependent response, concomitant with the peak of antibody production. The HPA axis contributes to the termination of the response and to the suppression of the late activation of non-specific clones.

The corticotrophin release inhibiting factor, present in the prepro-TRH gene, that inhibits ACTH secretion [48,49] may provide an explanation for the different response of the HPA axis to the inflammatory/T-cell-independent and the T-cell-dependent stimulation. Thus the TRH gene may also be responsible for coordinating PRL release and the inhibition of the HPA axis during this early activation phase and its elevation in the late phase, when TRH expression declines. In the LPS inflammatory/T-cell independent response, not only is the TRH/PRL activation not necessary for T-cell clonal expansion, but the regulation of inflammatory mediators (i.e. IL-1) by glucocorticoids is of immediate importance.

In conclusion, PRL and TRH play an important role during the induction (T-cell clonal expansion phase) of the primary immune response to T cell-dependent antigens. During the course of the T-cell dependent response the clonal expansion of T-lymphocytes is critically dependent on the coordinated response of the TRH/PRL axis. This is followed by the activation of the HPA axis, which is inhibitory.

ACKNOWLEDGEMENTS

This work was supported by grants from the University of Buenos Aires (UBA), the CONICET and Agencia Nacional de Promoción Científica y Tecnológica-Argentina.

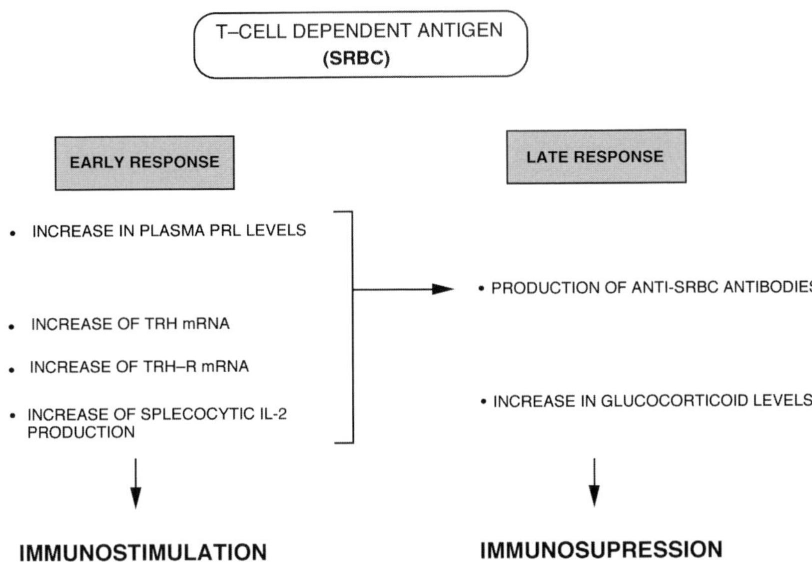

Figure 1. A model for the role of neuroendocrine changes and PRL levels occurring during the T-cell dependent antigen response is presented. Details are discussed in the text.

REFERENCES

1. Toni R, Jackson IM, Lechan RM. Thyrotropin-releasing-hormone-immunoreactive inervation of thyrotropin-releasing-hormone-tuberoinfundibular neurons in rat hypothalamus: anatomical basis to suggest ultrashort feedback regulation. Neuroendocrinology 1990;52:422–428.
2. Masuda M, Kanai S, Miyasaka K. Central somatostatin prevents vagal efferent nerve excitation produced by TRH but not by 2-deoxy-D-glucose. Am J Physiol 1997;272:G351–G356.
3. Raiden S, Polack E, Nahmod V, Labeur M, Holsboer F, Arzt E. TRH receptor on immune cells: *in vitro* and *in vivo* stimulation of human lymphocyte and rat splenocytes DNA synthesis by TRH. Journal Clin Immunol 1995;15:242–249.
4. Kruger TE, Smith LR, Harbour DV, Blalock JE. Thyrotropin: an endogenous regulator of the *in vitro* immune response. J Immunol 1989;142:744–747.
5. Komorowski J, Stepien H, Pawlikowski M. Increased interleukin-2 levels during standard TRH test in man. Nueropeptides 1994;27:151–156.
6. Pawlikowski M, Zerek-Melén G, Winczyk K. Thyroliberin (TRH) increases thymus cell proliferation in rats. Nueropeptides 1992;23:199–202.
7. Besedovsky H, Del Rey A, Sorkin E, Da Prada M, Burri R, Honegger C. The immune response evokes changes in brain noradrenergic neurons. Science 1983;221:564–565.
8. Gardier AM, Kachaner S, Shaghaghi EK, et al. Effects of primary immune response to T-cell dependent antigen on serotonin metabolism in frontal cortex: in vivo microdialysis study in freely moving Fischer 344 rat. Brain Research 1994;645:150–156.
9. Besedovsky H, Sorkin E, Kelier M, Muller J. Changes in blood hormone levels during the immune response. Proc Soc Exp Med 1975;150:466–502.

10. Perez Castro C, Peñalva R, Páez Pereda M, et al. Early activation of thyrotropin-releasing-hormone and prolactine plays a critical role during a T cell-dependent immune response. Endocrinology 1999;140:690–697.
11. Kakusca I, Romero L, Clark B, et al. Suppression of thyrotropin-releasing hormone gene expression by interleukin-1β in the rat: implications for nonthyroidal illness. Neuroendocrinol 1994;59:129–137.
12. Bernton E, Beach JE, Holaday JW, Smallridge RC, Fein HG. Release of multiple hormones by a direct action of interleuquin-1 on pituitary cells. Science 1987;238:519–521.
13. Arzt E, Buric R, Stelzer G, et al. Interleukin involvement in anterior pituitary cell growth regulation: effects of IL-2 and IL-6. Endocrinology 1993;132:459–467.
14. Arzt E, Sauer J, Buric R, Stalla J, Renner U, Stalla GK. Characterization of interleukin-2 (IL-2) receptor expression and action of IL-2 and IL-6 on normal anterior pituitary cell growth. Endocrine 1995;3:113–119.
15. Arzt E, Stalla G. Cytokines. autocrine and paracrine roles in the anterior pituitary. Neuroimmunomodulation 1996;3:28–34.
16. Arzt E, Páez Pereda M, Perez Castro C, Pagotto U, Renner U, Stalla GK. Pathophysiological role of the cytokine network in the anterior pituitary gland. Frontiers in Neuroendocrinology 1999;20:71–95.
17. Karanth S, McCann SM. Anterior pituitary hormone control by interleukin-2. Proc Natl Acad Sci USA 1991;88:2961–2965.
18. Karanth S, McCann MS. Influence of dopamine on the altered release of prolactin, luteinizing hormone and follicle stimulating hormone induced by interleukin-2 in vitro. Neuroendocrinology 1992;56:871–880.
19. Felten DL, Felten SY, Carlson SL, Olschowka JA, Livnat S. Noradrenergic and peptidergic innervation of lymphoid tissue. J Immunol 1985;135:755s–765s.
20. MacNeil BJ, Jansen AH, Greenberg AH, Nance DM. Activation and selectivity of splenic sympathetic nerve electrical activity response to bacterial endotoxin. Am J Physiol 1996;270:R264–R270.
21. Merrill J. Macroglia: neural cells responsive to lymphokines and growth factors. Immunology Today 1987;8:146–150.
22. Nagy E, Berczi I, Wren GE, Asa SL, Kovacs K. Immunomodulation by bromocriptine. Immunopharmacology 1983;6:231–243.
23. Besedovsky HO, Del Rey A, Schardt M, et al. Changes in plasma hormone profiles after tumor transplantation into syngeneic and allogeneic rats. Int J Cancer 1985;36:209–216.
24. Comsa J, Leonhardt H, Schwarz JA. Influence of the thymus-corticothropin-growth hormone interaction on the rejection of skin allografts in the rat. Ann NY Acad Sci 1975;249:387–401.
25. Prentice ED, Lipscomb H, Metcalf WK, Sharp JG. Effects of hypophysectomy on DNCB-induced contact sensitivity in rats. Scand J Immunol 1976;5:955–961.
26. Cross RJ, Markesbery WR, Brooks WH, Roszman TL. Hypothalamic-immune interactions: neuromodulation of natural killer activity by lesioning of the anterior hypothalamus. Immunology 1984;51:399–405.
27. Berczi I, Nagy E, Kovacs K, Horvath E. Regulation of humoral immunity in rats by pituitary hormones. Acta Endocrinol 1981;98:506–513.
28. Hartmann DP, Holaway JW, Bernton EW. Inhibition of lymphocyte proliferation by antibodies to prolactine. FASEB J 1989;3:2194–2202.
29. Clevenger CV, Russell DH, Appasamy PM, Prystowsky MB. Regulation of interleukin-2

driven T lymphocyte proliferation by prolactin. Proc Natl Acad Sci USA 1990;87:6460–6464.

30. Clevenger CV, Sillman AL, Hanley-Hyde J, Prystowsky MB. Requirement for prolactin during cell cycle regulated gene expression in cloned T-lymphocytes. Endocrinology 1992;130:3216–3222.
31. Clevenger C, Altmann SW, Prystowsky MB. Requeriment of nuclear prolactin for interleukin-2-stimulated proliferation of T lymphocytes. Science 1991;253:77–79.
32. Rycyzyn MA, Reilly SC, O'Malley K, Clevenger CV. Role of cyclophilin B in prolactin signal transduction and nuclear retrotranslocation. Mol Endocrinol 2000;14:1175–1186.
33. Clevenger CV, Rycyzyn MA. Translocation and action of polypeptide hormones within the nucleus. Relevance to lactogenic transduction. Adv Exp Med Biol 2000;480:77–84.
34. Clevenger CV, Freier DO, Kline JB. Prolactin receptor signal transduction in cells of the immune system. J Endocrinol 1998;157:187–197.
35. Pelligrini I, Lebrun JJ, Ali S, Kelly PA. Expression of prolactin and its receptor in human lymphoid cells. Mol Endocrinol 1992;6:1023–1031.
36. Matera L, Geuna M, Pastore C, Buttiglieri S, Gaidano G, Savarino A, Marengo S, Vonderhaar BK. Expression of prolactin and prolactin receptors by non-Hodgkin's lymphoma cells. Int J Cancer 2000;85:124–130.
37. Murkherjee P, Mastro AM, Hymer WC. Prolactin induction of interleukin-2 receptors on rat splenic lymphocytes. Endocrinology 1990;126:88–93.
38. Clevenger CV, Medaglia MV. The protein tyrosine kinase P59fyn is associated with prolactin (PRL) receptor and is activated by PRL stimulation of T-lymphocytes. Mol Endocrinol 1994;8:674–681.
39. Clevenger CV, Ngo W, Sokol DL, Luger SM, Gewirtz AM. Vav is necessary for prolactin-stimulated proliferation and is translocated into the nucleus of a T-cell line. J Biol Chem 1995;270:13246–13253.
40. Montgomery DW, Krumenacker JS, Buckley AR. Prolactin stimulates phosphorylation of the human T-cell antigen receptor complex and ZAP-70 tyrosine kinase: a potential mechanism for its immunomodulation. Endocrinology 1998;139:811–814.
41. Matsumoto A, Seki Y, Kubo M, et al. Suppression of STAT5 functions in liver, mammary glands, and T cells in cytokine-inducible SH2-containing protein 1 transgenic mice. 1999;19:6396–6407.
42. Belkowski SM, Levine JE, Prystowsky MB. Requirement of PI3-kinase activity for the nuclear transport of prolactin in cloned murine T lymphocytes. J Neuroimmunol 1999;94:40–47.
43. Gjorloff-Wingren A, Saxena M, Han S, et al. Subcellular localization of intracellular protein tyrosine phosphatases in T cells. Eur J Immunol 2000;30:2412–2421.
44. Luo G, Yu-Lee L. Stat5b inhibits NFkappaB-mediated signaling. Mol Endocrinol 2000;14:114–123.
45. Berchtold S, Volarevic S, Moriggl R, Mercep M, Groner B. Dominant negative variants of the SHP-2 tyrosine phosphatase inhibit prolactin activation of Jak2 (janus kinase 2) and induction of Stat5 (signal transducer and activator of transcription 5)-dependent transcription. Mol Endocrinol 1998;12:556–567.
46. Clevenger CV, Torigoe T, Reed JC. Prolactin induces rapid phosphorylation and activation of prolactin receptor-associated RAF-1 kinase in a T-cell line. J Biol Chem 1994;269:5559–5565.
47. Stenzel-Poore M, Vale WW, Rivier C. Relationship between antigen-induced immune

stimulation and activation of the Hypothalamic-pituitary-adrenal axis in the rat. Endocrinology 1993;132:1313–1318.
48. Redei E, Hilderbrand H, Aird F. Corticotropin release inhibiting factor is encoded within prepro-TRH. Endocrinology 1995;136:1813–1816.
49. Redei E, Hilderbrand H, Aird F. Corticotropin release-inhibiting factor is preprothyrotropin-releasing hormone-(178–199). Endocrinology 1995;136:3557–3563.

Growth and Lactogenic Hormones
Edited by L. Matera and R. Rapaport

Prolactin Regulates Macrophage and NK Cell Mediated Inflammation and Cytotoxic Response Against Tumor

UTPALA CHATTOPADHYAY and RATNA BISWAS

Department of Immunoregulation and Immunodiagnosis, Chittaranjan National Cancer Institute, 37, S.P.Mukherjee Road, Kolkata – 700 026, India

ABSTRACT

Prolactin (PRL) is known to stimulate lymphocyte growth and function, but its role in regulating innate immune response is not clear. We show that PRL, as a secondary signal, can induce tumoricidal macrophages in mice and augment tumor target killing by human monocytes and NK cells. PRL also induces IL-1, NO_2^-, and O_2^- release by the macrophages. Through release of IL-12 and IFN-γ, PRL regulates activation of monocytes/macrophages and NK cells. PRL mediated activation of monocytes/macrophages is inhibited by IL-4 and in malignancy. A defect in PRL-receptor expression by macrophages may occur in malignancy.

1. INTRODUCTION

Prolactin (PRL), a 24kDa single chain peptide, plays a key role in neuroendocrine regulation of immune functions [1–5]. Pituitary is the major source of the hormone, but PRL is also secreted by lymphocytes, decidua and placenta [6–15]. The difference in neucleotide sequence of the extra-pituitary PRL and pituitary PRL and regulation of expression of non pituitary and pituitary PRL-RNA transcripts suggest that a functional difference may exist between these two groups of PRL [15,16]. Autocrine regulation of PRL has been shown in pituitary derived GH3 cells [3,4,16,17]. It is not clear how the extra pituitary and pituitary PRL act in concert to regulate immune functions. The involvement of PRL in maintaining immune responses is well established from the classical observation of restoration of impaired immune functions in hypophysectomized or bromocriptine treated animals by PRL [1,18–22] and association between hyperprolactenimia and autoimmune disorders [23–26]. Synthesis of PRL by lymphoid cells, expression of PRL mRNA in IM-9 B lymphoblastoid cells and Jurkat leukemic cells [2,3,15–17,27], expression of PRL receptors (PRL-R) on T cells, B cells and monocytes/macrophages [28–34], and PRL induced regulation of proliferation, differentiation and/or apoptosis of immune cells, suggest that PRL may act as a cytokine in activation of immune cell functions [3–5,35–38]. The immunoregulatory role of PRL was questioned by the observation that in PRL -/- mice, rejection of allogenic tumor cells and other immune functions are not impaired [39]. However, the existence of cytokine receptors with homology to prolactin receptors (PRL-R) [40–42] and their reactivity with PRL suggest that PRL may affect immune cells even in the absence of PRL-R [39]. PRL received attention as a cytokine with the demonstration of PRL-R as a member of cytokine / haematopoetin receptor superfamily. PRL–R is present in various isoforms which differ in length and sequence of cytoplasmic tail and signaling property [43–48]. Like other cytokines, binding of PRL results in homodimerization of its receptors and PRL induced cellular signaling involves JAK/STAT pathway. JAK2/STAT5 is the major signaling cascade described for PRL-R [49–56]. Activation of common signaling molecules by PRL and IL-2 is correlated with the observed amino acid sequence homology between PRL-R and IL-2 R beta chain [56,57,40,41]. PRL induced IL-2 R expression in murine splenocytes, T lymphocytes, and human peripheral blood T and B lymphocytes [58–61] and the hormone mediated costimulation of IL-2 and mitogen induced proliferation [62] support the view that PRL regulates immune responses in concert with IL-2 . PRL treated T lymphocytes produce IL-2 in dose and time dependent fashion and IL-2 has been shown to translocate PRL-R complex to the nuclear periphery [35,63]. Both IL-2 and PRL activate STAT5 transcription factor and specific gene expression in Nb2, YT and C169 T lymphocytes [64,65,66]. STAT5 DNA recognition site is the same as interferon gamma activated site (GAS) in the IRF-1 gene [67–71]. GAS is necessary response to of both IL-2 and PRL. IRF-1 is a multifunctional transcription factor induced by many signals such as IL-2, IFN-γ and PRL [2]. Abrogation or delayed rejection of allogenic tumor in presence of anti IFN-γ [72] and PRL induced secretion of IFN-γ from peripheral blood mononuclear cells [73] indicated that PRL regulates immune functions through IFN-γ also. These observations suggest that PRL may work as a component of a network of TH-1 cytokines which regulates cellular immune responses in diseases including cancer.

2. PRL STIMULATES INNATE IMMUNE RESPONSES TO TUMORS

Endogenous hormones and cytokines are important regulators of tumor cell growth and survival. Distinct mechanisms that have been identified so far as anti-tumor responses of hosts, that are regulated by cytokines, include cytostasis, necrosis and apoptosis of tumor cells mediated by macrophages, NK and LAK cells, and generation of specific T cell immunity [74,75]. The TH1 cytokines play a major role in regulating cellular immune responses. In natural and innate immune responses, dominant TH-1 cytokine profile exists as NK cells produce IFN-γ while macrophages and dendritic cells (DC) produce IL-12. In malignancy a TH1/TH2 imbalance may occur, which results in faulty immune responses to the tumor [76]. PRL as a component of TH-1 cytokine network may play a role in regulating cellular immune responses to tumor and in this review we discuss the role of PRL in regulation of host's natural defence mechanisms against tumor mediated by monocytes/macrophages and NK cells.

2.1. PRL regulates monocyte/macrophage mediated anti-tumor responses

Monocytes/macrophages perform multiple functions contributing to different aspects of host defence against tumor [77]. These cells are involved in inflammatory responses through secretion of pro-inflammatory cytokines such as interleukin-1 (IL–1) and TNF [78,79]. Macrophages play a great role in tumor growth associated inflammatory reaction and presentation of tumor antigens to T cells. Classical macrophages require cell surface expression of IL-1 in order to activate T cells [79]. Through secretion of IL-12, the macrophages induce TH-1 cytokines, which activate NK cells and induce anti-tumor T lymphocyte response. The resting monocytes/macrophages are not cytotoxic but with appropriate stimulation become activated and kill tumor cells through release of TNF and other cytotoxic mediators such as nitric oxide (NO_2^-) and superoxide anion (O_2^-) [80].

2.1.1. *PRL differentially regulate IL-1 release by inflammatory and tumor associated macrophages*

PRL has been shown to blunt the inflammatory response associated with cell and organ depression. The hormone decreased the expression of IL-1 and TNF α in Kupffer cells [81] and reduced IL-1α release from Staphylococcal enterotoxin A activated murine splenocytes also [82]. However, PRL induced downregulation of IL-1 and TNFα secretion is a dose and time dependent phenomenon. Recently Kumar et al. [83] have shown that PRL (10–200ng/ml) enhanced IL-1 release by resident peritoneal macrophages and LPS activated macrophages. In contrast, we observed that resident peritoneal macrophages (RM), when treated with PRL (10–50 ng/ml) alone , thymocyte comitogenic activity of the culture supernatants remained similar to that of untreated control (Figure 1). This observation suggested that PRL is unable to induce IL-1 release by RM, which are resting cells. PRL mediated IL-1 release was also studied in tumor associated macrophages (TAM) in Erlichs ascitic carcinoma (EAC) bearing mice [84]. The TAM are the activated cells of monocytes-macrophage lineage and represent a major component of lympho-reticular infiltrate into the stroma of tumors [85]. PRL enhanced IL-1 release by the TAM isolated from 1 day old tumor (TAM-1) (Figure 1). Induction of IL-1 secretion by PRL in TAM, but not in RM, suggests that PRL provides second signal for activation of macrophages, and requires a stimulator to primarily activate the cell to induce cell signaling and gene activation associated with production of IL-1. PRL could not stimulate IL-1 release from TAM-3 in physiological doses but marginally stimulated the same at higher dose (50 ng/ml). The secretion of IL-1 by LPS activated macrophages was inhibited by PRL (Figure 1). This suppression may be due to hyperstimulation of macrophages resulting in anergy [86].

IL-1, the inflammatory cytokine, activates the hypothalamus-pituitary-adrenal axis resulting in secretion of glucocorticoid hormone (GC). GC in a feed back loop inhibit IL-1 release. It has been shown in hemorrhagic shock that corticosterone release is associated with the suppression of IL-1 and IL-6 release by the macrophages [87]. PRL treatment normalizes hemorrhage induced shock either by suppressing corticosterone release or by antagonising the corticosterone mediated effect on cytokine release. We have observed earlier that PRL reversed the GC mediated suppression of IL-1 secretion by LPS activated RM, TAM-1 and TAM-3 (Figure 1). Dexamethasone (Dex) induced suppression of IL-1 release by in vivo LPS treated mouse peritoneal macrophages was also reversed by PRL in a dose dependent manner (unpublished). However, in absence of LPS activation, PRL failed to reverse in vitro GC mediated inhibition of IL-1 secretion by the TAM (Figure 1). Earlier we have observed PRL induced reversal of GC mediated suppression of mitogenic proliferation of lymphocytes of normal mice but not of tumor bearing mice [84]. Most of the LPS induced cytokine share the 3' untranslated repetitive TTATTTAT consensus sequence [88]. This region may be responsible for post-transcriptional control of cytokine production and may be regulated by GC. Reversal of GC induced suppression of LPS induced IL-1 secretion by PRL may involve GC-R/STAT complex induced down regulation of the transcription factor [69,89]. The altered PRL response in TAM may be due to a malignancy associated defect in expression or function of PRL-R. In competitive binding assay [90] with ^{125}I-human growth hormone (hGH) and PRL, we observed a lesser lactogen binding with the microsomal membranes of TAM-3 as compared to that of RM, which indicated alteration in

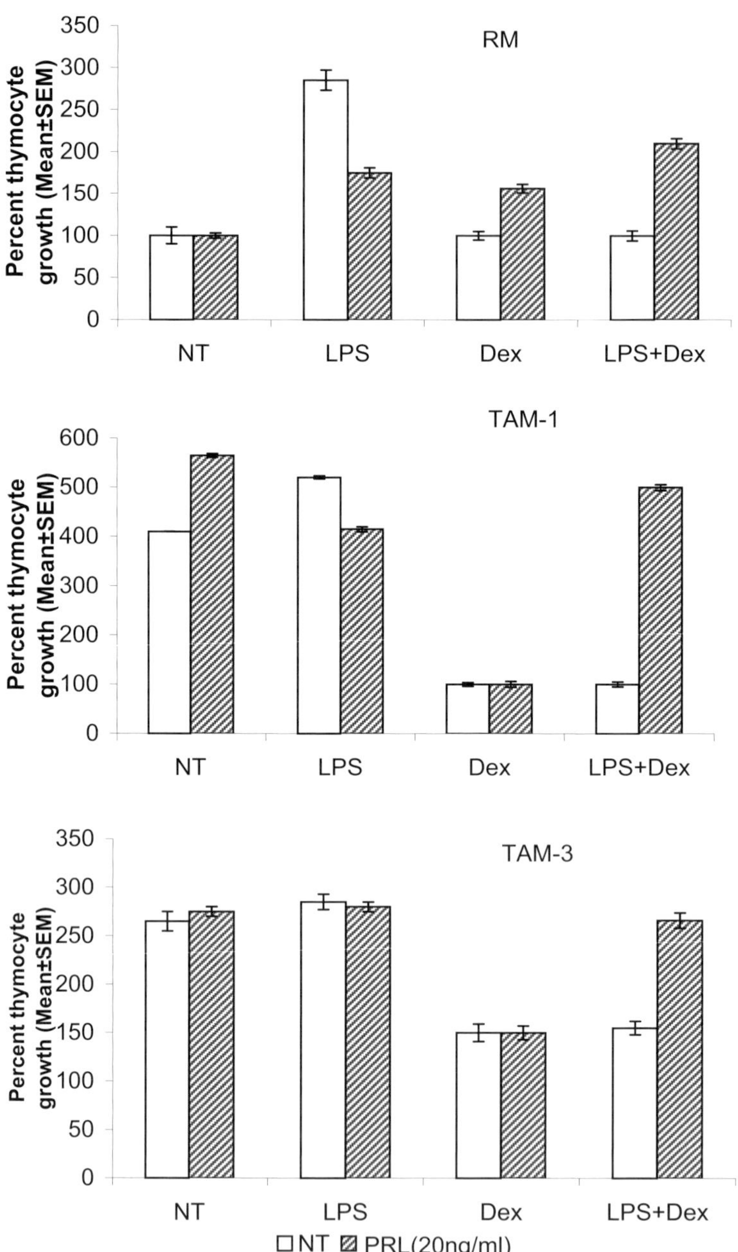

Figure 1. PRL differentially regulates GC induced suppression of IL-1 release in RM.

RM, TAM-1 and TAM-3 were cultured in medium RPMI-1640 supplemented with insulin transferrin selenium (1%) and charcoal stripped fetal bovine serum (2%) in presence or in absence of PRL, (20 ng/ml), LPS (1 μg/ml), Dex (10^{-7} M) alone or in combination for 18 hr. IL-1 bioactivity of the culture supernatants was determined by thymocyte comitogenic assay. Growth of thymocytes was determined by measuring 3H- thymidine incorporation into the cells. The data represent percent of thymocyte growth where growth, of thymocyte in presence of culture supernatant of untreated RM was taken as 100 (= 2500 cpm ± 230 / 2×10^6 cells). Each data represents the mean of 6 observations.

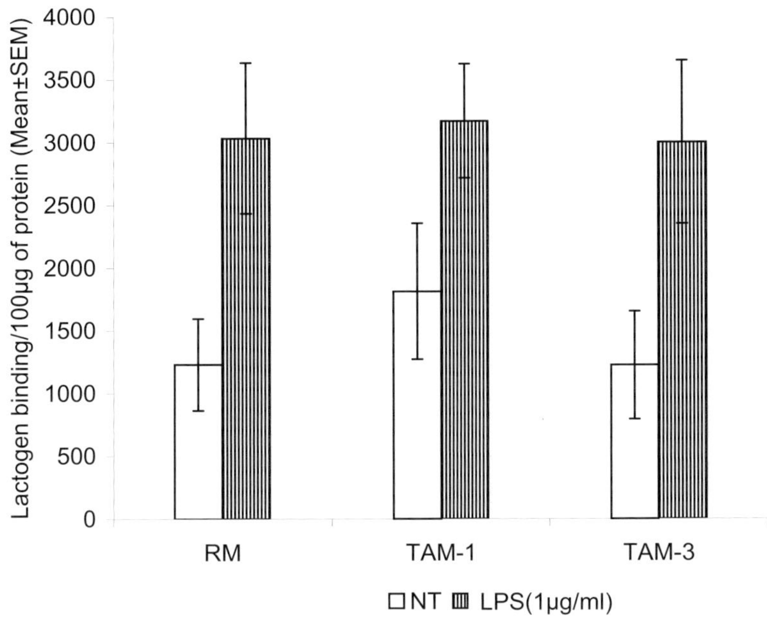

Figure 2. Binding of PRL to microsomal membranes of RM and TAM.
Specific binding of ^{125}I-hGH to microsomal membranes prepared from RM, TAM-1 and TAM-3 in presence or in absence of LPS (1µg/ml) was determined in presence or in absence of 10 mg of unlabelled ovine PRL for 4 hr at 37°. The hormone receptor complexes were separated from free hormone by precipitation with 0.5 ml of 25% (w/v) PEG 8000. Each experiment was performed minimum of 3 times.

number and/or affinity of PRL-R in TAM-3 in comparison to that in RM (Figure 2). However, in LPS treated TAM, ^{125}I-hGH binding was similar to that of LPS activated macrophages, which could be correlated with PRL response observed in LPS treated TAM-3 in respect to reversal of GC mediated suppression of IL-1 secretion (Figure 1).

2.1.2. *PRL does not regulate TNF release by monocytes/macrophages*

Treatment with PRL and TNF α protected mice lethally infected with *Toxoplasma gondii* from death. Similar protection was obtained when the mice were treated with PRL alone or TNF α alone. From these observations, the investigators suggested that PRL can regulate endogenous production of TNF α *in vivo* [91]. However, in *in vitro* studies we failed to see the induction of TNF secretion by elicited macrophages (EM) or TAM with PRL. The TNF bioactivity in the culture supernatants of peripheral blood monocytes (PBM) of cancer patients and TAM of EAC bearing mice was significantly higher than that of PBM of healthy individuals and RM. The PBM and TAM of tumor hosts are activated macrophages, and stimulation of these cells with PRL resulted in suppression of TNF secretion (unpublished).

2.1.3. *PRL regulates cytotoxic functions of monocytes/macrophages*

When appropriately activated, the cells of monocyte/macrophage lineage can kill tumor or microbial cells independent of conventional immune specificity. In resting condition the macrophages are not cytotoxic or tumoricidal. The activation of macrophages for tumoricidal activity is a complex process, which can be accomplished with bacterial products, like LPS or cytokines, such as IFN-γ [92]. We observed that the tumoricidal activity in PBM of healthy individuals and cancer patients with carcinoma of oral cavity (clinical stage II) can be induced in vitro by PRL (Figure 3B). In murine model PRL augmented tumor killing by both EM and TAM-1 in a dose dependent manner (Figure 3A). However, with progression of tumor, the TAM-3 became unresponsive to PRL. Correlation of PRL induced augmentation of cytotoxicity with progression of disease was not studied in cancer patients. To kill target cells, macrophages employ a selection of effector molecules including reactive nitrogen and oxygen species and TNFα. Since PRL is unable to influence TNF α secretion by monocytes/macrophages, the hormone induced cytotoxicity may involve other cytotoxic effector molecules released by the cells.

In vivo experiments in murine models demonstrated that administration of PRL primed peritoneal macrophages of hypophysectomised or normal mice to release elevated level of O_2^- and H_2O_2 [93,94] . PRL was also seen to activate human monocytes to release H_2O_2 in response to PMA [94]. In *in vitro* experiments we observed that treatment of EM or TAM with PRL resulted in a dose dependent increase in O_2^- release by these cells (Table I), which correlated with the PRL induced activation of tumoricidal activity of EM or TAM (Figure 3).

NO is one of the major effector molecules involved in destruction of tumor cells by activated macrophages. Meli et al. [95] reported that treatment of mice with PRL protected the animals against *Salmonella typhymurium* infection and involvement of NO_2^- induced phagocytic killing of the bacteria was demonstrated as the underlying mechanism for protection. It appears that like IFN-γ and LPS, PRL regulates cytotoxic activity of macrophages through induction of NO_2^-synthesis [96]. We observed PRL induced augmentation of NO_2^- release by EM and TAM-1 (Table I). The hormone failed to induce NO_2^- release from TAM-3. Defective iNOS expression in TAM of growing tumor is known [97]. In RT-PCR analysis we failed to detect PRL-R mRNA in TAM-3 (unpublished), which supported the view that malignancy associated defective expression of PRL-R in macrophages results in lack of PRL response of these macrophages.

2.2. PRL regulates NK cell activity

The NK cells express receptors for PRL, which suggests that PRL directly regulates NK cell functions [98]. A relationship between plasma levels of PRL and immune competence of experimental animals has been shown by many [99–101]. The disruption of tubero-infundibular dopaminergic system leads to uncontrolled secretion of pituitary PRL associated with impaired NK cell activity and increased tumor burden in mice [100].

Inhibition of NK activity in presence of high concentrations of circulating PRL was also suggested from the impaired NK activity observed in pregnant women and hyper-prolactenimic patients [3,102]. Cyclosporin-A (CSA), but not its analogue CSH, is known to inhibit PRL binding to NK cells [98], Nb2 cells, T cells and B cells [103]. Blocking of immunostimulatory activity of PRL may thus be associated with CSA induced immunosuppression .

The NK cells are capable of killing non-immunogenic tumor cells in a non-MHC restricted manner and are not dependent on prior immunization. Cytokines, particularly the TH-1 cytokines such as IL-2, IFN-γ and TNF stimulate NK cell cytotoxicity. Matera et al. [3] first demonstrated that the cytotoxic activity of purified NK cells displaying $CD16^+$ and $CD3^+$ phenotype plus LGL morphology could be augmented in vitro by physiological concentrations of PRL. Higher concentrations of PRL inhibited the NK cells. We studied the regulatory role of PRL on peripheral blood NK cell activity in peripheral blood lymphocytes (PBL) of young (near age 40 years) and

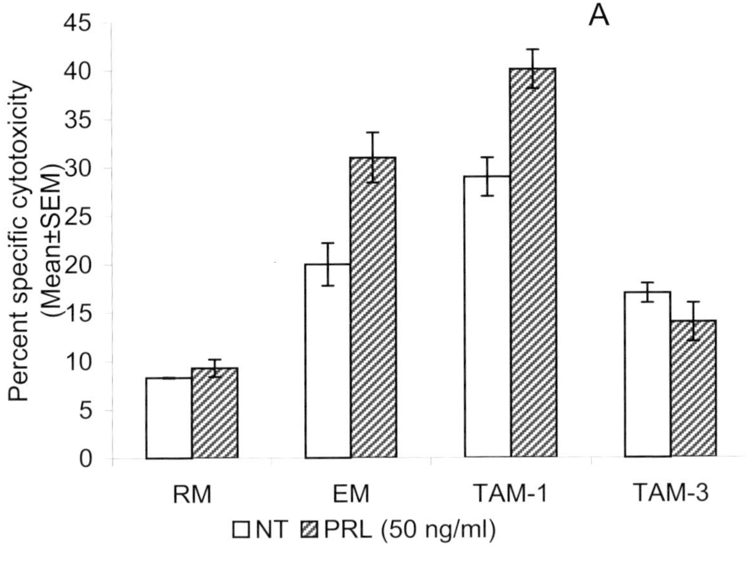

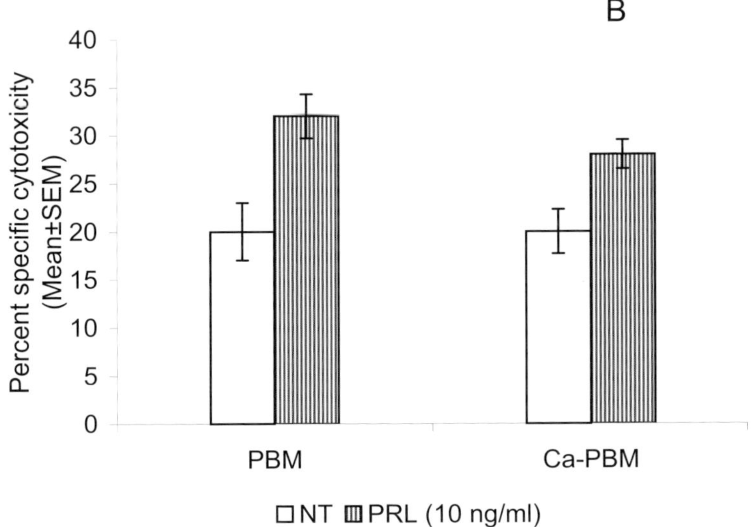

Figure 3. PRL augments tumor target killing by monocytes/macrophages. (A) Killing of P815 cells by murine macrophages. (B) Killing of P815 cells by the PBM of healthy individuals and patients with carcinoma of oral cavity.
RM, EM, TAM-1 and TAM-3 and PBM of healthy individuals and cancer patients were cultured in medium RPMI-1640 supplemented with insulin, transferin, selenium (1%), charcoal stripped fetal bovine serum (2%), in presence or absence of PRL for 18 hr. The cells were washed and co-cultured with ^{51}Cr-labelled P815 cells in an effector : target ratio of 40 : 1 and 25 : 1 for mouse macrophages and human monocytes, respectively. Cytotoxicity of the effector cells was determined by 4 hr ^{51}Cr-release assay. Each data represents mean ±SEM of 5 observations.

Table I Effect of IFN-γ and IL-4 on PRL induced tumor target killing and release of NO_2^- and O_2^- by EM and TAM-3.

Treatment	percent cytotoxicity (mean ± SEM)		μmole NO_2^- released/10^6 cells/18 hr (mean ± SEM)		nmole O_2^- released/10^6 cells/18 hr (mean ± SEM)	
	EM	TAM-3	EM	TAM-3	EM	TAM-3
NT	20 ± 2.2	17.3 ± 1.0	12.2 ± 0.6	7 ± 0.46	6 ± 0.69	3.3 ± 0.25
PRL (50 ng/ml)	31 ± 2.6	15.9 ± 1.36	17.2 ± 0.5	6.9 ± 0.2	9.6 ± 1.53	4.5 ± 0.18
IFN-γ (100 U/ml)	34 ± 1.9	17.1 ± 2.2	24.4 ± 1.01	10.4 ± 1.0	12.4 ± 1.14	7.1 ± 0.8
IFN-γ (100 U/ml) + PRL (50 ng/ml)	38 ± 0.02	17.9 ± 1.8	27.3 ± 1.12	8.4 ± 1.0	10.3 ± 0.89	7.1 ± 0.6
IL-4 (100 U/ml)	15 ± 2.3	14 ± 1.5	7.8 ± 0.4	6.4 ± 0.3	3.7 ± 0.26	4 ± 0.3
IL-4 (100 U/ml) + PRL (50 ng/ml)	15.3 ± 1.5	17 ± 1.0	7.8 ± 0.7	6.0 ± 0.4	4.3 ± 0.43	4.5 ± 0.25

The EM and TAM-3 were cultured in the medium RPM-1640 supplemented with Insulin-Transferrin-Selenium (1%), charcoal stripped fetal bovine serum (2%), penicillin (100 μg/ml), streptomycin (100 μg/ml) with PRL (50 ng/ml) alone, IFN-γ (100 U/ml) alone, IL-4 (100 U/ml) alone, PRL (50 ng/ml) + IFN-γ (100 U/ml) or PRL (50 ng/ml) + IL-4 (100 U /ml) for 18 hr at 37 °C in 5% CO_2 atmosphere. Lysis of P815 cells by the treated and washed macrophages was measured by co-incubation of the macrophages with the target cells at a ratio of 40 : 1 for 4 hr and assaying LDH released by the target cells into the medium. NO_2^- and O_2^- released in culture supernatant was measured by Griess reagent and by determining reduction of SOD inhibitable Ferricytochrome C. Each data represents mean ± SEM of 9 observations. NT represents untreated control.

elderly (mean age 68 years) healthy individuals and patients with carcinoma of oral cavity [104]. The peripheral blood NK cells were found to be depressed in oral cancer patients as compared to that of healthy individuals. Malignancy associated impairment of NK cell activity has been reported by many investigators. PRL produced a dose dependent stimulation of NK cell function of PBL from young healthy individuals and cancer patients [104]. At physiological dose PRL inhibited NK activity of PBL. Similar findings were reported by Matera et al. [3]. Stimulation of T suppressor cells in PBL by physiological doses of PRL might result in inhibition of NK cells. In the elderly groups less or no response of NK cells to low doses of PRL was evident [104].

Under serum free conditions PRL is more effective than IL-2 at stimulating NK cell growth [3]. PRL responsiveness of NK cells is much higher than that of T cells and B cells with reference to cell growth, which may be a reflection of presence of the higher number of PRL–R on NK cells. Matera et al. reported that in NK cell 660 PRL-R are present per cell with a kd of 3.0×10^{-10} M, whereas T and B cell expressed an average of 320 PRL-R per cell with kD of about 1.7×10^{-9} M [2,3]. We observed a marked inhibition of IL-2 induced stimulation of NK cells in presence of PRL [104]. Competitive binding of PRL to the IL-2R due to homology of IL-2R chain with PRL-R [40] may have resulted in the observed PRL mediated inhibition of IL-2 response of the NK cells. The release of PRL like peptide in response to IL-2 from the lymphocytes might also increase the actual concentration of PRL in the medium resulting in inhibition of NK functions. No inhibitory effect of PRL was observed on the IL-2 driven NK cell function of cancer patients [104], which suggested the presence of malignancy associated defect in PRL-R expression or PRL-R signaling in IL-2 driven NK cells.

We reported earlier that the PBL of both cancer patients and healthy individuals, when cultured for 5 days in presence of PRL, generated lymphokine activated killer (LAK) like cells [104]. These cells were functionally similar to the IL-2 driven LAK cells. The LAK cells share many properties with NK cells but are able to lyse a variety of fresh autologous, allogenic and xenogenic tumor targets which may be resistant to killing by NK cells. The progenitors of LAK cells are, like NK cells, $CD3^-/CD16^+/CD56^+$, although a small portion of $CD3^+$ T cells may develop into LAK cells. Generation of LAK activity by culturing NK cells in PRL containing medium was reported by Matera et al. [3]. They showed that PRL/IL-2 unresponsive lymphocytes could be stimulated to develop LAK activity by exposing the cells to both IL-2 and PRL from the onset of culture. We observed a deficient IL-2 and PRL response of NK cells but not of LAK progenitor cells in cancer patients. We also failed to demonstrate malignancy associated changes in biphasic inhibitory effect of PRL on LAK cells [104] which suggested that the LAK progenitor cells of the cancer patients might be non NK cells with no defect in either IL-2 or PRL-R expression. PRL generated LAK cells have been shown to kill non-Hodgkin's lymphoma (NHL) cell lines [105]. PRL-mRNA and PRL were detected in some of the target NHL cell lines, but their growth was independent of endogenous PRL.

2.3. IFN-γ costimulates and IL-4 down regulates PRL induced activation of macrophages

Activation of murine macrophages to a tumoricidal state requires more than one signal. IFN-γ along with IL-2 or LPS was shown to activate a macrophage cell line to kill tumor target cells [96]. We observed synergy of PRL with IFN-γ at inducing tumoricidal activity in EM, but not in TAM-3 (Table I). Tumor associated defect in PRL-R expression may relate to the lack of synergy of PRL with IFN-γ on TAM-3. Reactive nitrogen and oxygen species are important intermediates of macrophage mediated cytotoxicity. IFN-γ is a well known potentiator of NO_2^- and O_2^- release by macrophages [80]. Synergy between PRL and IFN-γ was also observed in inducing NO_2^- release by EM, but not by TAM-3 (Table I). Simultaneous induction of IRF-1 and subsequent expression of iNOS by both IFN-γ and PRL [2] may cause costimulation of NO_2^- production by these agents (Table I). No synergy of PRL with IFN-γ was observed at inducing O_2^- release by macrophages, though both stimulated O_2^- release in EM (Table I). Activation of divergent signaling molecules by PRL and IFN-γ leading to O_2^- release or hyperstimulation of the macrophages might lead to such observation. Anti-IFN-γ antibodies partially blocked IFN-γ as well as PRL mediated enhanced tumor target killing and release of NO_2^- by the macrophages (Figure 4). It appears that NO_2^- release by the macrophages might at least partially be a function of IFN-γ released by contaminating lymphocytes in response of PRL. Treatment of human PBMC with PRL was shown to enhance IFN-γ secretion [73]. Correction of a defect in altered lymphocyte proliferation and IFN-γ secretion in bromocriptine treated mice with PRL supplementation, and delayed or abrogated rejection of tumor in allogenic mice in presence of anti-IFN-γ antibody [3,72,73,106] further suggest PRL mediated regulation of IFN-γ secretion. We observed expression of IFN-γ mRNA in PRL treated thymocytes (unpublished). Matera et al. [107] observed IFN-γ mRNA expression and release of IFN-γ from NK cells following treatment with pituitary and recombinant human PRL. Anti-IFN-γ monoclonal antibody blocked PRL mediated augmentation of K562 target killing by NK cells. Further more, costimulation of IL-2 induced proliferation of lymphocytes and NK cells, and induction of IL-2 release from lymphocytes by PRL suggest that PRL through IFN-γ drives TH-1 responses or directly regulates

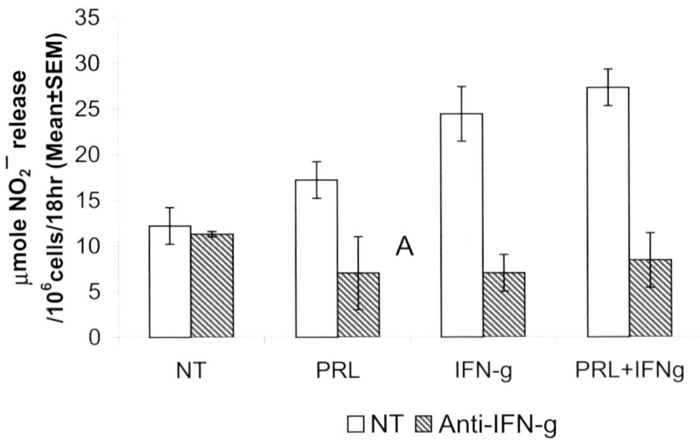

Figure 4. PRL mediated NO_2^- release by EM is inhibited in presence of anti IFN-γ antibodies.
EM was cultured in the medium RPMI-1640, supplemented with insulin, transferin, selenium (1%), charcoal stripped fetal bovine serum (2%), for 18 hr in presence or absence of PRL (50 ng/ml) alone, IFN-γ (100 U/ml) alone, anti IFN-γ antibodies (Hybridoma supernatant diluted 1 : 500) alone PRL (50 ng/ml) + IFN-γ (100 U/ml) and anti IFN-γ along with PRL, IFN-γ and PRL + IFN-γ. NO_2^- released by the cells in medium was measured with Griess reagent. Each data represents mean ±SEM of 5 observations.

immune functions by acting as a TH-1 cytokine. IFN-γ induced macrophage functions are down regulated by IL-4, a TH-2 cytokine, which inhibits transcriptional activation of IFN-γ inducible proteins [96]. We observed IL-4 induced suppression of PRL mediated activation of NO_2^-, O_2^- and target tumor cell killing by EM (Table I). Since these functions of PRL might be mediated through IFN-γ, IL-4 might antagonise PRL through suppression of IFN-γ inducible genes [108].

2.4. PRL induces release of IL-12 by monocytes/macrophages

IL-12 secreted by monocytes/macrophages and dendritic cells (DC), is the critical cytokine in induction of IFN-γ by $CD4^+$ T cells and thus plays a key role in driving TH-1 responses. Recently, PRL has been shown to stimulate GM-CSF driven differentiation of DC suggesting that PRL may play a role in regulation of IL-12 secretion by DC [109]. We observed that treatment of EM with PRL resulted in 30% increase in IL-12 release from the macrophages (5 pg/10^6 cells) compared to that of untreated control (3.7 pg/10^6 cells). A role of PRL in regulation of IL-12 release was suggested earlier by the observation that IRF-1 -/- mice lack the expression of IL-12 P40 [110]. PRL fails to induce IL-12 secretion in TAM-3 (3.7 pg/10^6 cells). We also observed PRL induced IL-12 production by the PBM of healthy individuals and lack of PRL response by PBM of cancer patients. In tumor bearing animals secretion of IL-12 and IFN-γ are negatively regulated. Tumor factor associated TH-2 phenotype may inhibit the production of these anti-tumor cytokines.

3. CONCLUDING REMARKS

From this overview it may be concluded that PRL plays an important role in generation of tumoricidal macrophages, LAK cell activity and augmentation of NK cell cytotoxicity. PRL induced tumor target killing is TNF independent and is correlated with augmented NO_2^- and O_2^- release. PRL regulates tumoricidal activity of macrophages through the release of IFN-γ. The hormone enhances the effect of IFN-γ in activating macrophages and NK cells also. IL-4 down regulates the PRL mediated activation of macrophages. PRL drives TH-1 responses and regulates immune responses against tumor through the release of IL-12 and IFN-γ. PRL is unable to stimulate TAM or NK cells in hosts with a growing tumor. Preponderance of TH-2 responses in malignancy and associated defects in PRL-R expression may be responsible for the lack of PRL response of these cells in tumor hosts .

ACKNOWLEDGEMENT

We thank the Department of Science and Technology, Government of India for sponsoring the work and financial support, Mr. Sumit Kr. Majumder for helping in preparation of the manuscript.

REFERENCES

1. Berczi I, Naggy E. The effect of prolactin and growth hormone on hemolympho poietic tissue and immune function. In Berczi I, Kovacs K, editors. Hormones and Immunity. Lancaster: MTP Press, 1987;145–171.
2. Yu-Lee L-Y. Molecular actions of prolactin in the Immune System. Proc Soc Expt Biol Med 1997;215:35–52.
3. Matera L, Bellone G, Cesano A. Prolactin and neuroimmune network. Adv Neuroimmunol 1991;1:158–172.
4. Kooijman R, Hooghe Peters EL, Hooghe R. Prolactin growth hormone and Insulin like growth factor –1 in the immune system. Adv Immunol 1996;63:377–454.
5. Savino W, Dardane M. Immune–neuroendocrine Interactions. Immunol Today 1995;16:318–322.
6. Montgomery DW, Shen GK, Ulrich ED, Steiner LL, Parrish PR, Zukoski CF. Human thymocytes express a prolactin-like mRNA and synthesize bioactive prolactin-like proteins. Endocrinology 1992;131:3019–3026.
7. O'Neal KD, Montgomery DW, Truong TM, Yu-Lee L-Y. Prolactin gene expression in human thymocytes. Mol Cell Endocrinol 1992;87:R19–R23.
8. Pellegrini I, Lebrun J-J, Ali S, Kelly PA. Expression of prolactin and its receptor in human lymphoid cells. Mol Endocrinol 1992;6:1023–1031.
9. Jurcovicova J, Day RN, MacLeod RM. Expression of prolactin in rat lymphocytes. Prog Neuro Endocrin Immunol 1992;5:256–263.
10. Sabharwal P, Glaser R, LaFuse W, et al. Prolactin synthesized and secreted by human peripheral blood mononuclear cells: An autocrine growth factor for lymphoproliferation. Proc Natl Acad Sci 1992;89:7713–7716.
11. Gala RR, Shevach EM. Evidence for the release of a Prolactin like substance by mouse

lymphocytes and macrophages. Proc Soc Exp Biol Med 1994;205:12–19.
12. Chao HS, Myers SE, Handwerger S. Endothelin inhibits basal and stimulated release of prolactin by human decidual cells. Endocrinology 1993;133:505–510.
13. Adler RA. The anterior pituitary-grafted rat: A valid model of chronic hyperprolactinemia. Endocr Rev 1986;7:302–313.
14. Gala RR, Rillema JA. Evaluation of prolactin-like activity produced by concanavalin-A stimulated mouse splenocytes. Life Sci 1995;57:1683–1700.
15. DiMattia GE, Gellersen B, Kuckworth ML, Friesen GH, Human prolactin gene expression: The use of an alternative noncoding exon in decidua and the IM-9-P3 lymphoblast cell line. J Biol Chem 1990;265:16412–16421.
16. Gellersen B, Kempf R, Telgmann R, DiMattia GE. Nonpituitary human prolactin gene transcriptin is independent of pit-1 and differentially controlled in lymphocytes and endometrial stroma. Mol Endocrinol 1994;8:356–373.
17. Gellersen B, Kempf R, Telgmann R, DiMattia GE. Pituitary-type transcription of the human prolactin gene in the absence of Pit-1. Mol Endocrinol 1995;9:887–901.
18. Baulieu EE, and Kelly PA. Hormones: From Molecules to Disease. Paris: Hermann/New York: Chapman and Hall, 1990:1–697.
19. Nagy E, Berczi I, Friesen HG. Regulation of immunity in rats by lactogenic and growth hormones. Acta Endocrinol 1983;102:351–357.
20. Nagy E, Berczi I. Pituitary dependence of bone marrow function. Br J Haemarol 1989;71:457–462.
21. Kelley KW, Brief S, Westly HJ, et al. GH_3 pituitary adenoma cells can reverse thymic aging in rats. Proc Natl Acad Sci 1986;83:5663–5667.
22. Li Ym, Brunke DL, Dantzer R, Kelley KW, Pituitary epithelial cell implants reverse the accumulation of $CD4^-CD8^-$ lymphocytes in thymus glands of aged rats. Endocrinology 1992;130:2703–2709.
23. Neidhart M, Fluckiger EW. Hyperprolactinaemia in hypophysectomized or intact male rats and the development of adjuvant arthritis. Immunology 1992;77:449–455.
24. Riskind PN, Massacesi L, Doolittle TH, Hauser SL. The role of prolactin in autoimmune demyelination: Suppression of experimental allergic encephalomyelitis by bromocriptine. Ann Neurol 1991;29:542–547.
25. McMurray R, Keisler D, Kanuckel K, Izui S, Walker SE. Prolactin influences autoimmune disease activity in the female B/W mouse. J Immunol 1991;147:3780–3787.
26. Walker SE, Allen SH, McMurry RW. Prolactin and autoimmune disease. Trends Endocrinol Metab 1993;4:147–151.
27. Gellersen B, DiMattia GE, Friesen HG, Bolnet HG. Regulation of prolactin secretion in the human B-lymphoblastoid cell line IM-9-P3 by dexamethasone but not other regulators of pituitary prolactin secretion. Endocrinology 1989;125:2853–2861.
28. Pellegrini I, Lebrun JJ, Ali S, Kelly PA. Expression of prolactin and its receptor in human lymphoid cells. Mol Endocrinol 1992;6:1023–1031.
29. Matera L, Muccioli G, Cesano A, Bellusi G, Gennazzani E. Prolactin dual regulation by Cyclosporin A. Brain behav Immunol 1988;2:1–10.
30. Fletcher-Chiappini SE, Comptom MM, La Voie HA, Day EB, Witorsch RJ. Glucocorticoid-prolactin interactions in Nb2 lymphoma cells: antiproliferative versus anticytolytic effects. Proc Soc Exp Biol Med 1993;202:345–352.
31. O'Neal KD, Schwarz LA, yu-Lee L-Y. Prolactin receptor gene expression in lymphoid cells. Mol cell Endocrinol 1991;82:127–135.

32. Dardenne M, Leite de Moraes M, Kelly PA, Gagnerault M-C. Prolactin receptor expression in human hematopoietic tissues analysed by flow cytoflourometry. Endocrinology 1994;134:2108–2114.
33. Gagnerault M-C, Touraine P, Savino W, Kelly PA, Dardenne M. Expression of prolactin receptors in murine lymphoid cells in normal and autoimmune situations. J Immunol 1993;150:5673–5681.
34. Gala RR, Shevach EM. Identification by analytical flow cytometry of prolactin receptors on immunocompetent cell populations in the mouse. Endocrinology 1993;133:1617–1623.
35. Clevenger CV, Sillman AL, Hanley Hyde J, Prystowsky MB. Requirement for prolactin during cell cycle regulated gene expression in cloned T-lymphocytes. Endocrinology 1992;130:3216–3222.
36. Hooghe R, Delhase M, Vergani P, Malur A, Hooghe-Peters EL. Growth hormone and prolactin are paracrine growth and differentiation factors in the haemopoietic system. Immunology Today 1993;14:212–214.
37. Murphy WJ, Rui H, Longo DL. Effects of growth hormone and prolactin in immune development and function. Life Sci 1995;57:1–14.
38. Matera L. Endocrine, paracrine and autocrine actions of prolactin on immune cells. Life Sci 1996;59:599–614.
39. Bouchard B, Ormandy J, Santo JP, Kelly PA. Immune system development and function in Prolactin receptor deficient mice. J Immunol 1999;163:576–582.
40. Bazan JF. A novel family of growth factor receptors: A common binding domain in the growth hormone, prolactin, erythropoietin and IL-6 receptors, and the p75 IL-2 receptor beta-chain. Biochem Biophys Res Commun 1989;164:788–795.
41. Cosman D. The hematopoietin receptor superfamily. Cytokine 1993;5:95–106.
42. Horseman ND, Yu-Lee L-Y. Transcriptional regulation by the helix bundle peptide hormones: Growth hormone, prolactin, and hematopoietic cytokines. Endocr Rev 1994;15:627–649.
43. Elberg G, Kelly PA, Djiane J, Binder L, Gertler A. Mitogenic and binding properties of monoclonal antibodies to the prolactin receptor in Nb_2 rat lymphoma cells. J Biol Chem 1990;265:14770–14776.
44. Kelly PA, Djiane J, Postel-Vinay M-C, Edery M. The prolactin/growth hormone receptor family. Endocr Rev 1991;12:235–251.
45. Davis JA, Linzer DI. Expression of multiple forms of the prolactin receptor in mouse liver. Mol Endocrinol 1989;3:674–680.
46. O'Neal KD, Yu-Lee L-Y. Differential signal transduction of the short, Nb2, and long PRL receptor: Activation of IRF-1 and cell proliferation. J Biol Chem 1994;269:26076–26082.
47. Nagano M, Kelly PA. Tissue distribution and regulation of rat prolactin receptor gene expression. J Biol Chem 1994;269:13337–13345.
48. Ali S, Pellegrini I, Kelly PA. A prolactin-dependent immune cell line (Nb2) expresses a mutant form of prolactin receptor. J Biol Chem 1991;266:20110–20117.
49. DaSilva L, Howard OM, Rui H, Kirken RA, Farrar WL. Growth signaling and JAK2 association mediated by membrane-proximal cytoplasmic regions of prolactin receptors. J Biol Chem 1994;269:18267–18270.
50. Chang W-P, Clevenger CV. Modulation of growth factor receptor function by isoform heterodimerization. Proc Natl Acad Sci 1996;93:5947–5952.
51. Rui H, Kirken RA, Farrar WL. Activation of receptor-associated tyrosine kinase JAK2 by prolactin. J Biol Chem 1994;269:5364–5368.

52. Campbell GS, Argetsinger LS, Ihle JN, Kelly PA, Rillema JA, Carter-su C. Activation of JAK2 tyrosine kinase by prolactin receptors in Nb2 cells and mouse mammary gland explants. Proc Natl Acad Sci 1994;91:5232–5236.
53. Lebrun J, Ali S, Sofer L, Ullrich A, Kelly PA. Prolactin-induced proliferation of Nb2 cells involves tyrosine phosphorylation of the prolactin receptor and its associated tyrosine kinase JAK2. J Biol Chem 1994;269:14021–14026.
54. Welte T, Garimorth K, Philipp S, Doppler W. Prolactin-dependent activation of a tyrosine phosphorylated DNA binding factor in mouse mammary epithelial cells. Mol Endocrinol 1994;8:1091–1102.
55. Gouilleux F, Wakao H, Mundt M, Groner B. Prolactin induces phosphorylation of Tyr694 of Stat5 (MGF), a prerequisite for DNA binding and induction of transcription. EMBO J 1994;13:4361–4369.
56. Gilmour KC, Reich NC. Receptor to nucleus signaling by prolactin and interleukin 2 via activation of latent DNA-binding factors. Proc Natl Acad Sci 1994;91:6850–6854.
57. Rao YP, Buckley DJ, Buckley AR. Rapid activation of mitogen-activated protein kinase and $p21^{ras}$ by prolactin and interleukin 2 in rat Nb2 node lymphoma cells. Cell Growth Differ 1995;6:1235–1244.
58. Mukherjee P, Mastro AM, Hymer WC. Prolactin induction of interleukin-2 receptors on rat splenic lymphocytes. Endocrinology 1990;126:88–94.
59. Gala RR. Influence of thyroxine and thyroxine with growth hormone and prolactin on splenocyte subsets of snell dwarf mice. Proc Soc Exp Biol Med 1995;210:117–125.
60. Lahat N, Miller A, Shtiller R, Touby E. Differential effects of prolactin upon activation and differentiation of human B-lymphocytes. J Neuroimmunol 1993;47:35–40.
61. Clevenger CV, Russell DII, Appasamy PM, Prystowsky MB. Regulation of interleukin 2-driven T-lymphocyte proliferation by prolactin. Proc Natl Acad Sci USA 1990;87:6460–6464.
62. Athreya BH, Pletcher J, Zulian F, Weiner DB, Williams WV. Subsetspecific effects of sex hormones and pituitary gonadotropins on human lymphocyte proliferation in vitro. Clin Immunol Immunopathol 1993;66:201–211.
63. Clevenger CV, Sillman AL, Prystowsky MB. Interleukin 2 driven nuclear translocation of prolactin in cloned T lymphocytes. Endocrinology 1990;127:3151–3159.
64. Gilmour KC, Pine R, Reich NC. Interleukin-2 activates STAT5 transcription factor (mammary gland factor) and specific gene expression in T lymphocytes. Proc Natl Acad Sci USA 1995;92:10772–10776.
65. Gouilleux F, Moritz D, Humar M, Moriggi R, Bechtold S, Groner B. Prolactin and interleukin-2 receptors in T-lymphocytes signal through an MGF-STAT5-like transcription factor. Endocrinology 1995;136:5700–5708.
66. Schwarz LA, Stevens AM, Hrachovy JA, Yu-Lee L-Y. Interferon regulatory factor-1 is inducible by prolactin, interleukin-2 and concanavalin A in T cells. Mol Cell Endocrinol 1992;86:103–110.
67. Stevens AM, Yu-Lee L-Y. The transcription factor interferon regulatory factor-1 is expressed during both early G1 and the G1/S transition in the prolactin-induced lymphocyte cell cycle. Mol Endocrinol 1992;6:2236–2243.
68. Harada H, Willison K, Sakakibara J, Miyamoto M, Fujita T, Taniguchi T. Absence of the type I IFN system in EC cells: transcriptional activator (IRF-1) and repressor (IRF-2) genes are developmentally regulated. Cell 1990;63:303–312.
69. Yu-Lee L-Y. Prolactin stimulates transcription of growth-related genes in Nb2 T lymphoma

cells. Mol Cell Endocrinol 1990;68:21–28.

70. Stevens AM, Yu-Lee L-Y. Multiple prolactin-responsive elements mediate G1 and S phase expression of the interferon regulatory factor-1 gene. Mol Endocrinol 1994;8:345–355.
71. Yu-Lee L-Y, Hrachovy JA, Stevens AM, Schwarz LA. Interferon-regulatory factor 1 is an immediate-early gene under transcriptional regulation by prolactin in Nb2 T cells. Mol Cell Biol 1990;10:3087–3094.
72. Landolfo S, Cofano F, Giovarelli M, et al. Inhibition of interferon gamma may suppress allograft reactivity by T lymphocytes in vitro and in vivo. Science 1985;229:176–179.
73. Cesario TC, Yousefi S, Carandang G, Sadati N, Le J and Vaziri N. Enhanced yields of gamma interferon in prolactin treated human peripheral blood mononuclear cells. Proc Soc Exp Biol Med 1994;205:89–95.
74. Balkwill FR. Cytokine therapy of cancer. The importance of knowing the context. Eur Cytokine Netw 1994;5:379–385.
75. Gutterman JU. Cytokine therapeutics: lessons from interferon α. Proc Natl Acad Sci 1994;91:1198–1205.
76. Lucey DR, Clerci M, Shearer GM. Type-1 and Type-2 cytokine dysregulation in human infectious, neoplastic and inflammatory diseases. Clinical Microbiology Review 1996:532–562.
77. Cline MJ, Lehrer RI, Territo C, Golde DW. Monocytes and macrophages: functions and diseases. Ann Intern Med 1978;88:78–88.
78. March CJ, Mosley B, Larsen A, et al. Cloning, sequence and expression of two distinct human interleukin-1 complementary DNAs. Nature 1985;315:641–647.
79. Dinarello CA, Cannon JG, Mier JW, et al. Multiple biological activities of human recombinant interleukin 1. J Clin Invest 1986;77:1734–1739.
80. Alprintis AO, Diez-Roux G, Mulder LCF, Zychlinsky A, Lang RA. Do macrophage kill through apoptosis. Immunol Today 1996;17:573–576.
81. Zhu XH, Zellweger R, Ayala A, Chaudry JH. Prolactin inhibits the increased cytokine gene expression in Kupffer cells following haemorrhage. Cytokine 1996;8:134–140.
82. Galdiero M, Donnarumma G, Cipollaro de l'Ero G, Marcatili A, Scarfogliero P, Sommese L. Effect of growth hormone, prolactin, and insulin on the release of IL-1-slpha, INF-gamma and IL-4 by staphylococcal enterotoxin A-stimulatee splenocytes. Eur Cytokine Netw 1995;6:187–194.
83. Kumar A, Singh SM, Sodhi A. Effect of Prolactin on nitric oxide and interleukin 1 production of murine peritoneal macrophages: role of Ca^{2+} and Protein Kinase C. Int J Immunopharmacol 1997;19:1–5.
84. Biswas R, Chattopadhyay U. Altered prolactin response of lymphocytes of tumor bearing mice. Int J Cancer 1992;50:93–98.
85. Moldawer LL, Lonnroth C, Mizel SB, Lundhoum C. Down regulation of interleukin-1 production by macrophages of tumor bearing mice. J Immunol 1987;138:4270–4274.
86. Erroi A, Fantvzzi G, Mengozzi M, et al. Differential regulation of cytokine production in Lipopolysaccharide tolerence in mice. Infect Immun 1993;61:4356–4359.
87. Zellweger R, Zhu X-H, Wich mann MW, Ayala A, Demaso CM, Chaudry IH. Parolactin administration following hemmorhogic shock improves macrophage cytokine release capacity and decreases mortality from subsequent sepsis. J Immunol 1996;157:5748–5754.
88. Caput D, Beutler B, Hartog K, Thayer R, Brown-Shimer S, Cerami A. Identification of common neucleotide sequence in 3^{-} untranslated region of mRNA molecules. Proc Natl

Acad Sci 1986;83:1670–1674 .
89. Djiane J, Clavser H, Kelly PA. Rapid down regulation of prolactin receptors in mammary gland and liver. Biochem. Biophys Res Conm 1979;90:1371–1379.
90. Biswas R, Vonderhaar BK. Antiestrogen inhibition of prolactin induced growth of Nb2 Rat lymphoma Rat lymphoma cell line. Cancer Research 1989;49:6295–6299.
91. Benedeho N, Fol gore A, Galdiero M, Meli R, Dicarlo R. Effect of prolactin r INF-gamma or r TNF-alpha in murine toxoplasmosis. Palhol Biol Paris 1995;43:395–400.
92. Wiltrout RH, Varesio L. Activation of macrophages for cytotoxic and suppressor effector functions. In: Oppenheim JJ, Shevach EM, editors. Immunophysiology: The role of cells and cytokines in immunity and inflammation. New York: Oxford Univ. Press, 1990:365–385.
93. Edward CK, Ghiasuddin SM, Libby MY, et al. In vivo administration of recombinant growth hormone or gamma interferon activates macrophages: enhanced resistance to experimental *Salmonella typhimurium* infection is correlated with generation of reactive oxygen intermediates. Infect Immun 1992;60:2514–2521.
94. Warwick DJ, Lowrie DB, Cole PJ. Growth hormone is a human macrophage activating factor: Priming of human monocytes for enhanced release of H_2O_2. J Immunol 1995;154:1909–1918.
95. Meli R, Raso GM, Bentizoglio C, Nuzzo I, Galdiero M, Di Carlo R. Recombinant human prolactin induces protection against Salmonella typhymurium in the mouse: role of nitaric oxide. Immunopharmacology 1996;34:1–7.
96. Cox GW, Mellilo G, Chattopadhyay U, Mullet D, Fretel RH, Varesio L. Tumor necrosis factor-A dependent production of reactive nitrogen intermediates mediates IFN-γ plus IL-2 induced murine macrophage tumoricidal activity. J Immunol 1992;149:3290–3296.
97. Dinapoli MR, Calderon CL, Lopez D. The altered tumoricidal capacity of macrophages isolated from tumor bearing mice is related to reduced expression of inducible Nitric oxide synthase gene. J Exp Med 1996;183:1323–1329.
98. Matera L, Muccioli G, Cesano A, Bellussi G, Genazzani E. Prolactin receptors on large granular lymphocytes: Dual regulation by cyclosporin A. Brain Behav Immun 1988;2:1–10.
99. Biodoni M, Belluardo N, Licciarello S, Marchese AE, Cicirata F. Growth of Yoshida ascite tumor in rat after radiofrequency destruction of the tuberoinfundibular region of the hupothalamus. Neuroendocrinology 1980;30:88–94.
100. Cross RJ, Markesbery WR, Brooks WH, Roszman TL. Hypothalamic immune interactions: neuromodulation of natural killer activity by lesioning of the anterior hypothalamus. Immunology 1984;51:399–408.
101. Forni G, Bindoni M, Santoni A, Belluardo N, Marchese AE, Giovarelli M. Radiofrequency destruction of the tuberoinfundibular region of the hypothalamus permanently abrogates NK cell activity in mice. Nature 1983;306:181–186.
102. Camanni F. Normal development of lymphokine activated killing (LAK) in peripheral blood lymphocytes from hyperprolactinemic patients. Int J Immunopharmacol 1992;14:1235–1240.
103. Matera L, Cesano A, Bellone G, Oberholtzer E. Modulatory effect of prolactin on the resting and mitogen-induced activity of T, B, and NK lymphocytes. Brain Behav Immun 1992;6:409–417.
104. Chakraborty A, Chakraborty NG, Chattopadhyay U. Prolactin response of NK cells but not LAK cells is deficient in patients with carcinoma of oral cavity and during aging. Int

J Cancer 1996;66:65–69.
105. Matera L, Geuna M, Pastore C, et al. Expression of prolactin and prolactin receptors by non-Hodgkin's lymphoma cells. Int J Cancer 2000;85:124–130.
106. Bernton EW, Meltzer MS, Holaday JW. Suppression of macrophage activation and T-lymphocyte function in hypoprolactenimic mice. Science 1988;239:401–404.
107. Matera L, Contarini M, Bellone G, Forno B, Biglino A. Up-modulation of interferon-gamma mediates the enhancement of spontanous cytotoxicity in prolactin-activated natural killer cells. Immunology 1999;98:386–392.
108. Larner AC, Pericoin EF, Nakagawa Y, Finbloom DS. IL-4 attenuates the transcriptional activation of both IFN γ and IFN γ induced cellular gene expression in monocytes and monocytic cell lines. J Immunol 1993;150:1944–1950.
109. Matera L, Galetto A, Geuna M, et al. Individual and combined effect of granulocyte-macrophage colony-stimulating factor and prolactin on maturation of dendritic cells from blood monocytes under serum-free conditions. Immunology 2000;100:29–36.
110. Yu-Lee L, Luo G, Moutoussamy S, Finidori J. Prolactin and growth hormone signal transduction in lymphohaemopoietic cells. Cell Mol Life Sciences 1998;54:1067–1075.

V. GLH AND DISEASE

Growth and Lactogenic Hormones
Edited by L. Matera and R. Rapaport

Acromegaly and Immune Function

ANNAMARIA COLAO[1], DIEGO FERONE[1,2], PAOLO MARZULLO[1] and GAETANO LOMBARDI[1]

[1]*Department of Molecular and Clinical Endocrinology and Oncology, Federico II University of Naples, Italy*
[2]*Department of Endocrine and Metabolic Sciences (Di.S.E.M.), University of Genova, Italy*

ABSTRACT

GH/IGF-I axes have long been supposed to play a major role in immunomodulation. They ensure ordered body growth and therefore are involved in complex interactions with most organ systems, tissues, and cell types. The homozygous Snell–Bagg dwarf mouse, GH-, PRL-, and thyroid hormones-deficient, has an associated poorly developed immune system including a marked spleen and thymus hypertrophy, a progressive loss of small lymphocytes in the thymic cortex, and a decreased number of peripheral blood lymphocytes. On the other hand, lymphoid cells themselves produce GH: 10% of unstimulated human peripheral blood mononuclear cells (PBMCs) were positive for GH, whereas after mitogen stimulation 20% were positive. A paracrine effect of thymocyte-secreted GH on human primary thymic epithelial cell (TEC) cultures has also been observed. Moreover, somatostatin and GH receptors are expressed in either human TEC or thymocytes. Both these types of receptor seem to be developmentally and differentially expressed on thymic cells, suggesting a crucial role of the local somatostatin-GH-IGF-I system in immune cell differentiation. Cytokines other than GH affect IGF-I synthesis in lymphoid tissues, e.g. in macrophages tumor necrosis factor-α (TNF-α) regulates IGF-I production, the colony-stimulating factors induce the expression of IGF-I mRNA, and the T cell-derived cytokine IFN-γ reduces IGF-I mRNA in a time- and dose-dependent manner. Therefore, the possibility that GH and IGF-I may have effects on immune responses in nonstressed, healthy animals that are compensated by the overlapping actions of other hormones or cytokines cannot be ruled out. Some changes in the lymphocyte subset pattern have been found in acromegaly, but whether these changes play a role in the increased prevalence of neo-plasms is still unknown.

1. INTRODUCTION

Acromegaly is a rare pituitary disorder with an estimated incidence of 3–4 cases per million population per year [1–3]. It is a severe systemic disease, because the GH/IGF-I excess causes impairment of cardiac and respiratory functions, which contribute to increased mortality and morbidity [1–3]. The clinical features of acromegaly develop insidiously and progressively over many years. Most signs and symptoms are the result of the longstanding overproduction of GH and/or IGF-I, but they may also be related to the pituitary lesion itself, which causes central nervous system (CNS) syndromes. In adults chronic GH hypersecretion causes acromegaly which is characterised by local bone overgrowth, while in children and adolescents it leads to gigantism because of the associated secondary hypogonadism which delays epiphysial closure, thus allowing continued acceleration of linear growth [1–3]. Additional features include hyperhydrosis, goiter, osteoarthitis, carpal tunnel syndrome and other peripheral neuropathies, fatigue, colonic polyps, sleep apnea and cardiomyopathy [3,4]. In acromegaly various neoplasms have also been reported to occur with a greater than expected incidence [5–7]. Particularly, a higher prevalence of tumors of the gastrointestinal tract was reported in acromegalics than in the healthy population [8–12], although data were not univocal [13]. The GH/IGF-I axis has long been supposed to play a major role in immunomodulation: acting together, they ensure ordered body growth and therefore are involved in complex interactions with most organ systems, tissues, and cell types. Due to these multiple effects, GH and IGF-I can affect diverse physiological processes, including immune function, in many ways, both directly and indirectly [14–16]. Binding of GH to its receptors on lymphocytes stimulates IGF-I production, which mediates mainly the effects of GH on cell proliferation, while other effects of GH on immune system seem to be direct or shared with IGF-I [17]. Despite these observations, and the fact that GH is produced and secreted by immunological tissues such as thymus and spleen, immune deficiency is not considered characteristic of GH deficiency in humans. Moreover, this issue has been poorly investigated in acromegaly. This study aims at reviewing the experimental basis for a role of GH/IGF-I in modulating immune response and the clinical impact of this modulation in patients with acromegaly. The role of GH and IGF-I on immune system has been recently reviewed [15,16].

2. GH/IGF-I AND THE IMMUNE SYSTEM: EXPERIMENTAL BASIS

Aging, stress, and nutrition affect blood concentrations of GH and IGF-I, which in turn modulate immune function [15]. The first available data on this topic are, indeed, very old. Hypophysectomy in the rat was shown to cause thymus regression as early as in 1930 [18]. Hypophysectomized rats were also shown to have a dramatic age-related decrease in both blood hemoglobin and white cell count, compared with normal animals [19]. Reduced antibody response to antigen was improved by GH and PRL treatment in these animals [20]. Moreover, the homozygous Snell-Bagg dwarf mouse, GH-, PRL-, and thyroid hormones-deficient, has an associated poorly developed immune system including a marked spleen and thymus hypertrophy, a progressive loss of small lymphocytes in the thymic cortex, and a decreased number of peripheral blood lymphocytes [21,22]. On the other hand, lymphoid cells themselves produce GH: 10% of unstimulated human peripheral blood mononuclear cells (PBMCs) were positive for GH, whereas after mitogen stimulation 20% were positive [23]. A plaque assay confirmed that the GH produced by human PBMCs is a biologically active GH [24]. By *in situ* hybridisation, immunocytochemistry and RT-PCR, GH (protein or mRNA) was shown to be expressed in human and rat bone marrow, spleen, thymus, lymph nodes, and in human tonsil [25,26]. The finding that GH is produced locally by lymphoid tissues has been given a functional significance by the evidence that lymphocytes also express the GH receptor. GH binding was first detected on a human B cell lymphoma (IM-9) lymphocyte cell line [27], and subsequently identified on human PBMCs [28]. In hypopituitary dwarf Weimaraner dogs, GH administration was shown to affect thymus size

[29]: thymic growth was observed in middle-aged, but not in aged, dogs and the blood level of thymic hormone increased [30]. The thymus glands of GH-treated dogs were described [31] as "resembling thymic tissue of young dogs". In rodents continuous infusions of GH was very effective in stimulating thymic growth [32]. In monkeys a depot GH, that chronically elevated blood GH levels gave higher serum IGF-I levels than did a comparable dose of GH given by daily injections [33], causing lymphoid tissue growth.

Moreover, thymulin plasma levels have been found increased in patients with acromegaly compared to controls [34]. Alternatively, it is possible that these effects of GH on lymphoid tissues may not be mediated by IGF-I. In fact, lymphocytes are exposed to endocrine IGF-I (human thymocytes express IGF-I receptors [35,36]), their own autocrine IGF-I, and perhaps most importantly, in lymphoid organs and bone marrow, a third source of IGF-I from epithelial cells [37] and stromal cells [38]. The proliferation of thymic epithelial cells can be stimulated *in vitro* by both GH and IGF-I [39] and the effect of GH is blocked by either an anti-IGF-I or an anti-IGF-I receptor antibody [34]. Human thymocytes synthesize and secrete GH and IGF-I, and GH functions as an autocrine/paracrine growth factor in the human thymus via locally synthesized IGF-I [40]. In fact, a paracrine effect of thymocyte-secreted GH on human primary thymic epithelial cell (TEC) cultures has been observed [40]. Moreover, it is intriguing that growth factors-stimulated TEC proliferation is significantly inhibited by the somatostatin analog octreotide [41]. On the other hand, somatostatin and GH receptors are expressed in either human TEC or thymocytes [41–43]. Interestingly, both these types of receptor seem to be developmentally and differentially expressed on thymic cells, suggesting a crucial role of the local somatostatin-GH-IGF-I system in immune cell differentiation [43,44]. Furthermore, cytokines other than GH affect IGF-I synthesis in lymphoid tissues, e.g. in macrophages tumor necrosis factor-α (TNF-α) regulates IGF-I production [45], the colony-stimulating factors induce the expression of IGF-I mRNA [46], and the T cell-derived cytokine IFN-γ reduces IGF-I mRNA in a time- and dose-dependent manner [47]. On bone marrow, IGF-I has two major effects on B cell development; it acts as a differentiation factor to potentiate pro-B to pre-B cell maturation [8], and as a B cell proliferation cofactor to synergize with IL-7 [49]. The mature B cell remains sensitive to IGF-I since immunoglobulin production is also stimulated by IGF-I *in vitro* and *in vivo* [50]. In the periphery, IGF-I enhances the proliferative response of lymphocytes to mitogens [51]. In the Snell dwarf mouse, treatment with bovine GH restores many measures of lymphocyte function, but pre-B cell numbers in bone marrow are not restored by bovine GH or ovine PRL [52]. The effects of IGF-I on T cell development are not as well characterized, although thymic T cell progenitors proliferate in response to IGF-I before they respond to any other known cytokine [53]. It is also clear that thymic epithelial cells produce IGF-I, thymocytes express IGF-I receptors [54], and the administration of IGF-I to animals affects the number of T cells in the thymus [33]. Information on the role of IGF-I in processes of positive or negative selection of thymocytes is limited: GH was shown to induce significant migration of resting and activated human T cells [55], suggesting that GH may play a role in normal lymphocyte recirculation by directly altering their adhesive and migratory capacities.

3. THE EFFECT OF GH AND IGF-I ADMINISTRATION ON IMMUNE FUNCTION

The evidence from experimental studies in animals that the entire pituitary, GH, and IGF-I affect hematopoietic and lymphoid tissues is generally accepted [14–16]. In humans, however, data are less clear. In particular, GH-deficient children are not clinically immunodeficient and therefore replacement therapy with hGH would not be expected to have significant effects on immune function [56,57]. However, cytokines, such as interleukin-6 (IL-6), which has a range of pleiotropic actions on T cell and B cell proliferation, IL-1, IL-2, IL-4, and IL-5 have important actions on the immune system, ensuring that homeostasis is maintained. Therefore, the apparent lack of effect of an endocrine GH deficiency should not be taken as evidence that GH has no effects on immune function in humans. In humans, attention has been focussed on investigating the immunological phenotype of GH-deficient patients [58,59]. In GH deficient patients the absolute number of total T lymphocytes and T-cell subsets (using monoclonal Ab as markers), Natural Killer cell activity (target K562) and response of lymphocytes to polyclonal mitogens (PHA, ConA, PWM) were all in the normal range and GH treatment had no effect [60]. In contrast, the absolute number of B lymphocytes was in the normal range before treatment while it was reported to drop significantly after 12 months of GH treatment. It should be mentioned, however, that no data are available on local GH, local IGF peptides, or receptors or local IGF-binding protein (IGFBP) status in GH deficient patients. One group of patients who may be immunologically impaired are patients with defective GH receptor function such as with the GH insensitivity syndrome [61]. The largest cohort yet described, that localized in Ecuador, has a significantly higher pediatric mortality [62], perhaps suggesting a deranged function of the immune system.

The administration of IGF-I has been shown to increase the size of lymphoid organs in several species. Increased spleen weight, due to a doubling in the number of both T- and B lymphocytes, thymic mass, due to a doubling in T lymphocytes, and increase in peanut agglutinin receptor binding (a marker for immature thymocytes), was shown in mice treated with rhIGF-1 for 7 or 14 days [32]. There were no changes in other blood cell numbers [32]. In rats and mice numerous studies report increases in lymphoid tissue mass with IGF-I administration [38]. Mice transgenic for GH or IGF-I have enlarged lymphoid organs [62]. An effect of endogenous IGF-I on lymphoid tissue growth was presumed in a large study [63] including lines of mice selected over many generations on the basis of high or low serum IGF-I levels. The high IGF-I line had spleen and thymus weights greater than the low IGF-I line and developmental patterns of thymus weight closely paralleled those of circulating IGF-I [63]. How IGF-I expands B and T cell number is unclear [15]; it could either act to potentiate differentiation or it to enhance survival, e.g. by reducing apoptosis. In fact, IGF-I has marked anti-apoptotic effects in many tissues and cell types, which may be important in normal growth and differentiation, in tumor growth, and for the protection of tissues from damage [64,65].

B cells seem to be preferentially responsive to rhIGF-1 and show an enhanced immunoglobulin production *in vitro* [15,31,66]. It is possible that if thymic growth and function could be stimulated by growth factors, such as rhIGF-1 or rhGH, then lymphocyte function might be more rapidly restored after chemotherapy, especially in adults. In the setting of combined chemotherapy and bone marrow transplantation, treatment with growth factors may be doubly valuable. The effect of such growth factors on tumor growth needs to be addressed, although initial short-term studies with IGF-I suggest that tumor growth is not enhanced [67]. The hypothesis that high-dose GH therapy may induce proinflammatory cytokines, which are implicated in septic shock has been recently investigated [68]. Cytokine levels were measured in patients undergoing laparoscopic cholecystectomy who were randomized to receive either high-dose GH therapy (13 IU/m^2·day) or placebo. GH did not play any effect on cell proliferation or the production of TNF-α, IL-6, or IFNγ. However, in patients undergoing laparoscopic cholecystectomy there was a time-related effect of surgery on cytokine levels: IL-6 rised while TNF-α decreased at 24 h after surgery [68]. High-dose GH therapy had no effect on the cytokine response, suggesting that high-dose GH therapy does not alter the proinflammatory cytokine response to surgery or endo-

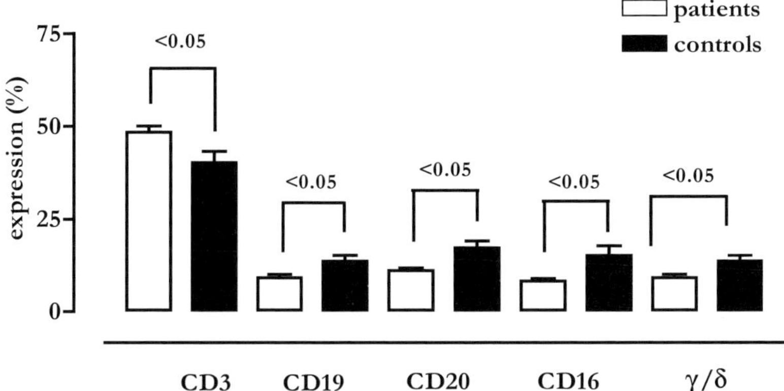

Figure 1. Lympocyte subset patter in the colonic lamina propria during colonoscopy in acromegalic patients as compared to controls. A significant increase of T cells (CD3), decrease of B cells (CD19, CD20), the Natural Killers (CD16), and the γ/δ ratio, was observed in patients with acromegaly.

toxin. On the other hand, overexpression of heterologous or homologous GH in transgenic mice was shown to induce significant stimulation of some parameters of immune function [69]. In fact, in metallothionein I (MT)-bGH transgenic mice with high peripheral levels of bovine GH, the absolute weight of the thymus and the spleen was significantly increased and the mitogenic responses of splenocytes to concanavalin A (ConA), lipopolysaccharide (LPS) and phytohemagglutinin (PHA) were enhanced, as compared to age-matched normal animals [69].

4. THE IMMUNE SYSTEM IN ACROMEGALY

Based on the foregoing, GH seems to be essential for the development, maintenance and regulation of immune function. Although an increased risk of developing neoplasms has been claimed in acromegaly [1,2,4–12], data concerning the immune function in patients with acromegaly are very limited. Colonic polyps, is one of the most frequent neoplasms reported in acromegaly [4–12]. In a previous study, the lymphocyte subset pattern was characterized in the colonic *lamina propria* of 34 acromegalics and 34 controls [11]. A significant decrease of the B cells (CD19, CD20), the Natural Killers (CD16), the γ/δ ratio, considered as index of intraepitelial cytotoxic activity [70], and an increase of the T cells as a whole (CD3) were observed in acromegalics when compared to controls (Figure 1). The changes of the lymphocyte subset pattern were found in patients with active disease while those with inactive disease had a pattern similar to that observed in healthy subjects. Since the study was performed at the mucosal level of the colonic *lamina propria* close to the polyp lesion, the decrease of the helper-inducer ($CD4^+$/leu8^-) lymphocytes suggested that the immunoglobulin synthesis mediated by these T cells at mucosal level is impaired [71] while the increase of the helper-suppressor ($CD4^+$/leu8^+) lymphocytes, that seem to mainly regulate the cell-mediated immune response at mucosal locoregional site, could be caused by the colonic polyp itself [11]. However, both patients with and those without polyps had a similar lymphocyte subset pattern ruling out a pathogenetic role of colonic polyps in the finding of altered lymphocyte subset pattern [11]. In another series of 100 patients with acromegaly compared to 200 sex-, age-matched healthy subjects, T-cell activity, measured as

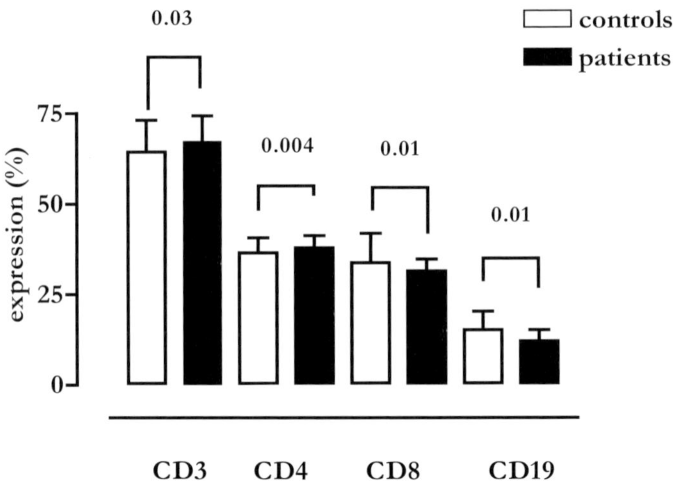

Figure 2. Lymphocyte whole population (CD3), T helpers cells (CD4), T suppressor cells (CD8), and B cell population as a whole (CD19) in 100 patients with acromegaly and 200 controls.

CD3 and CD4, was also shown to be enhanced (Figure 2). The existence of an altered immune function in acromegaly was also reported in a small series of acromegalic patients and controls: while NK cell activity, serum concentrations of immunoglobulins (IgG, IgM, IgA) and metabolic burst activity were within the normal range in both groups, a significantly enhanced phagocytic activity was observed in patients with acromegaly. Kotzmann et al. [72] reported also that surface markers on T lymphocytes (CD3, CD4, CD8), B lymphocytes (CD19) and NK cells (CD16/56) were normal in both groups, but in acromegalic patients, CD4+ and CD8+ cells showed a significant higher expression of transferrin receptors, indicating enhanced T-cell activity. Furthermore, aging has been shown to be accompanied by various changes in the lymphocyte subset distribution. In fact, a significant increase in both the absolute counts and the proportions of CD3 and CD57 was found with age, whereas the cytolytic T cell population showed less change with age [73]. In addition, soluble interleukin-2 receptor levels were found to increase significantly with age and correlated with certain NK cell subsets [73]. The functions of these changes is still unclear but the expansion of some lymphocyte subsets in the elderly was suggested to represent a remodeling of the immune system with ageing, with an increase in non-MHC-restricted cells likely to compensate the previously reported decline in T and B cells [73]. In the series of 100 patients with acromegaly, we observed that the age-related decline in T- and B-cells was also found in elderly acromegalics as compared to young and middle-aged patients [74], suggesting the persistence of this remodeling mechanism in patients with chronic GH/IGF-I hypersecretion (Figure 3).

Finally, a decrease in both directed migration and spontaneous migration was also found in patients with acromegaly when compared to controls, suggesting a putative direct or indirect GH effect on polymorphonuclear *cell* chemotaxis [75].

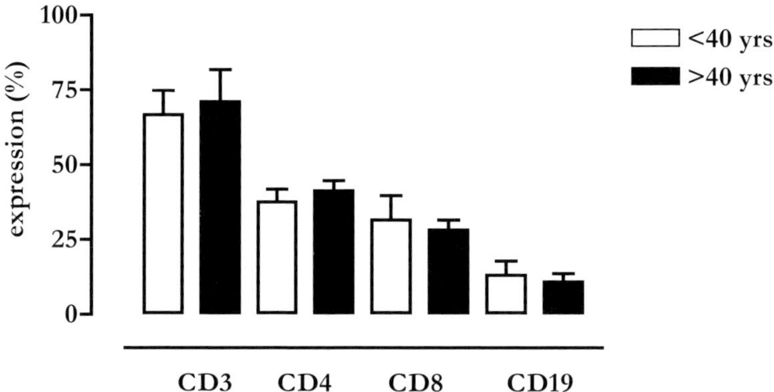

Figure 3. Lymphocyte whole population (CD3), T helpers cells (CD4), T suppressor cells (CD8), and B cell population as a whole (CD19) in 100 patients with acromegaly grouped on the basis of age below (n = 38) or above 40 yrs (n = 62).

CONCLUSIONS

It is conceivable that GH and IGF-I could play a role in lymphocyte development and function that was not detected because the range of assays employed is still too limited or not sensitive enough. The possibility that GH and IGF-I may have effects on immune responses in non-stressed, healthy animals that are compensated by the overlapping actions of other hormones or cytokines [76] cannot be ruled out. Some changes in the lymphocyte subset pattern have been found in acromegaly, but whether these changes play a role in the increased prevalence of neoplasms is still unknown.

REFERENCES

1. Melmed S, Ho K, Klibanski A, Reichlin S, Thorner MO. Recent advances in pathogenesis, diagnosis and management of acromegaly. J Clin Endocrinol Metab 1995;80:3395–3402.
2. Colao A, Merola B, Ferone D, Lombardi G. Extensive personal experience: Acromegaly. J Clin Endocrinol Metab 1997;82:2777–2781.
3. Lamberts SWJ, van der Lely AJ, de Herder WW. Clinical and medical diagnosis of acromegaly. Metabolism 1995;44(suppl.1):15–17.
4. Colao A, Lombardi G. GH and PRL excess. Lancet 1998;352:1455–1461.
5. Ezzat S, Melmed S. Are patients with acromegaly at increased risk for neoplasia? J Clin Endocrinol Metab 1991;72:245–249.
6. Rajasoorya C, Holdaway IM, Wrightson P, Scott DJ, Ibbertson HK. Determinants of clinical outcome and survival in acromegaly. Clin Endocrinol 1994;41:95–102.
7. Orme SM, McNally RJQ, Cartwright RA, Belchetz PE. Mortality and cancer incidence in acromegaly: a retrospective cohort study. J Clin Endocrinol Metab 1998;83:2730–2734.
8. Ituarte EA, Petrini J, Hershman JM. Acromegaly and colon cancer. Ann Int Med 1984;101:627–628.
9. Pines A, Rozen P, Ron E, Gilat T. Gastrointestinal tumor in acromegalic patients. Am J

Gastroenterol 1985;80:266–269.
10. Barzilay J, Heatley GJ, Cushing GW. Benign and malignant tumor in patients with acromegaly. Arch Intern Med 1991;151:1629–1632.
11. Colao A, Balzano A, Ferone D, et al. Increased prevalence of colonic polyps and altered lymphocyte subsets in the colonic lamina propria in acromegaly. Clin Endocrinol 1997;47:23–28.
12. Jenkins PJ, Frajese V, Jones AM, et al. Insulin-like growth factor I and the development of colorectal neoplasia in acromegaly. J Clin Endocrinol Metab 2000;85:3218–3221.
13. Renehan AG, Bhaskar P, Painter JE, et al. The prevalence and characteristics of colorectal neoplasia in acromegaly. J Clin Endocrinol Metab 2000;85:3417–3424.
14. Jones JI, Clemmons DR. Insulin-like growth factors and their binding proteins: biological actions. Endocr Rev 1995;16:3–34.
15. Clark R. The Somatogenic hormones and insulin-like growth factor-1: stimulators of lymphopoiesis and immune function. Endocr Rev 1997;18:157–179.
16. Dorshkind K, Horseman ND. The roles of prolactin, growth hormone, insulin-like growth factor-I, and thyroid hormones in lymphocyte development and function: insights from genetic models of hormone and hormone receptor deficiency. Endocr Rev 2000;21:292–312.
17. Geffner M. Effects of growth hormone and insulin-like growth factor I on T- and B-lymphocytes and immune function. Acta Paediatric Suppl 1997;423:76–79.
18. Smith P. The effect of hypophysectomy upon the involution of the thymus in the rat. Anat Rec 1930;47:119–143.
19. Nagy E, Berczi I. Immunodeficiency in hypophysectomized rats. Acta Endocrinol (Copenh) 1978;89:530–537.
20. Berczi I, Nagy E, Kovacs K, Horvath E Regulation of humoral immunity in rats by pituitary hormones. Acta Endocrinol (Copenh) 1981;98:506–513.
21. Pierpaoli W, Baroni C, Fabris N, Sorkin E. Hormones and immunological capacity. II. Reconstitution of antibody production in hormonally deficient mice by somatotropic hormone, thyrotropic hormone, and thyroxin. Immunology 1969;16:217–230.
22. Fabris N, Pieropaoli W, Sorkin E. Lymphocytes, hormones and ageing. Nature 1972;240:557–559.
23. Weigent DA, Baxter JB, Wear WE, et al. Production of immunoreactive growth hormone by mononuclear leukocytes. FASEB J 1988;2:2812–2818.
24. Warma S, Sabharwal P, Sheridan JF, Malarkey WB. Growth hormone secretion by human peripheral blood mononuclear cells detected by an enzyme-linked immunoplaque assay. J Clin Endocrinol Metab 1993;76:49–53.
25. Hooghe-Peters EL, Hooghe R, editors. Growth hormone, prolactin, and IGF-1 as lymphohemopoietic cytokines. Austin, TX: RG Landes Company, 1995;262.
26. Wu H, Devi R, Malarkey WB. Localization of growth hormone messenger ribonucleic acid in the human immune system: a clinical research center study. J Clin Endocrinol Metab 1996;81:1278–1282.
27. Lesniak MA, Gorden P, Roth J, Gavin Jr I. Binding of ^{125}I human growth hormone to specific receptors in human cultured lymphocytes. J Biol Chem 1974;249:1661–1667.
28. Kiess W, Butenandt O. Specific growth hormone receptors on human peripheral mononuclear cells: reexpression, identification and characterization. J Clin Endocrinol Metab 1985;60:740–746.
29. Roth JA, Kaeberle ML, Grier RL, Hopper JG, Spiegel HE, McAllister HA. Improvement

in clinical condition and thymus morphological features associated with growth hormone treatment of immunodeficient dwarf dogs. Am J Vet Res 1984;45:1151–1155.

30. Goff BL, Roth JA, Arp LH, Incefy GS. Growth hormone treatment stimulates thymulin production in aged dogs. Clin Exp Immunol 1987;68:580–587.
31. Monroe WE, Roth JA, Grier RL, Arp LH, Naylor PH. Effects of growth hormone on the adult canine thymus. Thymus 1987;9:173–187.
32. Clark R, Strasser J, McCabe S, Robbins K, Jardieu P. Insulin-like growth factor-1 stimulation of lymphopoiesis. J Clin Invest 1993;92:540–548.
33. Johnson OL, Cleland JL, Lee HJ, et al. A month-long effect from a single injection of microencapsulated human growth hormone. Nature Medicine 1996;2:795–799.
34. Timsit J, Savino W, Safieh B, Chanson P, Gagnerault MC, Bach JF. Growth hormone and insulin-like growth factor-I stimulate hormonal function and proliferation of thymic epithelial cells. J Clin Endocrinol Metab 1992;75:183–188.
35. Kooijman R, Scholtens LE, Rijkers GT, Zegers BJ. Type I insulin-like growth factor receptor expression in different developmental stages of human thymocytes. J Endocrinol 1995;147:203–209.
36. Kooijman R, Lauf JJ, Kappers AC, Rijkers G. Insulin-like growth factor induces phosphorylation of immunoreactive insulin receptor substrate and its association with phosphatidylinositol-3 kinase in human thymocytes. J Exp Med 1995;182:593–597.
37. Yamada M, Hato F, Kinoshita Y, Tominaga K, Tsuji Y. The indirect participation of growth hormone in the thymocyte proliferation system. Cell Mol Biol 1994;40:111–121.
38. Kelley KW, Arkins S, Li YM. Role of growth hormone and insulin-like growth factor I in immunoregulation. In: Bercu BB, Walker RF, editors. Growth Hormone II: Basic and Clinical Aspects. New York: Springer, 1994;87–103.
39. Tsuji Y, Kinoshita Y, Hato F, Tominaga K, Yoshida K. The *in vitro* proliferation of thymus epithelial cells stimulated with growth hormone and insulin-like growth factor-I. Cell Mol Biol 1994;40:1135–1142.
40. Sabharwal P, Varma S. Growth hormone synthesized and secreted by human thymocytes acts via insulin-like growth factor I as an autocrine and paracrine growth factor. J Clin Endocrinol Metab 1996;81:2663–2669.
41. Ferone D, van Hagen PM, van Koetsveld PM, et al. *In vitro* characterization of somatostatin receptors in the human thymus and effects of somatostatin and octreotide on cultured thymic epithelial cells. Endocrinology 1999;140:373–380.
42. Ferone D, van Hagen PM, Pivonello R, Colao A, Lamberts SW, Hofland LJ. Physiological and pathophysiological role of somatostatin receptors in the human thymus. Eur J Endocrinol 2000;143(Suppl 1):S27–34.
43. de Mello-Coelho V, Gagnerault MC, Souberbielle JC, et al. Growth hormone and its receptor are expressed in human thymic cells. Endocrinology 1998;139:3837–3842.
44. Ferone D, Pivonello R, Van Hagen PM, et al. Age-related decrease of somatostatin receptor number in the normal human thymus. Am J Physiol Endocrinol Metab 2000;279:E791–798.
45. Noble PW, Lake FR, Henson PM, Riches DWH. Hyaluronate activation of CD44 induces insulin-like growth factor-1 expression by a tumor necrosis factor-alpha dependent mechanism in murine macrophages. J Clin Invest 1993;91:2368–2377.
46. Arkins S, Rebeiz N, Brunke-Reese DL, Minshall C, Kelley KW. The colony-stimulating factors induce expression of insulin-like growth factor-I messenger ribonucleic acid during hematopoiesis. Endocrinology 1995;136:1153–1160.

47. Arkins S, Rebeiz N, Brunke-Reese DL, Biragyn A, Kelley KW. Interferon-gamma inhibits macrophage insulin-like growth factor-I synthesis at the transcriptional level. Mol Endocrinol 1995;9:350–360.
48. Landreth K, Narayanan R, Dorshkind K. Insulin-like growth factor-I regulates pro-B-cell differentiation. Blood 1992;80:1–6.
49. Gibson LF, Piktel D, Landreth KS. Insulin-like growth factor-1 potentiates expansion of interleukin-7-dependent pro-B cells. Blood 1993;82:3005–3011.
50. Robbins K, McCabe S, Scheiner T, Strasser J, Clark R, Jardieu P. Immunological effects of insulin-like growth factor-I: enhancement immunoglobulin synthesis. Clin Exp Immunol 1994;95:337–342.
51. Geffner ME, Bersch N, Golde DW. Insulin and IGF-I stimulate normal and virally transformed T-lymphocyte cell growth *in vitro*. Brain Behav Immun 1992;6:377–386.
52. Murphy WJ, Durum SK, Anver MR, Longo DL. Immunologic and hematologic effects of neuroendocrine hormones. Studies on DW/J dwarf mice. J Immunol 1992;148:3799–3805.
53. Gjerset RA, Yeargin J, Volkman SK, Vila V, Arya J, Haas M. Insulin-like growth factor-I supports proliferation of autocrine thymic lymphoma cells with a pre-T cell phenotype. J Immunol 1990;145:3497–3501.
54. Verland S, Gammeltoft S. Functional receptors for insulin-like growth factors I and II in rat thymocytes and mouse thyoma cells. Mol Cell Endocrinol 1989;67:207–216.
55. Taub DD, Tsarfaty G, Lloyd AR, Durum SK, Longo DL, Murphy WJ. Growth hormone promotes human T cell adhesion and migration to both human and murine matrix proteins *in vitro* and directly promotes xenogeneic. J Clin Invest 1994;94:293–300.
56. Auernhammer CJ, Strasburger CJ. Effects of growth hormone and insulin-like growth factor on the immune system. Eur J Endocrinol 1995;133:635–645.
57. Adler WH, Nagel JE, Harman SM, Corpas E, Blackman MR. Growth hormone and immune function in the elderly. In: Blackman MR, Roth J, Harman SM, Shapiro JR, editors. GHRH, GH and IGF-1: Basic and Clinical Advances. New York: Springer, 1995;277–290.
58. Bozzola M, Moretta A, Pagani S, Rappaport RG. Effects of growth hormone on natural killer cells. Acta Paediatr Suppl 1997;423:84.
59. Kiess W, Doerr H, Eisl E, Butenandt O, Belohradsky BH. Lymphocyte subsets and natural-killer activity in growth hormone deficiency. N Engl J Med 1986;314:321.
60. Spadoni GL, Rossi P, Ragno W, et al. Immune function in growth hormone-deficient children treated with biosynthetic growth hormone. Acta Paediatr Scand 1991;80:75–9.
61. Rosenfeld RG, Rosenblom AL, Guevara-Aguirre J. Growth hormone (GH) insensitivity due to primary GH receptor deficiency. Endocr Rev 1994;15:369–390.
62. D'Ercole J. Expression of insulin-like growth factor-1 in transgenic mice. Ann NY Acad Sci 1993;692:149–160.
63. Siddiqui RA, McCutcheon SN, Blair HT, et al. Growth allometry of organs, muscles and bones in mice from lines divergently selected on the basis of plasma insulin-like growth factor-I. Growth Dev Aging 1992;56:53–60.
64. Harrington EA, Bennett MR, Fanidi A, Evan GI. CMyc-induced apoptosis in fibroblasts is inhibited by specific cytokines. EMBO J 1994;13:3286–3295.
65. Baserga R. The insulin-like growth factor receptor: a key to tumor growth? Cancer Res 1995;55:249–252.
66. Kimata H, Yoshida A. Effect of growth hormone and insulin-like growth factor-I on

immunoglobulin production by and growth of human B-cells. J Clin Endocrinol Metab 1994;78:635–641.
67. Dong YL, Fleming RY, Huang KF, Herndon DN, Yan TZ, Waymack JP. Effect of insulin-like growth factor 1 on host response to tumor. J Surg Oncol 1993;53:121–127.
68. Zarkesh-Esfahani SH, Kolstad O, Metcalfe RA, et al. High-dose growth hormone does not affect proinflammatory cytokine (tumor necrosis factor-α, interleukin-6, and interferon-γ) release from activated peripheral blood mononuclear cells or after minimal to moderate surgical stress. J Clin Endocrinol Metab 2000;85:3383–3390.
69. Dialynas E, Brown-Borg H, Bartke A. Immune function in transgenic mice overexpressing growth hormone (GH) releasing hormone, GH or GH antagonist. Proc Soc Exp Biol Med 1999;221:178–183.
70. Viney JL, MacDonalds TT, Spencer J. Gamma/delta T cells in the gut epithelium. Gut 1990;31:841–844.
71. Kanof ME, Stober W, Fiocchi C, Zeitz M, James SP. CD4 positive Leu-8 negative helper-inducer T cells predominate in the human intestinal lamina propria. J Immunol 1988;14:3029–3036.
72. Kotzmann H, Koller M, Czernin S, et al. Effect of elevated growth hormone concentrations on the phenotype and functions of human lymphocytes and natural killer cells. Neuroendocrinology 1994;60:618–624.
73. McNerlan SE, Rea IM, Alexander HD, Morris TC. Changes in natural killer cells, the CD57/CD8 subset, and related cytokines in healthy aging. J Clin Immuno 1998;18:31–36.
74. Colao A, Ferone D, Marzullo P, Panza N, Pivonello R, Orio Jr F, Grande G, Bevilacqua N, Lombardi G. Lymphocyte subset pattern in acromegaly. J Endocrinol Invest 2002.
75. Fornari CM, Palacios FM, Diez RA, Intebi AD. Decreased chemotaxis of neutrophils in acromegaly and hyperprolactinemia. Eur J Endocrinol 1994;130:463–468.
76. Paul WE. Pleiotropy and redundancy: T cell-derived lymphokines in the immune response. Cell 1989;57:521–524.

Growth and Lactogenic Hormones
Edited by L. Matera and R. Rapaport

Growth Hormone and Insulin-Like Growth Factor-1 in Human Immunodeficiency Virus Infection

MITCHELL E. GEFFNER

Children's Hospital Los Angeles, Keck School of Medicine, University of Southern California, 4650 Sunset Blvd, Los Angeles, CA 90027, USA

ABSTRACT

The GH-IGF axis has been the focus of ongoing attention in both children with HIV infection and growth failure and in adults with HIV wasting. These studies have been propelled, in part, by the similarity in body composition seen in patients with HIV wasting and those with GH deficiency. In both HIV-infected children and growth failure and adults with wasting, variable derangements of GH secretion, circulating IGF-1 and IGFBP-3 levels, and GH sensitivity have been reported. Much of this variability likely stems from differences in study entry criteria as well as in subtle differences in the clinical status of participating subjects. Whether or not the observed abnormalities in the GH-IGF axis are intrinsically related to HIV infection or are secondary to nutritional, infectious, or other complications remains uncertain. There has been limited reported usage of GH in children with HIV infection and growth failure, although a multicenter study is soon to be commenced to evaluate efficacy and toxicity in this population. In contrast, a 12-week course of high-dose GH therapy for the treatment of AIDS wasting in adults appears to have short-term benefits in terms of body composition and is approved by the FDA for this purpose. To date, there have been limited data supporting the notion that either GH and/or IGF-1 therapy have immunological benefits in HIV-infected subjects. It also remains unknown whether GH has any sustained long-term benefits or possible late-presenting adverse effects in this setting.

1. INTRODUCTION

Dysfunction of the growth hormone (GH)-insulin-like growth factor-1 (IGF-1) axis has been reported in both children and adults with HIV infection. Observed abnormalities may involve disturbances of hypothalamic-pituitary function affecting production of GH, result from end-organ resistance to GH and/or its growth-promoting mediator, IGF-1, and/or be caused by associated conditions such as malnutrition. Although limited data support the use of GH to treat growth failure in HIV-infected children, short-term treatment with GH, capitalizing on its anabolic actions, is approved by the United States Food and Drug Administration (FDA) for use in HIV-infected adults with the wasting syndrome. There appear to be no unusual safety issues concerning GH usage in this population.

2. ABNORMALITIES OF GROWTH IN HIV-INFECTED CHILDREN

Growth failure, specifically failure-to-thrive (FTT), is a common and occasional presenting feature of pediatric HIV infection, acquired either by vertical transmission or by transfusion [1–7]. It is estimated to occur in 50% of children with symptomatic HIV disease [8]. As the infection becomes more advanced, growth failure may progress to a distinct wasting syndrome analogous to that seen in adults. In one unique at-risk population, seropositive hemophilic children, the onset of growth failure was an early sign of progression toward symptomatic HIV disease [9].

We studied the intrauterine and postnatal growth patterns (through 36 months of age) of children with transplacentally acquired HIV infection. Significantly lower mean standard deviation scores (SDS) were observed for birth weight in those infants who were subsequently found to be HIV-positive than in those who were uninfected, despite no difference in prevalence of prematurity. However, because of considerable overlap between groups, birth weight was not a useful predictor of ultimate HIV status. In this same cohort, there were no differences in either mean length or mean head circumference at birth between the two groups. Mean postnatal weight, length, and head circumference were significantly less in our HIV-positive infants than in uninfected infants at multiple time points during the first 36 months of life. Most of the observed postnatal growth delay occurred in HIV-infected infants who, during the first 3 years of life, shifted from an asymptomatic to a symptomatic state [10].

The etiology of growth failure in pediatric HIV infection appears to be multifactorial and likely includes effects from associated chronic disease, malnutrition, and opportunistic infection [3,11,12]. The non-specific effects of chronic disease may contribute by causing decreased intake (secondary to anorexia, esophagitis, and abdominal pain) and enteropathy (resulting in diarrhea and malabsorption). In addition, HIV-infected children manifest hypermetabolism (increased resting expenditure) and catabolism.

Hormonal abnormalities have also been implicated as the cause of growth failure at least in some HIV-infected children. While primary thyroid failure is rare, subtle central *hypothyroidism* in children has been reported to manifest by a failure to observe the expected rise in nocturnal TSH during serial sampling [3], as has elevated serum *thyroxine-binding globulin* levels [13,14]. Both central and primary *hypogonadism* has been reported commonly in HIV-infected adult males. Similarly delayed puberty, especially in boys, especially those with HIV secondary to hemophilia, has been reported frequently, and could also adversely impact growth during adolescence [13]. Slightly elevated basal and ACTH-stimulated serum *cortisol* levels have been described in some HIV-infected children that, in theory, could limit their linear growth [15]. The effects of drugs used to treat HIV infection or its complications could also interfere with the function of the endocrine system and normal growth, e.g., *ketoconazole* used to treat overwhelming fungal infections may reversibly inhibit *testosterone* biosynthesis [14].

3. POTENTIAL ROLE OF GH AND IGF-1 IN THE ETIOLOGY OF GROWTH FAILURE IN PEDIATRIC HIV INFECTION

GH deficiency has been reported infrequently in HIV-infected children using either pharmacological [3,10] or physiological [16] assessments and appears to occur less often than might be predicted based on the prevalence of AIDS encephalopathy in children [17] and pituitary infection and tumors in adults [18]. Furthermore, a direct effect of the HIV virus on GH secretion could occur, as the HIV envelope protein, *gp120*, is *neurotoxic* and its intracerebroventricular injection suppresses GH secretion in rats [19] and in culture [20]. Nonetheless, to date, only two children with HIV infection and GH deficiency following pharmacological stimulation have been described [21,22].

IGF-1 levels have been reported to range from variably reduced with some correlation to growth rate [5,10–12,16,21–23] to normal [3–10]. Low serum levels of the other GH-dependent surrogates, IGF binding protein-3 (IGFBP-3) and acid-labile sub-unit (ALS), have been found in poorly growing children with HIV infection. These proteins combine with IGF-1 and circulate as a three-part or ternary complex presumed to serve as the means by which hepatically-generated IGF-1 is delivered to its target tissues. These same children tend to have enhanced IGFBP-3 proteolysis and elevated levels of IGFBP-1, as seen in other hypercatabolic states [23]. Lower serum IGF-1 and IGFBP-3 levels were found in slowly growing HIV-infected children than in those growing better [24,25].

The possibility that GH and/or IGF-1 resistance may be present in HIV-infected subjects also needs to be considered in part because associated malnutrition has secondary effects on the GH-IGF-1 axis, both in animals and in humans. Rats, either fasted or fed a reduced protein diet, show decreased GH-receptor and IGF-1 mRNA levels in liver [26]. The low serum IGF-1 levels that ensued were not restored to normal following physiological GH replacement, but were comparable to GH-treated hypophysectomized rats [27]. That these malnourished animals also could have concomitant *IGF-1 resistance* was supported by a failure of administered IGF-1 to stimulate carcass growth [28,29]. In contrast, infusion of IGF-1 into well-nourished hypophysectomized rats at a 50% lower dose induced a significant increase in carcass growth. Furthermore, combined infusion of GH and IGF-1 also failed to stimulate significant carcass growth in the malnourished rats [30]. Similarly, increased serum GH and reduced serum IGF-1 levels have been reported in human fasting, kwashiorkor, and marasmus. In obese humans subjected to dietary restriction and treated with GH for 11 weeks a dissipation of an initially observed increase in nitrogen balance was observed, despite the maintenance of persistent elevations in serum IGF-1 concentrations [31]. This suggests that loss of the early anabolic action of GH may have resulted from development of IGF-1 resistance.

Our own study was aimed at assessing the possible roles of GH and/or IGF-1 resistance in the etiology of growth failure of HIV-infected children. We examined six asymptomatic normally-statured HIV-infected children (P1 patients); 10 symptomatic, short, but not acutely ill, HIV-infected children (P2 patients); and six equally short, normal children (controls) [10]. Mean weight-to-height SDS ratios were similar in all three groups, suggesting that nutritional status did not differ between groups. There were no significant differences between groups with respect to mean plasma levels of IGF-1 or other growth-promoting hormones. As an index of hormone sensitivity, we quantified *in vitro* colony formation of *erythroid progenitor cells*, isolated from peripheral blood of study subjects, in response to GH and IGF-1. Mean overall responsiveness of erythroid progenitor cells of P2 patients to both growth factors (studied separately) was reduced compared to that of both P1 subjects and controls. Thus, we found that more severe HIV infection in children is associated with *in vitro* presumably acquired resistance to the growth-promoting actions of GH and IGF-1 that is unrelated to stature, malnutrition, or overwhelming illness. Thus, we concluded that *resistance to GH and IGF-1*, intrinsic to HIV infection, could contribute to the poor *in vivo* growth seen in most symptomatic children.

The mechanism by which IGF-1 resistance develops in children with symptomatic HIV infection could be related to the increased production of cytokines that occurs in this population. Specifically, *tumor necrosis factor* (TNF) production is increased in some AIDS patients, particularly in the setting of opportunistic infections, progressive encephalopathy, and wasting [32,33]. Higher spontaneous and phytohemagglutinin-stimulated *interleukin-6* secretion by cultured peripheral blood mononuclear cells (PBMCs) was found in slowly growing HIV-infected children than in those growing better, consistent with a growth-inhibitory role for this cytokine [25]. While IGF-1 has been shown to promote human adult and embryonic erythropoiesis, TNF inhibits *in vitro* erythroid colony formation [34]. Although IGF-1 promotes *in vitro* cartilage matrix formation (proteoglycan synthesis) equally in the presence or absence of TNF-α [35], transient *in vitro* exposure of human myoblasts to TNF-α significantly inhibits IGF-1-stimulated protein synthesis [36]. Thus, the *in vitro* cellular resistance to IGF-1 could result from exaggerated production of TNF and/or other cytokines.

4. THE WASTING SYNDROME IN HIV-INFECTED ADULTS

The AIDS *wasting syndrome* is defined by the Centers for Disease Control and Prevention as follows: it is a profound, involuntary weight loss of greater than 10% of baseline body weight. Chronic diarrhea (≥2 loose stools per day) is present for more than 30 days, in the absence of a concurrent illness or condition (other than HIV infection) that could explain these symptoms (e.g., cancer, tuberculosis, cryptosporidiosis, or other specific enteritis) [37]. The lost weight consists of both lean body mass (in the form of muscle protein) and fat. The development of this *cachexia* represents a devastating complication of AIDS and, importantly, is associated with a poor prognosis [38–40]. The prevalence of the wasting syndrome appears to be decreasing among HIV-infected adults with advancements in therapy [41].

A specific etiology for the wasting syndrome in adults with HIV infection remains uncertain. Multiple contributing factors have been proposed [42], including decreased energy intake resulting from decreased appetite, depression and fatigue, painful lesions in the mouth and throat, and/or malabsorption; increased resting energy expenditure resulting from hypermetabolism and/or altered metabolism; and excess cytokine production (especially TNF-α, interleukin-1, and interleukin-6).

5. POTENTIAL ROLE OF GH AND IGF-1 IN THE PATHOGENESIS OF THE WASTING SYNDROME IN ADULT HIV INFECTION

GH deficiency is considered in the pathogenesis of the wasting syndrome, which is based on similarities in body composition between patients with AIDS wasting and those with GH deficiency. The preservation of fat mass and preferential loss of lean body mass [43] is characteristic, the latter leading to negative nitrogen balance. The data regarding GH secretion in response to pharmacological provocation in HIV-infected adults have been conflicting. In one study, HIV-infected adults who had no AIDS-defining illness and who were asymptomatic had a 50% reduction in GH secretion by Pulsar analysis, whereas those with AIDS were profoundly GH-deficient with an even greater reduction in GH secretion [44]. In contrast, in other studies, hypogonadal men with HIV infection secreted more GH than did age-matched control subjects [45,46] and maximal GH levels stimulated by GHRH were increased in men with AIDS [47]. Finally, in yet another study, no difference between GH secretory patterns of HIV-infected adults and controls was observed. Analysis of the circadian rhythm of physiological GH secretion in non-acutely ill, HIV-positive adults has also yielded conflicting results with one study demonstrating a normal and another study a blunted pattern of secretion [48,49].

With regard to circulating IGF-1 levels in HIV-infected adults, again variable results have been reported. Adults with AIDS often have plasma IGF-1 levels in the normal range [50–52], but, as the disease progresses and weight loss ensues, IGF-1 (and IGF-2) levels fall by between 50–75% [44,53]. Low IGF-1 levels may also reflect GH resistance as occurs in pediatric HIV infection [8]; however, studies in HIV-infected adults have yielded conflicting results. A study showing decreased *in vivo* GH-stimulated muscle protein synthesis has been reported in HIV-infected adults [54]. However, the same group subsequently found no difference in GH-stimulated serum IGF-1 levels among controls, HIV-positive subjects without weight loss, and HIV-positive subjects with >10% weight loss [50]. With regard to IGF binding proteins, basal IGFBP-3 levels are either reduced [54] or normal [50] in GH-deficient AIDS patients. The rise in GH-stimulated serum IGFBP-3 levels was similar in controls and in HIV-positive subjects without weight loss, but blunted in HIV-positive subjects with >10% weight loss [50]. As in children, many adult patients with HIV infections exhibited IGFBP-3 proteolysis [53]. Lastly, IGFBP-1 levels are increased (perhaps because of associated insulinopenia, hypercortisolemia, and/or elevated concentrations of cytokines such as TNF-α) and circulate in a free form devoid of IGFs [50,53].

6. GH AND IGF-1 THERAPY IN THE CONTEXT OF TREATMENT OF CHILDREN WITH HIV INFECTION

The mainstay of treatment for the growth failure seen in children with HIV infection involves a combination of nutritional supplementation and highly active anti-retroviral therapy (HAART), including the use of protease inhibitors [55]. In our own studies, protease inhibitor therapy was associated with a statistically significant increase in height velocity and in height SDS, but no significant increase in weight SDS. These responses were unrelated to virological or immunological status, or to degree of pretreatment growth failure [56]. Appetite stimulants have also been employed sporadically and appear to augment weight, but not height, in HIV-infected children [57]. Unfortunately, all of these approaches, when associated with improved growth and weight gain, preferentially increase fat and not lean body mass (muscle). Such a response in body composition, however, is not associated with improved disease outcome.

Hormonal therapy was reported for two HIV-infected children with growth failure and with apparent GH deficiency, one with moderate improvement in height velocity [21], and the other with substantial improvement in height velocity and correction of marked wasting [22] (Figures 1 and 2). Another reported patient failed to improve her growth velocity with a one-year trial of GH, but had a marked improvement in weight and height gain following the institution of treatment with protease inhibitors [58]. Hirschfeld et al. published a preliminary report of a large scale study. Anti-viral therapy was given to a group of HIV-infected children with growth failure for a 3-month run-in period, followed by a 6-month trial of either GH (at a dose of 40 μg/kg/day) or IGF-1 (at a dose of 90 μg twice daily). An improvement in linear growth rate and lean body mass resulted from such teratment. With both regimens, there were also increases in granulocyte number, stabilization of total lymphocyte number, increases in CD4 and CD8 lymphocyte numbers, humoral antibody responses to both T-cell-independent and -dependent antigens, and decreases in p24 antigen and HIV viral RNA [59,60]. A smaller pilot study of acceptability, tolerance, and efficacy was carried out on 5 non-GH-deficient children with HIV-associated growth failure. In response to GH treatment, given for 28 days at a dose of 0.067 mg/kg/day, the patients showed an increased weight (but not lean body mass), and serum levels of IGF-1 and IGFBP-3, with no adverse effect on viral load [61]. Two of these subjects were treated for a total of three years with sustained improvements in height standard deviation scores from –3.6 to –2.0 and –4.7 to –1.9, respectively. These preliminary findings, combined with the efficacy of short-term GH treatment of AIDS wasting in adults (see Section VII), have led to the initiation of a national multi-center trial of GH use in HIV-infected children under the auspices of the Pediatric AIDS Clinical Trial Group (PACTG).

7. GH AND IGF-1 THERAPY AS PART OF THE TREATMENT REGIMEN OF ADULTS WITH HIV INFECTION AND WASTING SYNDROME

The initial approach to the treatment of the wasting syndrome in HIV-infected adults has traditionally centered on improved nutrition (e.g., high-calorie supplements, enteral alimentation, and parenteral hyperalimentation), the use of appetite stimulants (e.g., megesterol acetate [Megace®], dronabinol, and corticosteroids), and HAART [62]. Agents that suppress cytokine activity (e.g., thalidomide) have also been employed. As in children, these regimens do not significantly increase lean body mass and, thus, do not alter disease outcome when the wasting syndrome is present [63,64]. Trials using anabolic agents, such as oxandrolone, nandrolone, and oxymethalone, have been associated with increased weight gain and, in the case of nandrolone, favorable improvements in body composition as well [65–67].

The rationale for consideration of GH (and/or IGF-1) treatment of the wasting syndrome is, in part, as previously noted, based on the fact that the body composition abnormalities resemble those seen in untreated GH-deficient adults. Additionally, GH administration to individuals with an array of catabolic disorders, including those with burns, cancer cachexia, and undergoing gastrointestinal surgery, has reversed negative nitrogen balance [51]. Furthermore, since both GH and IGF-1 cause proliferation of lymphocytes in vitro [68], the possibility of a beneficial effect in these immunodeficient subjects also made GH and IGF-1 potentially attractive therapies [69]. In a series of small trials of variable duration (all ≤12 weeks) and design, adults with HIV wasting were treated with high-dose GH regimens [70–74] and, in all studies, manifested, on average, decreased nitrogen excretion, increased lean body mass, and/or decreased fat mass. A larger randomized, double-blind, placebo-controlled multi-center study showed significant improvements in body weight after 6 weeks, but not 12 weeks, no significant changes in either median CD4 counts or in plasma HIV-RNA levels, and an improved quality-of-life [75]. The most prolific study evaluating GH treatment of AIDS wasting was a randomized, placebo-controlled Phase III trial (sponsored by Serono Laboratories, Inc.). GH was given (0.1 mg/kg/day) to 178 HIV-positive subjects (with an average pre-study weight loss of 14%), or placebo was administered for 12 weeks. The GH-treated subjects exhibited significant weight gains, a significant difference in lean body mass, a significant decrease in fat mass, and a significant increase in treadmill performance. No difference in quality-of-life as assessed by questionnaire, CD4, CD8 or lymphocyte counts, or plasma HIV RNA levels [76] occurred compared to those receiving placebo. A crossover extension to this study also suggested that AIDS-defining events, such as *Candida* esophagitis, lymphoma, and Kaposi sarcoma developed less frequently in subjects treated with GH than with placebo [77]. In each of these studies, however, there was little or no evidence of any clinically relevant improved functional correlate, e.g., improvement in muscle strength or endurance.

These results led to approval of the Serono GH product, known as Serostim®, by the FDA in 1997 for the treatment of AIDS wasting for 12 weeks, to be given subcutaneously on a daily basis concomitant with standard anti-retroviral therapy. The dosing instructions, according to the package insert, are as follows: 0.1 mg/kg/day (<35 kg), 4 mg/day (35–45 kg), 5 mg/day (45–55 kg), and 6 mg/day (>55 kg). The cost of such a regimen is approximately $20,000 per patient. The use of such high doses is presumed necessary to overcome associated GH resistance.

More recently, a two-week course of GH administration to HIV-positive patients concurrent with the presence of an opportunistic infection has been shown to improve protein metabolism and body composition [78]. The authors suggested that GH treatment may have been beneficial to the acute recovery of these individuals. Such a conclusion must be tempered with the reported increased mortality rate observed with GH therapy given to very ill patients hospitalized in intensive care units [79]. The explanation for the increased mortality is unclear, but, because of the increased rate of occurrence of sepsis, septic shock, uncontrolled infection, and multiple organ failure in the GH-treated group, a possible adverse effect of GH on the immune system must be considered. This effect could be mediated through cytokine generation, reduced levels of reactive oxygen species, enhanced susceptibility to endotoxin, reduced glutamine mobilization by white and red blood cells, and/or through secondary hyperglycemia [80]. In yet another short-term study of GH sensitivity in HIV-positive adults, McNurlan et al. found no difference in anabolic activity of bone, as measured by serum levels of propeptide of type 1 collagen (PICP) following a two-week course of high-dose GH treatment, between controls, subjects with asymptomatic HIV infection, subjects with AIDS, and subjects with AIDS and >10% weight loss. In the same study, however, GH-stimulated albumin synthesis, as measured isotopically, was markedly depressed in subjects with AIDS and weight loss [81]. The FDA-approved GH regimen has also been partially successful in reversing the fat redistribution syndrome (characterized by truncal adiposity and buffalo humps) seen in HIV-positive adults receiving HAART [82,83]. In one final study, short-term GH administration to 6 HIV-infected adults with weight loss resulted in a sustained improvement in body composition and sustained improved functionality [84].

Attempts to bypass possible GH resistance have led to several trials of IGF-1 therapy in adults with HIV wasting. In one such study of affected individuals fed a weight-maintaining diet, daily infusions of IGF-1 (at a dose of 4 μg/kg/hour) produced significant short-term, but not sustained, nitrogen retention [85]. Of note, a threefold higher IGF-1 infusion rate failed to induce any significant nitrogen retention, perhaps due to simultaneous increases in serum IGFBP-1 concentrations. In a subsequent pilot study reported in abstract form, IGF-1 administered at a daily dose of 90 μg/kg/day subcutaneously for 14 days to one HIV-negative and one HIV-positive individual resulted in significant and sustained nitrogen retention [86]. Overall, the observed nitrogen retention with IGF-1 was typically less than that seen with GH in this population. The use of IGF-1 in subjects with HIV-associated wasting must be balanced against its potential hypoglycemic action that may be heightened secondary to their low weight and limited energy stores, anorexia, malabsorption, and increased sensitivity to the metabolic actions of insulin.

As a result of this potentially unfavorable risk:benefit profile for IGF-1 therapy in this population, therapy with both GH and IGF-1 has been proposed. Further support of this concept comes from a study of combined therapy in healthy subjects consuming a hypocaloric diet, in which the degree of attained nitrogen retention exceeded that seen with IGF-1 therapy alone [87]. In a study of 60 subjects with HIV-associated wasting who were randomized to receive either GH (1.4 mg/day), IGF-1 (10 mg/day), both agents, or placebo for 12 weeks, the largest observed increase in lean body mass was found in the combination group [72]. However, this increase in lean body mass was not associated with any improvements in quality of life, muscle strength, or immune function. In another cohort, GH at a low dose of 0.7 mg/day combined with IGF-1 at a high dose of 10 mg/day was administered for 6 weeks to adults with HIV-associated wasting. This combination induced an increase in fat-free mass in the form of water, but not in true lean body mass or in nitrogen retention. This poor response was thought to be the result of use of a sub-optimal dose of GH [54,88]. In a separate study comparing immunodeficient HIV-infected adults treated with GH or IGF-1, or with both GH and IGF-1 for 12 weeks, no significant changes in CD4 counts, CD45 subsets, natural killer cell function, immunoglobulin levels, or *in vitro* IL-2 production in response to each of 5 HIV-envelope peptides were noted in any group [89].

8. SAFETY CONCERNS RELATED TO GH AND IGF-1 THERAPY IN HIV-INFECTED PATIENTS

That GH and/or IGF-1 treatment could adversely affect the clinical status in patients with HIV infection has been of theoretical concern. This concern is based on the fact that either GH or IGF-1 can stimulate T-cell proliferation, monocyte function, and cytokine production, and all of these are factors that could affect HIV replication. A clone of promonocytic cells *chronically* infected with HIV and susceptible to viral induction by various cytokines and protein kinase C activators, was exposed to GH, ranging between 10 to 500 ng/mL (including those induced in serum by GH treatment of humans). No direct effect on the HIV long terminal repeat was observed and GH did not upregulate the virus. However, viral replication was enhanced by GH in *acutely* infected PBMCs, but there was no blunting of the antiviral efficacy of co-administered zidovudine (AZT) [90]. In contrast, *in vitro* incubation of physiological concentrations of IGF-1 exerted significant inhibitory effects on chronically HIV-infected U937 cells [91] and variable effects on PHA-stimulated PBMCs [92]. Based on studies of mice with severe combined immunodeficiency disease treated with AZT, GH partially neutralized the myelosuppressive effects of AZT, which, in similarly treated human HIV infection, could be a beneficial side effect [93]. To date, there has been no direct association of GH and/or IGF-1 therapy with worsening of human HIV infection.

The more typical, presumably dose-related, side effects of GH that occur in GH-deficient adults without HIV infection, such as carpal tunnel syndrome and other complaints related to fluid retention, gynecomastia, insulin resistance, and glucose intolerance, have generally been of only mild-to-moderate degree and responded to symptomatic treatment, or to GH dose reduction.

9. CONCLUSION

The GH-IGF axis has been increasingly studied in both children and adults with HIV infection. Variable derangements have been observed in both cohorts. For the most part, GH secretion appears to be normal. Both low and normal IGF-1 and IGFBP-3 levels have been found, along with elevated IGFBP-1 levels. Insofar as GH action is concerned, depending on the method of assessment, both reduced and normal sensitivity has been described. Much of the variability probably stems from obvious differences in study entry criteria as well as subtle differences in the clinical status of study subjects. Whether the observed abnormalities in the GH-IGF axis are intrinsically related to HIV infection or are secondary to nutritional, infectious, or other issues, remains uncertain. In children with HIV infection and growth failure, routine treatment with GH is not indicated. However, a 12-week course of high-dose GH therapy for the treatment of AIDS wasting in adults appears to have short-term benefits in terms of body composition and is approved by the FDA for this purpose. Initial hopes that either GH and/or IGF-1 therapy would also have immunological benefits have not been realized. Furthermore, it remains unknown as to whether such an expensive modality of therapy has any sustained long-term benefits or possibly delays the adverse effects of infection [94].

ACKNOWLEDGEMENTS

This work was supported by grants from the Pediatric AIDS and Carolan Foundations, and from Pharmacia, Inc. I would also like to thank Dr. E. Richard Stiehm whose early guidance kindled my interest in this field and to Dr. David Golde for his unyielding support over the years. I would also like to acknowledge the assistance, advice, and collaboration of the following individuals with regard to some of the studies that form the basis for this review: Noelle Bersch, Robert Bailey, Ph.D., Daina Dreimane, M.D., Yvonne J. Bryson, M.D., Paul Krogstad, M.D., Karin Nielsen, M.D., Kerry Gallagher, M.D., Andrea Kovacs, M.D., Victor Israele, M.D., Diana Yeh, M.D., and Elliot Landaw, M.D., Ph.D.

REFERENCES

1. Falloon J, Eddy J, Weiner L, Pizzo PA. Human immunodeficiency virus in children. J Pediatr 1989;114:1–30.
2. Rapaport R, McSherry G, Connor E, et al. Growth and hormonal parameters in symptomatic human immunodeficiency virus (HIV) infected children. Pediatr Res 1989;25:187A (abstract).
3. Laue L, Pizzo PA, Butler K, Cutler Jr GB. Growth and neuroendocrine dysfunction in children with acquired immunodeficiency syndrome. J Pediatr 1990;117:541–45.
4. Grubman S, Conviser R, Oleske J. HIV infection in infants, children, and adolescents. In: Wormser GP, editor. AIDS and Other Manifestations of HIV Infection. New York, NY: Raven, 1992;201–16.
5. Rapaport R, Sills I, Figueroa W, Hoyt I, Mintz M. Growth failure (GF) in HIV infected children (HIV-IC). Pediatr Res 1993;82A (abstract).
6. Matarazzo P, Palomba E, Lala R, et al. Growth impairment, IGF I hyposecretion and thyroid dysfunction in children with perinatal HIV-1 infection. Acta Paediatr 1994;803:1029–34.

7. Laue L, Cutler Jr GB. Abnormalities in growth and development. In: Pizzo PA, Wilfert CM, editors. The Challenge of HIV Infection in Infants, Children and Adolescents, 2nd Ed. Baltimore, MD: Williams & Wilkins, 1994;575–90.
8. Arpadi SM. Growth failure in children with HIV infection. J Acquir Immune Defic Syndr 2000;25(suppl 1):S37–42.
9. Brettler DB, Forsberg A, Bolivar E, Brewster F, Sullivan J. Growth failure as a prognostic indicator for progression to acquired immunodeficiency syndrome in children with hemophilia. J Pediatr 1990;117:584–88.
10. Geffner ME, Yeh DY, Landaw EM, et al. *In vitro* insulin-like growth factor-I, growth hormone, and insulin resistance occurs in symptomatic human immunodeficiency virus-I-infected children. Pediatr Res 1993;34:66–72.
11. Schwartz LJ, St. Louis Y, Wu R, Wiznia A, Rubinstein A, Saenger P. Endocrine function in children with human immunodeficiency virus infection. Amer J Dis Child 1991;145:330–33.
12. Lepage P, Van de Perre P, Van Vliet G, et al. Clinical and endocrinological manifestations in perinatally human immunodeficiency virus type 1-infected children aged 5 years or older. Amer J Dis Child 1991;145:1248–1251.
13. Rapaport R. Endocrine alterations in human immunodeficiency virus infections. In: Lifshitz F, editor. Pediatric Endocrinology, 3rd Ed. New York, NY: Marcel Dekker, 1996;809–19.
14. Geffner ME. Endocrine, growth, and development: evaluation and therapy. In: Zeichner SL, Read JS, editors. Handbook of Pediatric HIV Care. Philadelphia, PA: Lippincott, Williams & Wilkins, 1999;433–39.
15. Oberfield SE, Kairam R, Bakshi S, et al. Steroid response to adrenocorticotropin stimulation in children with human immunodeficiency virus infection. J Clin Endocrinol Metab 1990;70:578–81.
16. Rondanelli M, Caselli D, Maccabruni A, et al. Involvement of hormonal circadian secretion in the growth of HIV-infected children. AIDS 1998;12:1845–50.
17. Wiznia AA, Nicholas SW. Organ system involvement in HIV-infected children. Pediatr Ann 1990;19:475–6, 479–81.
18. Ferreiro J, Vinters HV. Pathology of the pituitary gland in patients with the acquired immune deficiency syndrome (AIDS). Pathology 1988;20:211–15.
19. Buzy J, Brenneman DE, Pert CB, Martin A, Salazar A, Ruff MR. Potent gp120-like neurotoxic activity in the cerebrospinal fluid of HIV-infected individuals is blocked by peptide T. Brain Res 1992;598:10–18.
20. Mulroney SE, McDonnell KJ, Pert CB, et al. HIV gp120 inhibits the somatotropic axis: a possible GH-releasing hormone receptor mechanism for the pathogenesis of AIDS wasting. Proc Natl Acad Sci USA 1998;95:1927–32.
21. Jospe N, Powell KR. Growth hormone deficiency in an 8-year-old girl with human immunodeficiency virus infection. Pediatrics 1990;86:309–12.
22. Dreimane D, Gallagher K, Nielsen K, et al. Growth hormone exerts potent anabolic effects in an adolescent with HIV wasting. Pediatr Infect Dis J 1999;18:167–69.
23. Frost RA, Nachman SA, Lang CH, Gelato MC. Proteolysis of insulin-like growth factor-binding protein-3 in human immunodeficiency virus-positive children who fail to thrive. J Clin Endocrinol Metab 1996;81:2957–62.
24. Johann-Liang R, O'Neill L, Cervia J, et al. Energy balance, viral burden, insulin-like growth factor-1, interleukin-6 and growth impairment in children infected with human

immunodeficiency virus-positive children. AIDS 2000;14:683–90.

25. de Martino M, Galli L, Chiarelli F, et al. Interleukin-6 release by cultured peripheral blood mononuclear cells inversely correlates with height velocity, bone age, insulin-like growth factor-I, and insulin-like growth factor binding protein-3 serum levels in children with perinatal HIV-1 infection. Clin Immunol 2000;94:212–18.
26. Bornfeldt KE, Arnqvist HJ, Enberg B, Mathews LS, Norstedt G. Regulation of insulin-like growth factor-I and growth hormone receptor gene expression by diabetes and nutritional state in rat tissues. J Endocrinol 1989;122:651–56.
27. Thissen JP, Triest S, Underwood LE, Maes M, Ketelslegers JM. Divergent responses of serum insulin-like growth factor-I and liver growth hormone (GH) receptors to exogenous GH in protein-restricted rats. Endocrinology 1990;126:908–13.
28. Philipps AF, Persson B, Hall K, et al. The effects of biosynthetic insulin-like growth factor-I supplementation on somatic growth, maturation, and erythropoiesis on the neonatal rat. Pediatr Res 1988;23:298–305.
29. Thissen JP, Underwood LE, Maiter D, Maes M, Clemmons DR, Ketelslegers JM. Failure of insulin-like growth factor-I (IGF-I) infusion to promote growth in protein-restricted rats despite normalization of serum IGF-I concentrations. Endocrinology 1991;128:885–90.
30. Ketelslegers JM, Maiter D, Maes M, Underwood LE, Thissen JP. Nutritional regulation of insulin-like growth factor-I. Metabolism 1995;44 (suppl 4):50–57.
31. Snyder DK, Clemmons DR, Underwood LE. Treatment of obese, diet-restricted subjects with growth hormone for 11 weeks: effects on anabolism, lipolysis, and body composition. J Clin Endocrinol Metab 1988;67:54–61.
32. Mintz M, Rapaport R, Oleske JM, et al. Elevated serum levels of tumor necrosis factor are associated with progressive encephalopathy in children with acquired immunodeficiency syndrome. Am J Dis Child 1989;143:171–74.
33. Remick DG. Significance of *in vivo* detection of tumor necrosis factor. Lab Invest 1991;65:259–61.
34. Migliaccio AR, Migliaccio G. Human embryonic hemopoiesis: control mechanisms underlying progenitor differentiation *in vitro*. Devel Biol 1988;125:127–34.
35. Tyler JA. Insulin-like growth factor 1 can decrease degradation and promote synthesis of proteoglycan in cartilage exposed to cytokines. Biochem J 1989;260:543–48.
36. Frost RA, Lang CH, Gelato MC. Transient exposure of human myoblasts to tumor necrosis factor α (TNF) selectively inhibits serum and insulin-like growth factor (IGF) stimulated protein synthesis. Endocr Soc Prog Abstr 1997;100 (abstract).
37. Council of State and Territorial Epidemiologists: AIDS Program, Center for Infectious Diseases, Centers for Disease Control and Prevention. Revision of the CDC surveillance case definition for acquired immunodeficiency syndrome. MMWR Morb Mortal Wkly Rep 1987;36(suppl 1):3S-15S.
38. Chlebowski RT. Significance of altered nutritional status in acquired immunodeficiency syndrome (AIDS). Nutr Cancer 1985;7:85–91.
39. Von Roenn JH, Murphy RL, Wegener N. Megesterol acetate for treatment of anorexia and cachexia associated with human immunodeficiency virus infection. Semin Oncol 1990;176:13–16.
40. Grunfeld C, Feingold KR. Metabolic disturbances and wasting in the acquired immunodeficiency syndrome. N Engl J Med 1992;3275:329–37.
41. Abrams DL. Potential interventions for HIV/AIDS wasting: an overview. J Acquir Immune Defic Syndr 2000;25(suppl 1):S74–80.

42. Mulligan K, Schambelan M. Growth hormone in AIDS. In: Smith RG, Thorner MO, editors. Human growth hormone: research and clinical practice. Totowa, NJ: Humana Press, 2000;285–95.
43. Bengtsson BÅ, Edén S, Lönn L, et al. Treatment of adults with growth hormone (GH) deficiency with recombinant human GH. J Clin Endocrinol Metab 1993;76:309–17.
44. Frost RA, Fuhrer J, Steigbigel R, et al. Wasting in the acquired immunodeficiency syndrome is associated with multiple defects in the serum insulin-like growth factor system. Clin Endocrinol 1996;44:501–14.
45. Grinspoon S, Corcoran C, Lee K, et al. Loss of lean body and muscle mass correlates with androgen levels in hypogonadal men with acquired immunodeficiency syndrome. J Clin Endocrinol Metab 1996;81:4051–58.
46. Grinspoon S, Corcoran C, Stanely T, Katznelson L, Klibanski A. Effects of androgen administration on the growth hormone-insulin like growth factor I axis in men with acquired immunodeficiency syndrome. J Clin Endocrinol Metab 1998;83:4251–56.
47. Wilson LD, Truong MPM, Barber AR, et al. Anterior pituitary and pituitary target organ function in men infected with human immunodeficiency virus. Metabolism 1996;145:738–46.
48. Heijligenberg R, Sauerwein HP, Brabant G, Endert E, Hommes MJ, Romijn JA. Circadian growth hormone secretion in asymptomatic human immunodeficiency virus infection and acquired immunodeficiency syndrome. J Clin Endocrinol Metab 1996;81:4028–32.
49. Rondanelli M, Solerte SB, Fioravanti M, et al. Circadian secretory pattern of growth hormone, insulin-like growth factor type I, cortisol, adrenocorticotropic hormone, thyroid-stimulating hormone, and prolactin during HIV infection. AIDS Res Hum Retroviruses 1997;13:1243–49.
50. Minarcik DC, Frost RA, Lang CH, et al. Insulin-like growth factor system in patients with HIV infection: effect of exogenous growth hormone administration. J Acquir Immune Defic Syndr 1999;22:49–55.
51. Frost RA, Lang CH, Gelato MC. Growth hormone/insulin-like growth factor axis in human immunodeficiency virus-associated disease. The Endocrinologist 1997;7:23–31.
52. Mulligan K, Grunfeld G, Hellerstein MK, Neese RA, Schambelan M. Anabolic effects of recombinant human growth hormone in patients with wasting associated with human immunodeficiency virus infection. J Clin Endocrinol Metab 1993;77:956–62.
53. Lee PDK, Pivarnik JM, Bukar JG, et al. A randomized, placebo-controlled trial of combined insulin-like growth factor I and low dose growth hormone therapy for wasting associated with human immunodeficiency virus infection. J Clin Endocrinol Metab 1996;81:2968–75.
54. Gelato MC, Frost RA, DiCristofara K, Garlick PJ, Lang CH. Growth hormone (GH) deficiency and diminished anabolic response in muscle of HIV infected adults. Program of the Annual Meeting of the Endocrine Society, San Francisco, CA, 1996;253 (abstract).
55. Nadal D, Steiner F, Cheseaux JJ, Rudin C. Ritonavir promotes increased growth in HIV-infected children. AIDS 1998;12:2356–57.
56. Dreimane D, Nielsen K, Devakis A, Bryson YJ, Geffner ME. Effect of protease inhibitors combined with standard antiretroviral therapy on liner growth and weight gain in HIV-1-infected children. Pediatr Infect Dis J 2001;20:315–16.
57. Clarick RH, Hanekom WA, Yogev R, Chadwick EG. Megestrol acetate treatment of growth failure in children infected with human immunodeficiency virus. Pediatrics 1997;99:354–57.

58. Zeitler PS, Travers S, Kappy MS. Advances in the recognition and treatment of endocrine complications in children with chronic illness. Adv Pediatr 1999;46:101–47.
59. Hirschfeld S. Use of human recombinant growth hormone and human recombinant insulin-like growth factor-I in patients with human immunodeficiency virus infection. Horm Res 1996;46:215–21.
60. Hirschfeld S, Tudor-Williams G, Zafonte M, Jacobsen F, Pizzo PA. HIV-infected children with growth failure respond to growth hormone of IGF-1. Program of the Annual Meeting of the Endocrine Society, San Francisco, CA, 1996;766 (abstract).
61. Pinto G, Blanche S, Thiriet I, Souberbielle JC, Goulet O, Brauner R. Growth hormone treatment of children with human immunodeficiency virus-associated growth failure. Eur J Pediatr 2000;159:937–38.
62. Gilden D. No attrition in research on wasting therapies. GMHC Treat Issues 1997;11:12–16.
63. Kotler DP, Tierney AR, Pierson Jr RN. Magnitude of body-cell-mass depletion and the timing of death from easting in AIDS. Am J Clin Nutr 1989;50:444–47.
64. Suttmann U, Ockenga J, Selberg O, Hoogestraat L, Deicher H, Muller MJ. Incidence and prognostic value of malnutrition and wasting in human immunodeficiency virus-infected patients. J Acquir Immune Defic Syndr 1995;9:20–25.
65. Berger JR, Pall L, Hall CD, Simpson DM, Berry PS, Dudley R. Oxandrolone in AIDS-wasting myopathy. AIDS 1996;10:1657–62.
66. Gold J, High HA, Li Y, et al. Safety and efficacy of nandrolone decanoate for treatment of wasting in patients with HIV infection. AIDS 1996;10:745–52.
67. Mulligan K, Tai VW, Schambelan M. Use of growth hormone and other anabolic agents in AIDS wasting. J Parenter Enteral Nutr 1999;23:S202–09.
68. Geffner ME. Effects of growth hormone and IGF-I on T- and B-lymphocytes and immune function. Clin Pediatr Endocrinol 1999;8(suppl 12):19–28.
69. Chappel S. Growth hormone in immune reconstitution. J Acquir Immune Defic Syndr Hum Retrovirol 1999;20:423–31.
70. Mulligan K, Grunfeld C, Hellerstein MK, Neese RA, Schambelan M. Anabolic effects of recombinant human growth hormone in patients with wasting associated with human immunodeficiency virus infection. J Clin Endocrinol Metab 1993;77:956–62.
71. Mulligan K, Tai VW, Schambelan M. Effects of chronic growth hormone treatment on energy intake and resting energy metabolism in patients with human immunodeficiency virus-associated wasting. J Clin Endocrinol Metab 1996;83:1542–47.
72. Krentz AJ, Koster FT, Crist DM, et al. Anthropometric, metabolic, and imunological effects of recombinant human growth hormone in AIDS and AIDS-related complex. J Acquir Immune Defic Syndr 1993;6:245–51.
73. Waters D, Danska J, Hardy K, et al. Recombinant human growth hormone, insulin-like growth factor I, and combination therapy in AIDS-associated wasting: a randomized, double-blind, placebo-controlled trial. Ann Intern Med 1996;125:865–72.
74. Klaude S, Fruhauf L, Michels B, Malkowski JP, Helm EB, Althoff PH. HIV-associated wasting syndrome – therapy with mammalian cell-derived recombinant human growth hormone (r-hGH[m]). XI International Conference on AIDS. Vancouver, British Columbia, Canada. July 7–12, 1996;1384 (abstract).
75. Berger DS, LaMarca A, Landy H, Kauffman RS, Breitmeyer A. A Phase III study of recombinant human growth hormone (mammalian cell-derived) in patients with AIDS wasting. XI International Conference on AIDS, Vancouver, British Columbia, Canada,

July 7–12, 1996;422.
76. Schambelan M, Mulligan K, Grunfeld G, et al. and the Serostim Study Group. Recombinant human growth hormone in patients with HIV-associated wasting. Ann Intern Med 1996;125:873–82.
77. Daar ES, La Marca A, Schambelan M, Grunfeld C, Mulligan K, Breitmeyer J. Effect of continuous growth hormone therapy on lean body mass (LBM) and AIDS-defining events in HIV-associated wasting (HIV-W). Third Conference on Retrovirus and Opportunistic Infections, Washington DC, January 28–February 1, 1996;125 (abstract).
78. Paton NI, Newton PJ, Sharpstone DR, et al. Short-term growth hormone administration at the time of opportunistic infections in HIV-positive patients. AIDS 1999;13:1195–1202.
79. Takala J, Roukonen E, Webster NR, et al. Increased mortality associated with growth hormone treatment in critically ill patients. N Engl J Med 1999;341:785–92.
80. Ruokonen E, Takalaa J. Dangers of growth hormone therapy in critically ill patients. Ann Med 2000;32:317–22.
81. McNurlan MA, Garlick PJ, Frost RA, et al. Albumin synthesis and bone collagen formation in human immunodeficiency virus-positive subjects: differential effects of growth hormone administration. J Clin Endocrinol Metab 1998;83:3050–55.
82. Torres RA, Unger KW, Cadman JA, Kassous JY. Recombinant human growth hormone improves truncal adiposity and ‘buffalo humps’ in HIV-positive patients on HAART. AIDS 1999;13:2479–81.
83. Wanke C, Gerrior J, Kantaros J, Coakley E, Albrecht M. Recombinant human growth hormone improves the fat redistribution syndrome (lipodystrophy) in patients with HIV. AIDS 1999;12:2099–2103.
84. Nemechek PM. Short-term human growth hormone therapy for HIV-associated wasting. AIDS Patient Care STDS 1999;7:391–93.
85. Lieberman SA, Butterfield GE, Harrison D, Hoffman AR. Anabolic effects of recombinant insulin-like growth factor-I in cachectic patients with the acquired immunodeficiency syndrome. J Clin Endocrinol Metab 1994;78:404–10.
86. Mulligan K, Schambelan M. Growth hormone and other pharmacologic interventions. In: Gorbach SL, Miller TD, editors. Nutritional Aspects of HIV Infection. Philadelphia, PA: Thomson Science, 1999;169–78.
87. Kupfer AR, Underwood LE, Baxter RC, Clemmons DR. Enhancement of the anabolic effects of growth hormone and insulin-like growth factor I by use of both agents simultaneously. J Clin Invest 1993;91:391–96.
88. Ellis KJ, Lee PDK, Pivarnik JM, Bukar JG, Gesundheit N. Changes in body composition of human immunodeficiency virus-infected males receiving insulin-like growth factor I and growth hormone. J Clin Endocrinol Metab 1996;81:3033–38.
89. Nguyen B-Y, Clerici M, Venzon DJ, et al. Pilot study of the immunologic effects of recombinant insulin-like growth factor in HIV-infected patients. AIDS 1998;12:895–904.
90. Laurence J, Grimison B, Gonenne A. Effect of recombinant human growth hormone on acute and chronic human immunodeficiency virus infection *in vitro*. Blood 1992;79:467–72.
91. Germinario RJ, DeSantis T, Wainberg MA. Insulin-like growth factor 1 and insulin inhibit HIV Type 1 replication in cultured cells. AIDS Res Human Retroviruses 1995;11:555–61.
92. Sharma UK, Song H, Casella SJ, Schwartz DH. Failure of insulin-like growth factor I to suppress HIV type 1 in adult or umbilical cord mononuclear blood cells. AIDS Res Human Retroviruses 1997;13:105–10.
93. Murphy WJ, Tsarfaty G, Longo DL. Growth hormone exerts hematopoietic growth-

promoting effects *in vivo* and partially counteracts the myelosuppressive effects of azidothymidine. Blood 1992;80:1443–47.

94. Frindisch PA, Papatheofanis FJ, Matuszewski KA. Recombinant human growth hormone for AIDS-associated wasting. Ann Pharmacother 1998;32:437–45.

Growth and Lactogenic Hormones
Edited by L. Matera and R. Rapaport

Human Prolactin as an Immunohematopoietic Factor: Implications for the Clinic

SUSAN M. RICHARDS

Immunology Laboratory, Cell and Protein Therapeutics R&D, Genzyme Corporation, Framingham, MA 10701, USA

ABSTRACT

Human prolactin is increasingly recognized as a hormone capable of influencing the neuroendocrine, immune and hematopoietic systems. Prolactin has been shown to affect various cell lineages, resulting in both direct and indirect effects that influence and possibly accelerate immune and hematopoietic parameters. Both *in vitro* and *in vivo* studies suggest that prolactin can enhance cell function, accelerate lymphoid and myeloid reconstitution and promote restoration of deficient hematopoiesis. The hormone appears to play this role primarily under conditions of immunohematologic deficiency rather than steady-state hematopoiesis. Therefore prolactin is implicated as a "stress-hormone", involved in restoring immune and hematopoietic homeostasis under conditions of impaired function. Use of such an immunohematopoietic factor for certain systemic conditions, such as drug-induced myelosuppression, immune reconstitution, or hematopoietic cell transplants may provide clinical benefit.

1. INTRODUCTION

Human prolactin (PRL) is a single polypeptide of 199 amino acid residues that is synthesized and secreted by the anterior pituitary as well as by extrapituitary tissues. In adult females, plasma PRL concentrations are generally <20 ng/ml, and increase markedly during pregnancy. In adult males, plasma PRL concentrations are generally <10 ng/ml [1]. The higher PRL levels in women reflect the role of estrogen in regulating the secretion of PRL. Plasma levels are under negative control by dopaminergic neurons residing in the hypothalamus.

2. PROLACTIN SECRETION

Normal circadian secretion of PRL occurs periodically in a distinctive pattern. In normal human subjects, there are about fourteen pulses of PRL secretion in 24 hours, approximately one every 95 minutes [2]. This is overlaid on a continuous PRL secretion pattern that is bimodal; i.e. peaks during mid-sleep and is minimal around noon [2,3]. The need for these secretary patterns is not well understood.

The pulsatile and diurnal mechanisms of PRL secretion appear to be independent of each other and can be affected individually by disease or stress. For example, patients with prolactinomas continue to have pulsatile PRL secretion but the circadian variation is abolished [4]. In addition, a subset of systemic sclerosis patients has been reported as presenting with a shift in diurnal PRL rhythm [5]. Neuroendocrine changes during stress responses can also differ in the acute and chronic phase of critical illness [6]. Alterations in the hypothalamic dopaminergic tone can also occur. A study in HIV-positive individuals suggested that induced PRL release, versus basal serum concentrations, may be impaired in these patients [7]. PRL was also one of the first hormones identified to show increased serum concentrations in response to acute physical or psychological stress [8], a rise that may be mediated by vasoactive intestinal polypeptide, oxytocin, dopaminergic pathways or other still uncharacterized factors. The basis of these alterations and their impact on homeostasis vs. disease warrants further investigation.

In addition to prolactin that is secreted by the anterior pituitary gland, PRL is also synthesized at other sites where it mediates local function [9]. Extra-pituitary prolactin produced by human lymphocytes is believed to have effects on the immune system, much like a cytokine [10–12]. Local production of prolactin has also been reported to occur in the bone marrow environment [13,14]. Receptors for prolactin are expressed on diverse bone marrow-derived cell types [11,14,15] including CD34+ human hematopoietic progenitor cells [16]. The multiple sites of PRL production as well as the widespread expression of prolactin receptors imply a more generalized physiologic role for the molecule. In elucidating the effect of this hormone, consideration needs to be given to the influence of systemic circulating levels of PRL vs. local production, where PRL concentrations may be significantly higher and have an autocrine/paracrine role.

3. RATIONALE FOR USE OF PRL IN THE CLINIC

The importance of PRL to the development of blood and immune cell lineages was initially suggested by the association of defects in hematopoiesis with deficiencies in prolactin production or secretion. Hypophysectomized or bromocriptine-treated rats [17,18] and DW/J dwarf mice, which lack prolactin and other pituitary hormones [19] typically exhibit immunologic and hematologic defects. These animals present with an atrophied thymus and peripheral lymphoid organs as well as a wide range of immunologic abnormalities, including suppressed numbers of splenic hematopoietic progenitor cells. Pituitary transplants or administration of exogenous prolactin was able to restore these functions. An immunohematologic role is further implicated by the observation that a combination of hypophysectomy and immunoneutralization using anti-prolactin antibodies resulted in severe anemia and death from hematological failure [20].

Numerous *in vitro* and *in vivo* studies have provided evidence that prolactin effects multi-lineage cell compartments of both the immune and hematopoietic system. *In vitro* studies have demonstrated that prolactin can enhance responses of activated T cells, B cells, macrophages, antigen presenting cells and NK cells [21–24]. Prolactin can also act in concert with other cytokines. It has been reported that PRL may function with IL-2 as a progression factor for T cell proliferation [25,26] and can synergize with low-dose IL-2 to activate LAK cells and enhance their function [27,28]. Prolactin has been reported to up modulate INF-γ production [29,30]. Prolactin can also increase IL-2 induced INF-γ production by NK cells and act cooperatively

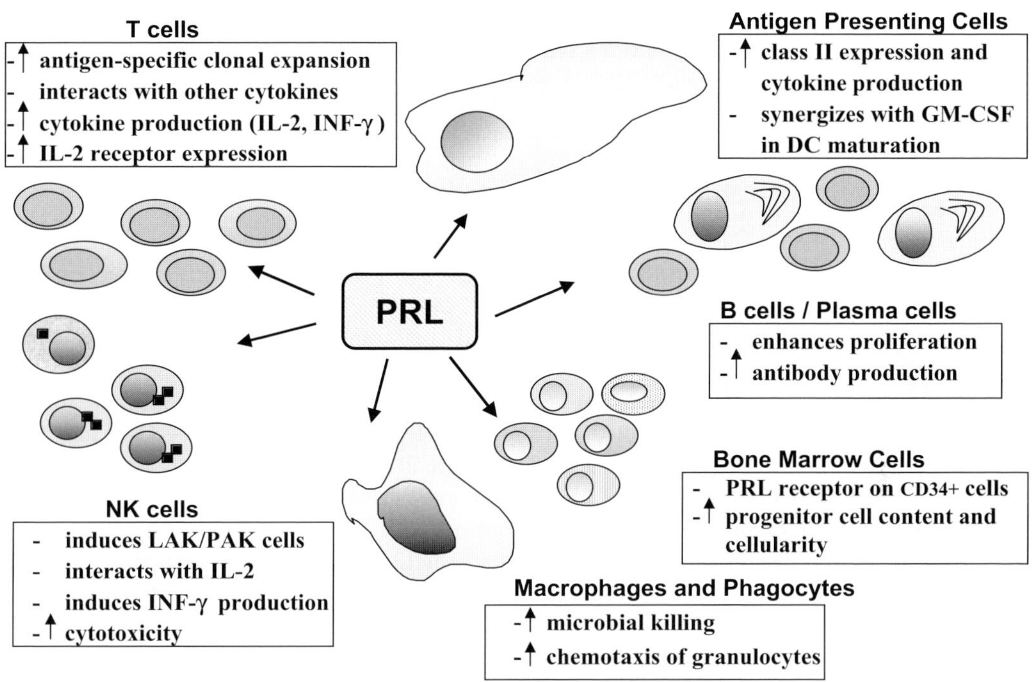

Figure 1. PRL effect on the Immunohematopoietic System. PRL can influence the development, number, and function of various cells in the bone marrow as well as lymphoid organs. It has direct and indirect actions, which can affect hematopoiesis and immune function.

with IL-12 to enhance INF-γ release by T cells [31]. With respect to B cells, PRL can increase proliferation of B cell hybridomas in response to IL-4, IL-5, and IL-6, resulting in an overall increase in antibody production [32].

In vivo immunomodulatory studies have demonstrated that PRL can enhance human T cell engraftment in HuPBL-SCID mice and promote antibody development in vaccine adjuvant studies [23]. Evidence for prolactin having hematopoietic activity was seen in studies of AZT-induced myelosuppression [33] as well as in a murine syngeneic bone marrow transplantation model where administration of exogenous prolactin promoted multi-lineage hematopoietic reconstitution and accelerated hematopoiesis [34,35]. Collectively these studies suggest that prolactin may have both direct and indirect effects in potentiating progenitor cells and enhancing immune and hematopoietic function (Figure 1).

4. POTENTIAL CLINCAL APPLICATIPON

Prolactin, unlike other cytokines, has the potential of providing broad multi-lineage effects with limited toxicity after systemic administration. Such a factor could be of use in several clinical settings. Disease-related situations such as HIV, cancer and opportunistic infections often present with aberrant or suboptimal immune function. States of generalized systemic trauma such as surgery, burns or hematopoietic reconstitution, where multiple cellular defects are detected, would benefit from accelerated kinetics of restored immunohematopoietic function. Potential clinical indications for PRL are further detailed below.

4.1. Immune compromised conditions

Preclinical studies support the concept that prolactin can play a role in restoring depressed immune function. The disease progression seen with HIV infection provides a clinical indication where this hypothesis could be evaluated. HIV infection is a very complex interaction between the virus and the immune system. Latently infected CD4+ T cells or infectious antigen presenting cells (APC) release virus in transient lytic bursts thereby infecting other immunologically co-activated cells. This results in a significant decrease in T cell number, reduced supply of memory lymphocytes, an imbalance in the T-cell repertoire, and overall reduced immune activation. Initiation of highly active antiretroviral therapy (HAART) during primary HIV infection significantly suppresses viral replication but does not lead to complete eradication of the virus [36]. While early initiation of HAART does help to preserve CD4+/CD8+ T-cell homeostasis, differences between HIV-infected patients and controls with regard to T cell and T-cell receptor gene repertoires becomes increasingly greater over time [37,38]. Immune-enhancing strategies that prevent the disruption of homeostasis in the immune system may provide clinical benefit. Therapeutic administration of PRL holds promise for this indication.

Prolactin has been shown to affect the in vitro growth of CD34+ human hematopoietic progenitor cells [16]. Human CD34+ cells express prolactin receptors; therefore the molecule may act as a growth factor or lineage-specific differentiation factor for these cells. Preclinical studies have also demonstrated that PRL has an apparent effect on multi-lineage progenitor cells [23,33–35]. These findings suggest that treatment with PRL could potentially ameliorate the depleted T cell repertoire seen in HIV-infected patients [37,39] as well as provide improved general immune function [40].

Prolactin can also antagonize TGF-β's suppressive activity on proliferation [32]. This may have clinical relevance since TGF-β is implicated as a host factor in the regulation of HIV-replication. TGF-β has an enhancing effect upon the rate of virus production in monocyte-derived macrophages [41]. TGF-β has also been reported as being over expressed in B cells of HIV-infected individuals, resulting in B cell dysfunction [42]. Prolactin may play a role in minimizing TGF-β induced immune suppression in these patients.

Burn patients represent another patient population with a dysfunctional immune system, which contributes to their risk for infection. These patients have acquired defects in multiple cell types including macrophages, neutrophils, T cells and B cells. Defective neutrophil function has been observed in several functional assays, including chemotaxis, phagocytosis, and intracellular killing of bacteria or fungi [43,44]. Generalized decreased immunoresponsiveness occurs, including cutaneous anergy, decreased lymphoproliferative responses, decreased cytotoxicity, alterations in cytokine induction and increased apoptosis in CD4+ T lymphocytes [44–46]. B cell abnormalities are also observed. Decreased serum immunoglobulins levels and depressed antibody responses have been reported [47]. In addition, suppressive serum factors, such as TGF-β, prostaglandins, endotoxin, and corticosteriods are also detected in burn patient sera [48]. The extent of immune dysfunction after burn injury depends both on the severity of the burn and time after injury.

Prolactin's ability to affect various immune cell types and compensate for depressed immune function makes it a candidate for consideration in this patient population. In vitro and in vivo studies indicate PRL broadly influences immune function. In addition to its ability to modulate T and B cell function, PRL has also been shown to influence function of macrophages. This has been largely studied in infectious disease models in which macrophage function is crucial to the clearance of infection and survival of the host. PRL increases survival of *Salmonella typhimurium*-infected mice [49,50] and in combination with INF-γ or TNF-α increased survival of mice infected with *Toxoplasma gondii* [51]. PRL restored the release of IL-1β and IL-6 by macrophages and increased survival of sepsis after hemorrhage [52,53]. Phagocytosis of *Candida albicans* by macrophages is increased by PRL treatment [52]. The action of PRL on macrophage function may be mediated indirectly though T cells that release cytokines that in turn activate macrophages [53].

4.2. Immunomodulation

Several cancers have been shown to respond to immunomodulation. The immune response to tumor cells involves the participation of several distinct leukocyte populations including natural killer (NK) cells and cytotoxic T lymphocytes. NK cells are large granular lymphocytes that use cytoplasmic granules containing perforins and granzymes to kill the target cells in the absence of any deliberate immunization. They make up about 15% of the peripheral blood lymphocytes and about 3–4% of the splenic lymphocytes. Appreciable numbers of NK cells are also found in the lung interstitium, the intestinal mucosa and the liver. The cytolytic activity of NK cells is profoundly affected by cytokines [54]. This is illustrated by the *in vitro* induction of lymphokine activated killer (LAK) cells generated in the presence of high, nonphysiologic doses of IL-2 [55]. Currently patients with metastatic renal cell carcinoma and melanoma are undergoing treatment with IL-2. The antineoplastic effects of IL-2 appear to be potentially dose dependent having side effects associated with induction of secondary cytokines including the proinflammatory cytokines IL-1, IL-6, IL-8, and TNF-α [56,57].

Prolactin can induce *in vitro* NK cell proliferation and up-regulate INF-γ production, which results in enhanced cytotoxicity [27,30]. Prolactin has also been shown to synergize with suboptimal concentrations of IL-2 to induce the maturation of NK cells to fully competent LAK effectors [27]. Generation of LAK activity in six-day PBMC cultures stimulated with IL-2 is markedly reduced by a neutralizing anti-prolactin antiserum. Further, it has been demonstrated that PRL acts as a paracrine factor during IL-2 induced LAK differentiation of NK cells [21]. The mechanism of action of PRL on NK activity may, in part, be due to the amplification of the IL-2 signal. The upregulation of IL-2 and the IL-2 receptors by PRL has been demonstrated [58,59]. It is also possible that PRL may activate IL-2 target genes relevant to LAK maturation. Activation of common genes has been demonstrated after binding of PRL and IL-2 to their cognate receptor [60,61].

Current evidence indicates that the effects of PRL on NK cells are dependent on the dose of the hormone. Suboptimal concentrations of the hormone synergize with IL-2 to augment in vitro activation [27,28], but supraphysiological levels may be inhibitory. Studies using patient derived tumor cells have demonstrated that PRL can influence the sensitivity of these cells to killing by effector cells. Susceptibility of acute myeloid leukemia cells to cell-mediated cytotoxicity by NK/LAK cells was increased in the presence of PRL [62]. Prolactin alone can also induce PRL activated lymphocytes (PAK) as demonstrated in studies using NK susceptible cell lines and endemic Burkitt's lymphoma cells [63]. Different sensitivities to the three types of Burkitt's lymphoma cells were seen with PRL or the combination of PRL and IL-2, pointing out the necessity of testing each tumor type. Prolactin alone may offer a non-toxic way of inducing the activation of NK cell for tumor immunomodulation in specific cancer types. Prolactin can also enhance antigen-specific clonal expansion of T cells and may be an essential competence signal for naïve T cells to mature to antigen-specific cells [23]. This may provide a means to propagate cytotoxic T lymphocytes. PRL could also be used in combination with lower dose of IL-2, thereby providing similar clinical benefit as higher IL-2 doses but with less toxicity.

4.3. Hematologic/Immunologic reconstitution

Cancer chemotherapy can be looked upon as a continuum, from conventional chemotherapy to hematopoietic stem cell transplantation (HSCT). Patients may be cytopenic for days to weeks after conventional chemotherapy, pending on the type of cancer being treated. Following hematopoietic stem cell transplant, there is a prolonged period of profound immune deficiency [64]. The etiology of the immune defect is multi-factorial. Thymopoietic defects resulting in decreased ability to generate new T cells after HSCT are important since complete immune reconstitution depends on the generation of new T cells from hematopoietic stem cells, just as long-term myeloid and erythroid reconstitution depends on human stem cell engraftment. Transfer of committed progenitors or mature donor-derived T cells may permit short-term immune function.

Preclinical data suggest that the use of PRL to enhance reconstitution warrants further study. Prolactin was shown to accelerate hematopoiesis in a syngeneic murine bone marrow transplantation model [34,35]. Significant increases in hematopoietic progenitor cell (CFU-GM and BFU-e) content and cellularity was measured in both the bone marrow and spleen. Administration of PRL to these animals also increased B cell progenitor cell content in the spleen and bone marrow as well as increased T cell progenitor content in the thymus and mature T cell content in the spleen and lymph nodes. Functionality of mature cells such as B, T, and NK cells was also improved.

The SCID (severe combined immune deficiency) mouse provides another model to assess in vivo immunomodulatory activity. SCID mice have a genetic defect resulting in their lacking T or B cells. Human peripheral blood lymphocytes (PBL) can be administered to these mice to study their function. HuPBL-SCID mice receiving PRL exhibited significantly increased number of human CD3+ T cells and human CD19+ B cells in spleen, lymph node and thymus [23]. These results indicate that administration of PRL can enhance the engraftment of human lymphoid cells into appropriate lymphoid organs. In addition, the functional activity of these cells was also improved.

A recent study evaluated plasma prolactin and its drug-induced modulation in 20 women with breast cancer undergoing high-dose chemotherapy and autologous blood stem-cell transplantation [65]. Elevated PRL levels were observed during conditioning and within thirty days after transplant. Prolactin plasma levels were further elevated during antiemetic therapy, such as metoclopramide and phenothiazines. Patients showing higher PRL levels while receiving PRL elevating drugs appeared to have the most favorable course after transplant. Further follow-up of these patients (median, 3 years) indicated that patients remaining in continuous complete remission after transplant had higher prolactin levels compared with those obtaining only partial remission or ensuing early relapse [65]. These findings suggest that PRL may improve immunotherapeutic strategies after HSCT. Larger studies will be needed to validate this approach. However, as with all ablative strategies, immunohematopoietic reconstitution with PRL would require that the growth of the underlying tumor is not promoted, and that the chance of developing graft-versus host disease is not enhanced .

5. SAFETY CONSIDERATIONS

A general concern for any molecule having immunoregulatory function is whether it would exacerbate autoimmune disease or accelerate tumor growth. A causal link between elevated PRL levels and clinical autoimmune disease has not been conclusively established. Elevated PRL levels were first noted in male patients with systemic lupus erythematosus [66] and a correlation was made between hyperprolactinemia and active systemic lupus erythematosus [66–68]. Other studies have documented that although elevated PRL levels were associated with the detection of autoantibodies, there was no correlation between the hyperprolactinemia and disease activity in these patients [69–72]. The current literature would suggest that although hyperprolactinemia may contribute to the development of autoantibodies, there is no clear indication this is associated with the development of clinical disease or of exacerbating preexisting disease. This, however, does not eliminate the possibility that there may be a subset of patients who prove to be sensitive to prolactin.

The development or exacerbation of cancer in susceptible patients due to hyperprolactinemia is controversial. Prolactin receptors have been identified on different cancers including breast [73], prostate [74], and osteosarcoma [75]. Prolactin can also be synthesized in different tissues, such as the mammary gland [9]. Consequently, paracrine/autocrine effects of the locally produced hormone may be more significant than circulating PRL. Recently an isoform of the prolactin receptor that does not bind PRL was identified on normal and neoplastic breast tissue [76]. This may provide a mechanism to regulate the expression of the receptor gene transcript in accordance to the microenvironment. It is important to note that cancer cells are responsive to many factors including cytokines and estrogen. Biologically available estradiol is a risk factor for subsequent development of breast cancer [77,78]. The effect of PRL relative to estrogen, cytokines and other growth factors, as well as their inter-relationship, needs further study.

Results of two long-term prospective studies argue against a causal relationship between elevated PRL levels and cancer. In a prospective 20-year cohort study of 5100 women living in Great Britain on the Island of Guernsey (chosen because of its stable population), 145 women developed breast cancer. At ten years when the statistics were evaluated, the authors initially reported there might be a relationship between breast cancer in postmenopausal women and prolactin. [79]. However, when re-examined at twenty years there was no significant relationship between the development of breast cancer and prolactin in either premenopausal or postmenopausal women [80]. In the second study, a cohort of 9156 schizophrenic patients on their first admission to a psychiatric hospital was evaluated. The results indicated that although these patients had chronically elevated prolactin levels secondary to neuroleptic medication, they did not have an increased incidence of cancer, including breast cancer. The overall incidence of cancer was found to be reduced, especially in males, with decreased rates of testicular cancer and melanoma [81]. At the present time, there is no epidemiological evidence to indicate that elevated systemic levels of PRL can inherently induce or exacerbate cancer.

6. CONCLUDING REMARKS

Prolactin has the ability to function as a hormone and a cytokine. The ability of PRL to accelerate hematopoietic recovery, influence progenitor cells, normalize peripheral blood counts, and activate both myeloid and lymphoid mature blood cells could be utilized in several clinical indications. The hormone appears to have a compensatory role during states of immunohematopoietic dysregulation and can influence multi-cell lineages. Physiologic versus pharmacologic levels of PRL could have different effects, therefore clinical studies will need to establish the appropriate therapeutic doses. Further studies are also needed to understand the relationship between circulating PRL levels and those achieved in the tissue microenvironment. Lastly, a significant body of literature exists on the clinical effects of hyperprolactinemia. These data come from patients with prolactinomas, patients receiving PRL elevating drugs and women both during pregnancy and lactation. The resulting side effects of elevated PRL levels are well understood and appear less toxic than those reported with several of the cytokines. Understanding how to utilize this molecule in the clinic may provide a way to "tune up" and normalize immunohematopoietic function with minimal toxicity.

REFERENCES

1. Aziz D. Use and Interpretation of Tests in Endocrinology. Specialty Laboratories, 1997.
2. Van Cauter E, L'Hermite M, Copinschi G. Qauntitative analysis of spontaneous variations of plasma prolactin in normal man. Am J Physiol 1981;241:355–363.
3. Sassin J, Frantz A, Kapen S, Weitzman E. The nocturnal rise of human prolactin is dependent on sleep. J Clin Endocrinol Metab 1973;37:436–440.
4. Samuels M, Kleinschmidt-Demasters B, Lillehei K, Ridgway E. Pulsatile prolactin secretion in hyperprolactinemia due to presumed pituitary stalk interruption. J Clin Endocrinol Metab 1991;73:1289–1293.
5. Hilty C, Bruhlmann P, Sprott H, et al. Altered diurnal rhythm of prolactin in systemic sclerosis. Journal of Rheumatology 2000;27:2160–2165.
6. Van den Berghe G. Novel insights into the neuroendocrinology of critical illness. European Journal of Endocrinology 2000;143:1–13.
7. Para A, Ramirez-Peredo J, Countina B, Coutino B. Impaired metoclopramide-induced pituitary prolactin release in men with human immunodeficiency virus infection. J Lab Clin Med 1999;133:70–74.
8. Noel G, Suh H, Stone S, Frantz A. Human prolactin and growth hormone release during surgery and other conditions of stress. J Clin Endocrinol Metab 1972;35:840–851.
9. Ben-Jonathan N, Mershon J, Allen D, Steinmetz R. Extrapituitary prolactin: distribution, regulation, functions, and clinical aspects. Endocrine Reviews 1996;17:639–669.
10. O'Neal K, Montgomery D, Truong T, Yu-Lee L. Prolactin gene expression in human thymocytes. Molecular and Cellular Endocrinology 1992;87:19–23.
11. Pellegrini I, Lebrun J-J, Ali S, Kelly P. Expression of prolactin and its receptor in human lymphoid cells. Molecular Endocrinology 1992;6:1023–1031.
12. Sabharwal P, Glaser R, Lafuse W, et al. Prolactin synthesized and secreted by human peripheral blood mononuclear cells: an autocrine growth factor for lymphoproliferation. Proc Natl Acad Sci USA 1992;89:7713–7716.
13. Bellone G, Astarita P, Artusio E, et al. Bone marrow stroma-derived prolactin is involved in basal and platelet-activating factor-stimulated in vitro erythropoiesis. Blood 1997;90:21–27.
14. Kooijman R, Gerlo S, Coppens A, Hooghe-Peters E. Growth hormone and prolactin expression in the immune system. Annals of the New York Academy of Sciences 2000;917:534–540.
15. Dogusan Z, Book M, Verdood P, Yu-Lee L, Hooghe-Peters E. Prolactin activates interferon

regualtory factor-1 expression in normal lympho-hemopoietic cells. European Cytokine Network 2000;11:435–442.

16. Bellone G, Geuna M, Carbone A, et al. Regulatory action of prolactin on the in vitro growth of CD34+ve human hemopoietic progenitor cells. Journal of Cellular Physiology 1995;163:221–231.
17. Berczi I. Pituitary Function and Immunity. Boca Raton, FL: CRC Press, 1986.
18. Berczi I, Nagy E. The effect of prolactin and growth hormone on hemolymphopoietic tissue and immune function. In: Hormones and Immunity. Lancaster: MTP Press, 1987.
19. Murphy W, Durum S, Anver M, Longo D. Immunologic and hematologic effects of neuroendocrine hormones. Studies on DW/J dwarf mice. J Immunol 1992;148:3799–3805.
20. Nagy E, Berczi I. Hypophysectomized rats depend on residual prolactin for survival. Endocrinology 1991;128:2776–2784.
21. Matera L. Endocrine, paracrine and autocrine actions of prolactin on immune cells. Life Sciences 1996;59:599–614.
22. Murphy W, Durum S, Longo D. Differential effects of growth hormone and prolactin on murine T cell development and function. The Journal of Experimental Medicine 1993;178:231–236.
23. Richards S, Murphy W. Use of human prolactin as a therapeutic protein to potentiate immunohematopoietic function. J Neuroimmunology 2000;109:56–62.
24. Matera L, Galetto A, Geuna M, et al. Individual and combined effect of granulocyte-macrophage colony-stimulating factor and prolactin on maturation of dendritic cells from blood monocytes under serum-free conditions. Immunology 2000;2100:29–36.
25. Clevenger C, Altman S, Prystowsky M. Requirement of nuclear prolactin for interleukin-2 stimulated proliferation of T lymphocytes. Science 1991;253:77–79.
26. Clevenger C, Russell D, Appasamy P, Prystowsky M. Regulation of interleukin 2-driven T-lymphocyte proliferation by prolactin. Proc Natl Acad Sci USA 1990;87:6460–6464.
27. Cesano A, Oberholtzer E, Contarini M, Bellone G, Matera L. Independent and synergistic effects of interleukin-2 and prolactin on development of T-and NK-derived LAK effectors. Immunopharmacology 1994;28:67–75.
28. Matera L, Bellone G, Lebrun J-J, et al. Role of prolactin in the in vitro development of interleukin-2-driven anti-tumoural lymphokine-activated killer cells. Immunology 1996;89:619–626.
29. Cesario T, Yousefi S, Carandang G, Sadati N, Le J, Vaziri N. Enhanced yields of gamma interferon in prolactin treated human peripheral blood mononuclear cells. Proceedings of the Society for Experimental Biology and Medicine 1994;205:89–95.
30. Matera L, Contarini M, Bellone G, Forno B, Biglino A. Up-modulation of interferon-gamma mediates the enhancement of spontanous cytotoxicity in prolactin-activated natural killer cells. Immunology 1999;98:386–392.
31. Matera L, Mori M. Cooperation of pituitary hormone prolactin with interleukin-2 and interleukin-12 on production of interferon-gamma by natural killer and T cells. Annals of the New York Academy of Sciences 2000;917:505–513.
32. Richards S, Garman R, Keyes L, Kavanaugh B, McPherson J. Prolactin is an antagonist of TGF-beta activity and promotes proliferation of murine B cell hybridomas. Cell Immunology 1998;184:85–91.
33. Woody M, Welniak L, Sun R, et al. Prolactin exerts hematopoietic growth-promoting effects in vivo and partially counteracts myelosuppression by azidothymidine. Exp

Hematol 1999;27:811–816.

34. Woody M, Welniak L, Richards S, et al. Use of neuroendocrine hormones to promote reconstitution after bone marrow transplantation. Neuroimmunomodulation 1999;6:69–80.
35. Welniak L, Tian Z, Sun R, et al. Effects of growth hormone and prolactin on hematopoiesis. Leukemia Lymphoma 2000;38:435–445.
36. Grossman Z, Polis M, Feinberg M, et al. Ongoing HIV dissemination during HAART. Nature Medicine 1999;5:1099–1104.
37. Schito A, Vittinghoff E, Hecht F, et al. Longitudinal analysis of T-cell receptor gene use by CD8(+) T cells in early human immunodeficiency virus infection in patients receiving highly active antiretroviral therapy. Blood 2001;97:214–220.
38. Pitcher C, Quittner C, Peterson D, Koup R, Maino V, Picker L. HIV-1 specific CD4+ T cells are detectable in most individuals with active HIV-1 infection, but decline with prolonged viral suppression. Nature Medicine 1999;5:518–525.
39. Gea-Banacloche J, Weiskopf E, Hallahan C, et al. Progression of human immunodeficiency virus disease is associated with increasing disruptions within the CD4+ T cell receptor repertoire. Journal of Infectious Diseases 1998;177:579–585.
40. Para A, Peredo-Ramirez J. The uncoupled couple? Prolactin and CD4 lymphocytes in HIV infection. Medical Hypothesis 1999;53:425–428.
41. Lazdins J, Klimkait T, Alteri E, et al. TGF-β upregulator of HIV replication in macrophages. Res. Virol 1991;142:239–242.
42. Kekow J, Wachsman W, McCutchan J, et al. Transforming growth factor-β and suppression of humoral immune responses in HIV infection. American Society for Clinical Investigation 1991;87:1010–1016.
43. Peterson V, Ambruso D. Postburn defects in phagocyte function. In: Phagocyte Production and Function Following Burn Injury. Austin: RG Landes Co, 1994:93–151.
44. Ogle C, Alexander W, Nagy H, et al. A long-term study and correlation of lymphocyte and neutrophil function in the patient with burns. Journal of Burn Care & Rehabilitation 1990;11:105–111.
45. Ayala A, Chung C, Xu Y, Evans T, Redmond K, Chaudry I. Increased inducible apoptosis in CD4+ T lymphocytes during polymicrobial sepsis is mediated by fas ligand and not endotoxin. Immunology 1999;97:45–55.
46. Deveci M, Sengezer M, Bozkurt M, Eski M, Inal A. Comparison of lymphocyte population in cutaneous and electrical burn patients: a clinical study. Burns 2000;26:229–232.
47. Hansbrough J, Field TJ, Gadd M, Soderberg C. Immune response modulation after burn injury: T cells and antibodies. Journal of Burn Care & Rehabilitation 1987;8:509–512.
48. Ninneman J. Clinical and immune status of burn patients. Antibiot Chemother 1987;39:16–25.
49. Di Carlo R, Meli R, Galdiero M, Nuzzo I, Bentivoglio C, Carratelli C. Prolactin protection against lethal effects of salmonella typhimurium. Life Science 1993;53:981–989.
50. Meli R, Raso G, Bentivoglio C, Nuzzo I, Galdiero M, Di Carlo R. Recombinant human prolactin induces protection against salmonella typhimurium infection in the mouse: role of nitric oxide. Immunopharmacology 1996;34:1–7.
51. Benedetto N, Folgore A, Galdiera M, Meli R, Di Carlo R. Effect of prolactin, rINF-γ or TNF-α in murine taxoplasmosis. Pathol Biol 1995;43:395–400.
52. Chen Y, Johnson A. *In vivo* activation of macrophages by prolactin from young and aging mice. J Immunopharmacol 1993;15:39–45.

52. Zellweger R, Zhu X-H, Wichmann M, Ayala A, De Maso C, Chaundry I. Prolactin administration following hemorrhagic shock improves mortality from subsequent sepis. J Immunol 1996;157:5748–5754.
53. Bernton E, Meltzer M, Holaday J. Suppression of macrophage activation and T-lymphocyte function in hypoprolactinemic mice. Science 1988;239:401–404.
54. Robertson MJR. Biology and clinical relevance of human natural killer cells. Blood 1990;76:2421–2438.
55. Grimm E, Robb R, Roth J, et al. Lymphokine-activated killer cell phenomenon. III. Evidence that IL-2 is sufficient for direct activation of peripheral blood lymphocytes into lymphokine-activated killer cells. J Exp Med 1983;155:1356–1361.
56. Mier J, Atkins M. Mechanisms of action and toxicity of immunotherapy with cytokines. Current opinion in oncology 1993;5:1067–1072.
57. Gemlo B, Palladino M, Jaffe H, Espevik T, Rayner A. Circulating cytokines in patients with metastatic cancer treated with recombinant interleukin-2 and lymphokine-activated killer cells. Cancer Research 1988;48:5864–5867.
58. Mukherjee P, Mastro A, Hymer W. Prolactin induction of interleukin-2 receptors on rat splenic lymphocytes. Endocrinology 1990;126:88–94.
59. Viselli S, Stanek E, Mukherjee P, Hymer W, Mastro A. Prolactin induced mitogenesis of lymphocytes from ovarectomized rats. Endocrinology 1991;129:983–990.
60. Gilmour K, Reich N. Receptor to nucleus signaling by prolactin and interleukin-2 via latent DNA-binding factors. Proc Natl Acad Sci USA 1994;91:6850–6854.
61. Horseman N, Yu-Lee L. Transcriptional regulation by the helix bundle peptide hormones: Growth hormone, prolactin, and hematopoietic cytokines. Endocrine Reviews 1994;15:627–649.
62. Oberholtzer E, Contarini M, Veglia F, et al. Prolactin increases the susceptibility of primary leukemia cells to NK and LAK effectors. Advances in Neuroimmunology 1996;6:233–247.
63. Gaidano G, Contarini M, Pastore C, Saglio G, Matera L. AIDS-related Burkitt's-type lymphomas are a target for lymphokine-activated killers induced by interleukin-2 and prolactin. Experimental Biology & Medicine 1996;213:196–205.
64. Parkman R, Weinberg K. Immunological reconstitution following hematopoietic stem cell transplantation. In: Thomas E, Blume K, editors. Hematopoietic Cell Transplantation. Oxford: Blackwell Science, 1999;704–711.
65. Hinterberger-Fischer M, Ogris E, Kier P, et al. Elevation of plasma prolactin in patients undergoing autologous blood stem-cell transplantation for breast cancer. Am J Clin Oncol 2000;23:325–329.
66. Lavelle C, Loyo E, Paniagua R, et al. Correlation study between prolactin and androgens in male patients with systemic lupus erythematosus. J Rheumatol 1987;14:268–273.
67. Quezada L, Graef A, Lavalle C. Prolactin and gonadal hormones during pregnancy in systemic lupus erythematosus. J Rheumatol 1991;18:349–353.
68. McMurray R, Allen S, Braun A, Rodriguez F, Walker S. Longstanding hyperprolactinemia associated with systemic lupus erythematosus: Possible hormonal stimulation of an autoimmune disease. J Rheumatol 1994;21:843–850.
69. Buskilla D, Berezin M, Gur H, et al. Autoantibody profile in the sera of women with hyperprolactinemia. Journal of Autoimmunity 1995;8:415–424.
70. Buskilla D, Lorber M, Neumann L, Flusser D, Shoenfeld Y. No clinical correlation between prolaction levels and clinical activity in patients with systemic lupus erythematosus. J

Rheumatol 1996;23:629.

71. Pauzner R, Urowitz M, Gladman D, Gough J. Prolactin in systemic lupus erythematosus. J Rheumatol 1994;21:2064–2067.
72. Ostendorf B, Fischer R, Santen R, Schmitz-Linneweber B, Specker C, Schneider M. Hyperprolactinemia in systemic lupus erythematosus. Scand J Rheumatol 1996;25:97–102.
73. Turoct-Lemay L, Kelly P. Prolactin receptors in human breast tumors. Journal of the National Cancer Institute 1982;68:381.
74. Nevalalnen M, Valve E, Ingleton P, Nurmi M, Martlkainen P, Harkonen P. Prolactin and prolactin receptors are expressed and functioning in human prostate. J Clinical Invest 1997;99:618–627.
75. Simoneau-Bataille N, Gerland K, Chappard D, Basle M, Mercier L. Expressions of prolactin receptors in human osteosarcoma cells. Biochemical and Biophysical Research Communications 1996;229:323–328.
76. Laud K, Gourdou I, Belair L, Peyrat J, Djiane J. Characterization and modulation of a prolactin receptor mRNA isoform in normal and tumoral human breast tissues. International Journal of Cancer 2000;85:771–776.
77. Kabuto M, Akiba S, Stevens RJ, Neriishi K, Land C. A prospective study of estradiol and breast cancer in Japanese women. Cancer Epidemiology, Biomarkers & Prevention 2000;9:575–579.
78. Abu-Bedair F, El-Gamal B, Ibrahim N, El-Aaser A. Hormonal profiles and estrogen receptors in Egyptian female breast cancer patients. Tumori 2000;86:24–29.
79. Kwa H, Cleton F, Wang D, et al. A prospective study of plasma prolactin levels and subsequent risk of breast cancer. Int J Cancer 1981;28:673–676.
80. Wang D, De Stavola B, Bulbrook R, et al. Relationship of blood prolactin levels and the risk of subsequent breast cancer. International Journal of Epidemiology 1992;21:214–221.
81. Mortensen P. The occurrence of cancer in first admitted schizophrenic patients. Schizophrenia Research 1994;12:185–194.

Growth and Lactogenic Hormones
Edited by L. Matera and R. Rapaport

Effectiveness of Bromocriptine in the Treatment of Autoimmune Diseases

SARA E. WALKER

The University of Missouri-Columbia, Harry S. Truman Memorial Veterans' Hospital, Research, 800 Hospital Drive, Columbia, MO 65201, USA

ABSTRACT

The rationale for treating autoimmune diseases with bromocriptine is based upon the observation that prolactin acts as a cytokine. It was reasoned that bromocriptine, which specifically suppresses pituitary secretion of prolactin and results in decreased concentration of circulating prolactin, would suppress immune responses and ameliorate autoimmune illness. Bromocriptine has been shown to alter the expression of a number of induced and spontaneous autoimmune states in experimental animals. Bromocriptine treatment, however, was not uniformly beneficial in animal models of autoimmune insulin-dependent diabetes mellitus, or in humans with rheumatoid arthritis, diabetes mellitus, or multiple sclerosis. Preventive treatment with bromocriptine was effective in the F_1 NZB/NZW mouse model of systemic lupus erythematosus. NZB/NZW mice with established disease also responded to bromocriptine therapy. Three separate trials have provided evidence that bromocriptine is a promising drug for treatment of systemic lupus eythematosus.

1. PROLACTIN

The lactotropic polypeptide hormone, prolactin, regulates a number of biochemical processes required for cell growth and division, activates protein kinase C [1], and stimulates growth-related gene mRNA [2]. Prolactin receptors are included in the novel cytokine/growth hormone/prolactin receptor family, comprising at least 10 receptors and including receptors for interleukin (IL)-2 beta, IL-3, IL-4, IL-6, IL-7, growth hormone, and erythropoietin [3]. Prolactin is traditionally thought to be a pituitary hormone, but also functions as a cytokine. Prolactin has structural motifs that resemble cytokines, is synthesized in multiple sites including lymphocytes, has similar receptor structures and signal transduction pathways, and is an important lymphocyte growth factor [4]. Prolactin receptors are distributed throughout the immune system and have been identified on thymic epithelium [5–7]. In rodents, prolactin influences T-cells, B-cells, macrophages, and natural killer cells. High levels of circulating prolactin stimulate immune responses and low levels result in immune suppression in experimental animals (reviewed in

Ref. [8]).

2. BROMOCRIPTINE

Bromocriptine is an ergot derivative with potent dopamine receptor agonist activity which selectively inhibits secretion of prolaction from the anterior pituitary. Bromocriptine is relatively safe and is used widely. In humans the usual dose of bromocriptine is 5.0 to 7.5 mg/day, given orally. This dose typically causes a marked fall in serum prolactin concentration. Bromocriptine was approved for use in the United States in 1978, and is used for treatment of microprolactinoma, acromegaly, symptomatic hyperprolactinemia that results in amenorrhea or galactorrhea or female infertility, and Parkinsonism. Undesirable side effects include nausea, orthostatic hypotension, headache, fatigue, abdominal cramps, nasal congestion, and constipation [9,10]. Serious adverse events have been almost completely limited to the use of bromocriptine in suppressing postpartum lactation, or in treating Parkinson's disease. Bromocriptine given as postpartum treatment has been associated with myocardial infarction [11], puerperal hypertension, stroke, and seizures [12]. Pleural effusions, pleural thickening, and parenchymal lung disease were described in men treated with bromocriptine for Parkinson's disease. The subjects were usually in the sixth or seventh decade of life and had taken high doses of bromocriptine (15–100 mg/day) for periods of 2 months to 3 years. Typically, the pleuropulmonary involvement improved after the drug was discontinued [13,14].

2.1. Immunosuppressive effects of bromocriptine

The mechanism whereby bromocriptine induces immunosuppression is a subject of debate. A number of reports have suggested that the drug suppressed immune responses by inhibiting secretion of pituitary prolactin. Bromocriptine inhibited antibody formation and cell mediated immune responses, and this suppression was reversed by treatment with exogenous prolactin [15,16]. Bernton [17] reported that macrophages from bromocriptine-treated mice had impaired production of interferon-gamma, and this defect was abrogated by treating the animals with ovine prolactin.

Other investigators have proposed that bromocriptine has immunosuppressive effects that are separate from its effects on the pituitary. Bromocriptine inhibited proliferation and immunoglobulin production in mitogen-stimulated cultures of human tonsillar B-cells. Bromocriptine was also associated with reversible inhibition of early stages of human T-cell activation, and the inhibitory effect of bromocriptine appeared to be independent of its effect on the concentration of extracellular prolactin [18,19]. The findings of Neidhart [20], however, contradicted the theory that direct suppression of lymphocyte function accounted for immunosuppressive properties of bromocriptine. Increasing concentrations of bromocriptine were added to mitogen-stimulated mouse spleen cell cultures. The lowest concentration of bromocriptine that was required to suppress *in vitro* lymphocyte proliferation was 200 ng/ml. The serum concentration of bromocriptine was tested in autoimmune F_1 hybrid NZB/NZW mice that responded to bromocriptine treatment with reduced severity of renal disease. In these animals, serum bromocriptine concentrations were only 2 to 6 ng/ml. Substantial suppression of autoimmune disease had occurred in the presence of relatively low levels of bromocriptine, and it was concluded that bromocriptine suppressed immunity in treated animals through suppression of pituitary secretion of prolactin, and not through direct suppression of cells of the immune system [20].

Table I Effective treatment of animal models of autoimmune disease with bromocriptine.

Induced autoimmune disease	
Prevention of disease	
Experimental autoimmune uveitis, Lewis rat	[21, 22*]
Experimental allergic encephalomyelitis, Lewis rat	[23]
Adjuvant arthritis, Fisher rat	[24]
Adjuvant arthritis, Sprague-Dawley rat	[22, 25]
Treatment of disease	
Experimental allergic encephalomyelitis, Lewis rat	[23, 26, 27]
Systemic lupus erythematosus, BALB/c mouse	[28]
Primary anti-phospholipid syndrome, BALB/c mouse	[28]
Spontaneous autoimmune disease	
Prevention of disease	
Diabetes mellitus, BB rat	[22*, 29*]
Diabetes mellitus, NOD mouse	[30]
Polyarteritis nodosa, aged Sprague-Dawley rat	[22]
Systemic lupus erythematosus, F1 NZB/NZW mouse	[32, 33]
Treatment of disease	
Systemic lupus erythematosus, F1 NZB/NZW mouse	[34]

* Bromocriptine was effective when given together or sequentially with cyclosporine A.

2.2. Bromocriptine treatment of animal models with autoimmune disease

Table I describes experiments in which bromocriptine treatment was effective in preventing or delaying the onset of autoimmune disease that had been induced in animal models. Bromocriptine, given as a single drug, did decrease the incidence of experimental autoimmune uveitis (EAU) in female Lewis rats [21] but was effective in males only if it was given with low-dose cyclosporine A [22]. Riskind [23] reported that treating female Lewis rats with bromocriptine before immunization with spinal cord homogenate reduced the incidence and severity of experimental allergic encephalomyelitis (EAE). Subcutaneous injections of bromocriptine also inhibited the development of adjuvant arthritis (AA) in female Fisher rats [24]. In male Sprague-Dawley rats, injections of long-acting bromocriptine microcapsules suppressed serum prolactin concentrations throughout the day. If treatment was given 3 days before complete Freund's adjuvant was injected, there was significant inhibition of AA-related hind limb swelling and systemic manifestations of disease [22,25].

In some experimental models of autoimmune disease, bromocriptine treatment was effective after clinical signs appeared. Lewis rats that received bromocriptine either one week after immunization or after the development of clinical signs had suppression of EAE [23,26,27]. Bromocriptine was effective in treating BALB/c mice with 2 separate induced autoimmune states: systemic lupus erythematosus (SLE) that was induced by injecting a human anti-ds DNA monoclonal antibody, or anti-phospholipid syndrome induced by injecting monoclonal mouse anti-cardiolipin antibody. In both instances, daily intraperitoneal injections of bromocriptine were given 2 months after disease induction when serum antibody titers were elevated but clinical findings of disease had not appeared. The SLE mice did have decreased glomerular deposition of immunoglobulin. The mice with induced anti-phospholipid syndrome had improvement in 2 manifestations: the activated partial thromboplastin time was decreased from 78 to 23 seconds, and fetal resorption was decreased from 57 to 5 percent of fetuses [28].

The spontaneously diabetic BB rat had decreased severity of disease following treatment with both bromocriptine and cyclosporine A [22,29]. Nonobese diabetic (NOD) mice had varying responses to bromocriptine therapy. Early, intensive treatment with bromocriptine, 200 micrograms injected daily from the age of 21 days, reduced the incidence of diabetes [30]. In contrast, long term treatment with a higher dose (10 mg/kg, estimated to be 300 micrograms, injected intraperitoneally 5 days a week) accelerated the onset of diabetes in male NOD mice and increased islet inflammation in males and females [31].

Bromocriptine also suppressed the severity of spontaneous periarteritis in aged Sprague-Dawley rats. Monthly injections of bromocriptine microcapsules were given over the life span of the rats and the animals were necropsied when they died spontaneously or were sacrificed at the age of 2 years. Severity of arteritis was graded on a scale of 0 (normal) to 5 (severe histological changes with over half the vesels involved). Rats that received doses of either 1 or 10 mg/kg had significant decrease of the severity score compared with untreated rats [22].

2.3. Bromocriptine treatment of NZB/NZW mice

Several studies have shown that early treatment with bromocriptine ameliorates the severity of spontaneous SLE in female NZB/NZW mice. McMurray [32] treated NZB/NZW females with daily subcutaneous injections of 300 micrograms of bromocriptine. This dose had been shown to suppress serum prolactin to 8 ng/ml in swim-stressed NZB/NZW females, compared to 76 ng/ml in stressed female controls. Treatment began at the age of 5 weeks, before the appearance of overt autoimmune disease, and continued through the life spans of the mice. At the age of 24 weeks, anti ds-DNA antibodies were decreased in treated mice compared to controls and this difference persisted through 30 weeks of age. Survival was prolonged significantly by bromocriptine treatment, and 25 weeks after the study began, 90% of bromocriptine-treated mice and 57% of control mice were alive [32]. The protective effects of bromocriptine in NZB/NZW female mice were confirmed by Elbourne [33]. Oophorectomized mice that received bromocriptine, 10 mg/kg/day from the age of 6 weeks, developed autoimmune disease of intermediate severity and their disease was not accelerated by concordant treatment with high-dose estrogen. It was concluded that prolactin was a more important stimulant of acceleration of disease than estrogen in NZB/NZW mice.

Neidhart [34] reported that bromocriptine was beneficial in mature NZB/NZW females. Treatment was begun at the age of 36 weeks, when the mice were expected to havc clinical SLE. Bromocriptine, 5 mg/kg/day, was injected daily and serum prolactin was suppressed to undetectable levels. Autoantibodies were not suppressed, but bromocriptine did suppress proteinuria and histological evidence of glomerulonephritis.

Table II Effective treatment of human autoimmune disease with bromocriptine.

Iridocyclitis	[35]
Reactive arthritis	[38]
Type 2 diabetes mellitus	[39]
Systemic lupus erythematosus	[46-52]

2.4. Bromocriptine treatment of autoimmune diseases in humans

Table II lists reports of effective treatment of autoimmune illnesses in humans. Hedner [35] was the first to report success with bromocriptine, in patients with inflammatory eye disease. Four individuals with ankylosing spondylitis and iridocyclitis, reactive arthritis associated with iritis, or idiopathic unilateral iridocyclitis were treated with bromocriptine in doses of 2.5 to 5.0 mg/day. The bromocriptine was given because these patients developed hyperprolactinemia, galactorrhea, or Parkinson's disease. The patients had resolution of eye disease that, at the time of the report, had lasted for 4 to 12 months. Fourteen patients whose severe, sight-threatening uveitis was resistant to corticosteroids received bromocriptine, 7.5–10.0 mg/day, in combination with cyclosporine A (starting dose 4 mg/kg). Serum prolactin levels were suppressed to less that 2 ng/ml in all patients. There was marked improvement in visual acuity in 8 patients, and it was concluded that results of combination therapy were similar to the outcome of treatment with high-dose cyclosporine alone [36]. In a double-blind study of 15 patients with recurrent anterior uveitis, 7 subjects were randomized to receive bromocriptine, 5 mg/day, and 8 subjects received placebo for 48 weeks. Two of the bromocriptine-treated patients stopped treatment because of recurrences, and 5 patients in the placebo group left the study because of recurrences. A conclusion was not reached concerning the effectiveness of bromocriptine therapy [37].

Dramatic response to bromocriptine treatment was reported in 4 men with severe reactive arthritis following enteritis caused by *Salmonella typhi* or *Shigella flexerni*. Doses of either 2.5 or 5.0 mg/day resulted in decreased numbers of affected joints after 1 to 8 days of treatment [38].

Trials of bromocriptine in other immune-mediated diseases have given inconclusive results or have shown no efficacy. In obese individuals with type 2 diabetes mellitus, bromocriptine was given to 15 subjects in a 16-week double-blind study and 7 subjects received placebo. Bromocriptine therapy was associated with favorable response in the form of small but significant decreases of hemoglobin A1c and fasting plasma glucose. There were, however, no changes in body composition, body fat distribution, oral glucose tolerance, insulin-mediated glucose disposal, or endogenous glucose production [39]. Other investigators suggested that bromocriptine was not beneficial for obese type 2 diabetic patients [40]. Atkinson [41] found no advantage in combining bromocriptine with cyclosporine-A in treatment of newly diagnosed insulin dependent diabetics.

The beneficial effects of bromocriptine in treating EAE in Lewis rats [23,26,27] suggested that bromocriptine treatment would help humans with multiple sclerosis [42]. Results of an open label study, however, were not encouraging. Eighteen patients with either the relapsing-remitting or chronic progressive form of multiple sclerosis were treated with bromocriptine, 5 mg/day. Disease manifestations — including clinical relapses, new lesions on magnetic resonance imaging (MRI) of the brain and brainstem, and increased visual or auditory evoked responses — showed progression in 14 of the 15 subjects who completed 1 year of treatment [43].

Two open label studies (reviewed in Ref. [8]) provided evidence that prolactin-lowering therapy did not consistently produce benefits in patients with rheumatoid arthritis. Only 4 of 9 women treated with bromocriptine (mean dose 19.7 mg/day) for 90 days had improvement according to American College of Rheumatology criteria [44]. Another dopamine agonist, quinagolide, was given to 9 patients with rheumatoid arthritis for 24 weeks and serum prolactin concentrations were suppressed to undetectable levels. Two patients were judged to have moderate response to this treatment [45].

2.5. Bromocriptine treatment of human SLE

Several early reports described successful use of bromocriptine to treat active SLE. A patient who was not reported to be hyperprolactinemic had central nervous system lupus and was resistant to conventional therapy. She improved after treatment was started with bromocriptine and intravenous immunoglobulin. In 2 instances, the disease flared after bromocriptine was discontinued [46, L. Schanberg, personal communication].

Four women with symptomatic hyperprolactinemia had mild SLE characterized by photosensitivity, malar rash, arthralgias, and positive FANA tests with titers ranging from 1:320 to 1:640. Two had antibodies to ds-DNA. None had renal involvement. Three of these patients had pituitary microadenomas, all had normal serum concentrations of 17-beta estradiol, and 2 had suppressed serum testosterone. In 2 cases, SLE flared after bromocriptine was stopped [47].

Funauchi [48] reported concordance between serum prolactin concentrations and SLE disease activity in a 31-year-old woman with SLE who had a prolactinoma. Her disease manifestations included polyarthralgia, facial erythema, low grade fever, thrombocytopenia, and antinuclear antibodies. She became amenorrheic 12 years after the onset of signs and symptoms of SLE. The plasma prolactin was 39 ng/ml (normal concentration 2.9 to 30.5), and MRI of the brain showed a hyperintense lesion in the pituitary gland. The patient was treated with bromocriptine 2.5 followed by 5 mg/day. Once, bromocriptine was discontinued and circulating prolactin increased and the patient experienced a lupus flare. Over a period of 6 years, 29 serial measurements of serum prolactin had positive correlation with levels of anti-DNA antibodies.

2.6. Bromocriptine trials in SLE

McMurray [49] treated 7 SLE patients who had mild to moderately active disease with bromocriptine in an open label study. Patients with rapidly progressive disease, life threatening complications, or significant hematologic, renal, or central nervous system lupus were excluded. At entry, 6 patients had normal serum prolactin concentrations and one had a microprolactinoma and borderline hyperprolactinemia (20.6 ng/ml; normal = 2 to 18.5). Each patient received a dose of bromocriptine (3.75 to 7.5 mg/day) that was adjusted to keep serum prolactin concentrations less than 3 ng/ml. Treatment continued for 6 to 9 months, and results of the first 6 months of treatment were analyzed.

The SLE Activity Measure (SLAM) and the Toronto SLE Disease Activity Index (SLEDAI) were suppressed significantly and constitutional symptoms, skin involvement, Raynaud's phenomenon, arthralgias, and arthritis improved. Patients also experienced improvement in mood states as measured by the Symptom Questionnaire survey. Total distress scores improved, and this improvement did correlate positively with improvement in both the SLAM and the SLEDAI.

The anger-hostility measure decreased during bromocriptine treatment. Improvement in this measure did not correlate with improved disease activity, and this observation raised the question that bromocriptine may have exerted favorable psychotropic effects [50]. Side effects were mild and included nasal stuffiness (1), insomnia (1), and nausea (2 patients). Three of 4 patients who took prednisone at entry were able to reduce the dose while they received bromocriptine, but the prednisone dose reduction was not significant. After bromocriptine was stopped, subjects were followed during a 5 month washout period. Five patients became hyperprolactinemic and all patients had increased lupus activity, and 6 patients required changes in medication to control lupus activity during the follow up period.

In a double-blind study, Alvarez-Nemeguei [51] showed that treatment with a fixed low dose of bromocriptine (2.5 mg/day) reduced lupus flares. Thirty-six of 66 consecutive SLE patients were randomized to receive bromocriptine, 2.5 mg/day, and 30 controls received placebo. Patients were allowed to enter the study without regard to disease activity, although those with organ failure were excluded. Modest hyperprolactinemia was found at entry in 51% of subjects in the bromocriptine treatment group and 40% of the controls. Subjects were followed prospectively for 2 to 17 months (mean 12.5 months) and evaluated at intervals of 30 to 45 days.

Serum prolactin concentrations were reduced significantly in the bromocriptine treatment group after 3, 6, 9, and 12 months of treatment. After 1 year, serum prolactin in the bromocriptine group was (mean ± standard deviation) 5.8 ng/ml ± 9 vs. 20.3 ± 14 in controls ($p = 0.001$). The normal value for serum prolactin was less than 20 ng/ml. The SLEDAI score on the fifth protocol visit was significantly smaller in patients who received bromocriptine (0.9 ± 1.4) compared with the control group (2.6 ± 4.5; $p = 0.05$). The mean number of flares/patient/month in bromocriptine-treated patients was reduced to 0.08 ± 0.1 vs. 0.18 ± 0.2 in controls ($p = 0.03$). Five subjects in the bromocriptine treatment group quit the study because of headache and nausea which were attributed to the drug. The prednisone dose did not differ between the treatment and control groups at any period during the study.

In a separate study, bromocriptine was compared to hydroxychloroquine in treatment of active but not life threatening SLE. Patients randomized to one group received bromocriptine, in a dose designed to suppress serum prolactin to a level less than 1 ng/ml, and the other group received hydroxychloroquine, 6 mg/kg. Treatment was given for one year. Preliminary data from 24 patients were reported recently [52]. Bromocriptine produced significant reduction of serum prolactin concentrations. In 11 bromocriptine-treated patients, the SLAM score decreased from (mean ± SEM) 14.0 ± 1.1 at entry to 7.5 ± 0.8 ($p < 0.05$) after one year. Thirteen SLE patients were randomized to receive hydroxychloroquine. The mean SLAM score decreased from 13.4 ± 1.3 at entry to 9.0 ± 1.4 ($p < 0.001$) at the end of the study. Prednisone doses, numbers of patients who started and stopped prednisone, and numbers of patients who left the study were similar in both treatment groups. Continued analyses of the complete data set following completion of the study have confirmed that SLE activity did improve in both the bromocriptine treatment group and the hydroxychloroquine treatment group (S.E. Walker, Unpublished data).

ACKNOWLEDGEMENTS

This research was supported by the Office of Research and Development, Medical Research Service, Department of Veterans Affairs, and by grants from the American Autoimmune Related Diseases Association, Inc. and The Kettering Family Foundation.

REFERENCES

1. Russell DH. New aspects of prolactin and immunity: a lymphocyte-derived prolactin-like product and nuclear protein kinase C activation. Trends Pharmacol Sci 1989;10:40–44.
2. Berczi I, Nagy E, de Toledo SM, Matusik RJ, Friesen HG. Pituitary hormones regulate c-myc and DNA synthesis in lymphoid tissue. J Immunol 1990;146:2201–2206.
3. Thoreau E, Petridou B, Kelly PA, Djiane J, Mornon JP. Structural symmetry of the extracellular domain of the cytokine/growth hormone/prolactin receptor family and interferon receptors revealed by hydrophobic cluster analysis. FEBS Letters 1991;282:26–31.
4. Yu-Lee L-Y. Molecular actions of prolactin in the immune system. Proc Soc Exp Biol Med 1997;215:35–52.
5. Russell DH, Kibler R, Matrisian L, Larson DF, Boulos B, Magun BE. Prolactin receptors on human T and B lymphocytes: Antagonism to prolactin binding by cyclosporine. J Immunol 1985;134:3027–3031.
6. Pellegrini I, Lebrun J, Ali S, Kelly PA. Expression of prolactin and its receptor in human lymphoid cells. Molec Endocrinol 1992;6:1023–1021.
7. Dardenne M, Kelly PA, Bach J, Savino W. Identification and functional activity of prolactin receptors in thymic epithelial cells. Proc Natl Acad Sci USA 1991;88:9700–9704.
8. Walker SE, McMurray RW, Houri JM, Allen SH, Keisler D, Sharp GC, Schlechte JA. Effects of prolactin in stimulating disease activity in systemic lupus erythematosus. Proc NY Acad Sci 1998;840:762–772.
9. Parkes D. Bromocriptine. N Engl J Med 1979;301:873–878.
10. Vance ML, Evans WS, Thorner MO. Bromocriptine. Diagnosis and treatment. Drugs five years later. Ann Intern Med 1984;100:78–91.
11. Larrazet F, Spaulding C, Lobreau HJ, Weber S, Guerin F. Possible bromocriptine-induced myocardial infarction. Ann Intern Med 1993;118:199–200.
12. Katz M, Kroll D, Pak I, Osimoni A, Hirsch M. Puerperal hypertension, stroke, and seizures after suppression of lactation with bromocriptine. Obstet Gynecol 1985;66:822–824.
13. McElvaney NG, Wilcox PG, Churg A, Fleetham JA. Pleuropulmonary disease during bromocriptine treatment of Parkinson's disease. Arch Intern Med 1988;148:2231–2236.
14. Kinnunen E, Viljanen A. Pleuropulmonary involvement during bromocriptine treatment. Chest 1988;94:1034–1036.
15. Hiestand PC, Mekler P, Nordmann R, Greider A, Permmangkol C. Prolactin as a modulator of lymphocyte responsiveness provides a possible mechanism of action for cyclosporine. Proc Natl Acad Sci USA 1986;83:2599–2603.
16. Nagy E, Berczi I, Wren GE, Asas SL, Kovacs K. Immunomodulation by bromocriptine. Immunopharmacology 1983;6:231–243.
17. Bernton EW, Meltzer MS, Holaday JW. Suppression of macrophage activation and T-lymphocyte function in hypoproalctinemic mice. Science 1988;239:401–404.
18. Morikawa K, Oseko F, Morikawa S. Immunosuppressive property of bromocriptine on human B lymphocyte function *in vitro*. Clin Exp Immunol 1993;93:200–205.
19. Morikawa K, Oseko F, Morikawa S. Immunosuppressive activity of bromocriptine on human T lymphocyte function *in vitro*. Clin Exp Immunol 1994;95:514–518.
20. Neidhart M. Bromocriptine has little direct effect on murine lymphocytes, the immunomodulatory effect being mediated by the suppression of prolactin secretion. Biomed & Pharmacother 1997;51:118–125.

21. Palestine AG, Muellenberg-Coulombre CG, Kim MK, Gelato MC, Nussenblatt RB. Bromocriptine and low dose cyclosporine in the treatment of experimental autoimmune uveitis in the rat. J Clin Invest 1987;79:1078–1081.
22. Neidhart M. Synergism between long-acting bromocryptine microcapsules and cyclosporine A in the prevention of various autoimmune diseases in rats. Experientia 1996;52;892–899.
23. Riskind PN, Massacesi L, Doolittle TH, Hauser SL. The role of prolactin in autoimmune demyelination: Suppression of experimental allergic encephalomyelitis by bromocriptine. Ann Neurol 1991;29:542–547.
24. Berczi B, Nagy E, Asa SL, Kovacs K. The influence of pituitary hormones on adjuvant arthritis. Arthritis Rheum 1984;27:682–688.
25. Neidhart M. Bromocriptine microcapsules inhibit ornithine decarboxylase activity induced by Freund's complete adjuvant in lymphoid tissues of male rats. Endocrinology 1989;125:2846–2852.
26. Dijkstra CD, Rouppe van der Voort E, De Groot CJA, Uitdehaag BMJ, Polman CH, Berkenbosch F. The therapeutic effect of bromocriptine on acute and chronic experimental allergic encephalomyelitis. Ann Neurol 1992;31:450–451.
27. Dijkstra CD, Rouppe van der Voort E, De Groot CJA, et al. Therapeutic effect of the D^2-dopamine agonist bromocriptine on acute and relapsing experimental allergic encephalomyelitis. Psychoneuroendocrinology 1994;19:135–142.
28. Blank M, Karuae I, Buskila D, et al. Bromocriptine immunomodulation of experimental SLE and primary anti-phospholipid syndrome via induction of nonspecific T suppressor cells. Cell Immunol 1995;162:114–122.
29. Mahon JL, Gunn HC, Stobie K, Gibson C, Garcia B, Dupre J, Stiller CR. The effect of bromocriptine and cyclosporine on spontaneous diabetes in BB rats. Transplant Proc 1998;20 (suppl):197–200.
30. Hawkins TA, Gala RR, Dunbar JC. Prolactin modulates the incidence of diabetes in male and female NOD mice. Autoimmunity 1994;18:155–162.
31. Durant S, Alves V, Coulaud J, El Hasnaoui A, Dardenne M, Homo-Delarche F. Attempts to pharmacologically modulate prolactin levels and type 1 autoimmune diabetes in the non-obese diabetic (NOD) mouse. J Autoimmunol 1995;8:875–885.
32. McMurray R, Keisler D, Kanuckel K, Izui S, Walker SE. Prolactin influences autoimmune disease activity in the female B/W mouse. J Immunol 1991;147:3780–3787.
33. Elbourne KB, Keisler D, McMurray RW. Differential effects of estrogen and prolactin on autoimmune disease in the NZB/NZW F1 mouse model of systemic lupus erythematosus. Lupus 1998;7:420–427.
34. Neidhart M. Bromocriptine has little direct effect on murine lymphocytes, the immunomodulatory effect being mediated by the suppression of prolactin secretion. Biomed & Pharmacother 1997;51:118–125.
35. Hedner LP, Bynke G. Endogenous iridocyclitis relieved during treatment with bromocriptine. Amer J Ophthalmol 1985;100:618–619.
36. Palestine AG, Nussenblatt RB, Gelato M. Therapy for human autoimmune uveitis with low-dose cyclosporine plus bromocriptine. Transpl Proc 1988;20:131–135.
37. Schaaf L, Zierhut M, Baur EM, et al. Bromocriptine in patients with chronic autoimmune-associated disorders. Klin Wochenschr 1991;69:943.
38. Bravo G, Zazueta B, Lavalle C. An acute remission of Reiter's syndrome in male patients treated with bromocriptine. J Rheumatol 1992;19:747–750.

39. Pijl H, Ohashi S, Matsuda M, et al. Bromocriptine: a novel approach to the treatment of type 2 diabetes. Diabetes Care 2000;23:1154–1161.
40. Wasada T, Kawahara R, Iwamoto Y. Lack of evidence for bromocriptine effect on glucose tolerance, insulin resistance, and body fat stores in obese type 2 diabetic patients. Diabetes Care 2000;23:1040.
41. Atkison PR, Mahon JL, Dupre J, et al. Interaction of bromocriptine and cyclosporine in insulin dependent diabetes mellitus: results from the Canadian open study. J Autoimmun 1990;3:793–797.
42. Draca S, Levic Z. The possible role of prolactin in the immunopathogenesis of multiple sclerosis. Medical Hypotheses 1996;47:89–92.
43. Bissay V, De Klippel N, Herroelen L, et al. Bromocriptine therapy in multiple sclerosis: an open label pilot study. Clin Neuropharmacol 1994;17:473–476.
44. Figueroa FE, Carrion F, Martinez ME, et al. Bromocriptine induces immunological changes related to disease parameters in rheumatoid arthritis. Br J Rheumatol 1997;36:1022–1023.
45. Finkelstein JS, Whitcomb RW, Longcope C, et al. Sex steroid control of gonadotropin secretion in the human male. I. Effects of testosterone administration in normal and gonadotropin-releasing hormone-deficient men. J Clin Endocrinol Metab 1991;73;609–620.
46. Rabiniovich CE, Schanberg LE, Kredich DW. Intravenous immunoglobulin and bromocriptine in the treatment of refractory neuropsychiatric systemic lupus erythematosus. Arthritis Rheum 1990;33 (Suppl.):R22 (abstract).
47. McMurray RW, Allen SH, Braun AL, Rodriguez F, Walker SE. Longstanding hyperprolactinemia associated with systemic lupus erythematosus: Possible hormonal stimulation of an autoimmune disease. J Rheumatol 1994;21:843–850.
48. Funauchi M, Ikoma S, Enomoto H, et al. Prolactin modulates the disease activity of systemic lupus erythematosus accompanied by prolactinoma. Clin Exp Rheumatol 1998;16:479–482.
49. McMurray RW, Weidensaul D, Allen SH, Walker SE. Efficacy of bromocriptine in an open label therapeutic trial for systemic lupus erythematosus. J Rheumatol 1995;22:2084–2091.
50. Walker SE, Smarr KL, Parker JC, Weidensaul DN, Nelson W, McMurray RW. Mood states and disease activity in patients with systemic lupus erythematosus treated with bromocriptine. Lupus 2000;9:527–533.
51. Alvarez-Nemeguei J, Cobarrubias-Cobbs A, Escalante-Triay F, Sosa-Munoz J, Miranda JM, Jara LJ. Bromocriptine in systemic lupus erythematosus: A double-blind, randomized, placebo-controlled study. Lupus 1998;7:414–419.
52. Walker SE, Reddy GH, Miller D, et al. Treatment of active systemic lupus erythematosus (SLE) with the prolactin (PRL) lowering drug, bromocriptine (BC): Comparison with hydroxychloroquine (HC) in a randomized, blinded one-year study. Arthritis Rheum 1999;42:S282 (abstract).

Growth and Lactogenic Hormones
Edited by L. Matera and R. Rapaport

The Pathogenic Role of Prolactin in Patients with Rheumatoid Arthritis

NOBORU SUZUKI

Departments of Immunology and Medicine, St. Marianna University School of Medicine, 2-16-1, Sugao, Miyamae-ku, Kawasaki, Kanagawa 216-8511, Japan

LIST OF ABBREVIATIONS

PRL: prolactin; RA: rheumatoid arthritis; PRLR: PRL receptor; MMP: matrix metalloproteinase; TIMP: tissue inhibitor of metalloproteinase; STAT: signal transduction and activation of transcription; BRC: bromocriptine; TNF: tumor necrosis factor; IL: interleukin; Th: T helper; HLA: human leucocyte antigen.

ABSTRACT

Defects in the hypothalamus-pituitary-adrenal axis have been observed in patients with rheumatoid arthritis (RA). Prolactin (PRL) levels are often elevated in patients with RA. To elucidate roles of PRL in the pathological responses occurring within the affected joints in patients with RA, we have studied PRL production and PRL receptor (PRLR) expression in RA synovium and its effects on RA synovial cell functions.

We found the accumulation of PRL producing lymphocytes and PRLR bearing synovial cells in RA synovial tissue. We also found that *in vitro* treatment with PRL induces enhanced proliferation of RA synovial cells. PRL treatment provoked excessive production of proinflammatory cytokines and of the tissue destructive proteolytic enzyme, matrix metalloproteinase (MMP), by RA synovial cells. In addition, PRL inhibited tissue inhibitor of metalloproteinase (TIMP)-1 production by the synovial cells. Thus, total collagenase activity in the joints may be upregulated in case of excessive PRL secretion in the joints. PRLR was exclusively expressed on fibroblast like synovial cells and lymphocytes infiltrating into the synovium in patients with RA. Both synovium infiltrating T lymphocytes and fibroblast like synovial cells synthesized PRL, suggesting that PRL acts as a paracrine as well as autocrine activator of RA synovial cell functions. Stimulation by PRL of synovial cells induced rapid translocation of signal transduction and activation of transcription (STAT)-5 from cytoplasm into nuclei of RA synovial cells, suggesting transcriptional regulation involving STAT-5 by PRL of RA synovial cell functions. Bromocriptine (BCR), an inhibitor of PRL release, inhibited proliferation of RA synovial cells. BCR inhibited proinflammatory cytokine and collagenase production by RA synovial cells. We want to

Table I Evidences for the involvement of PRL in the local inflammation of RA joints.

Effects of PRL on RA synovial cell function
1. Enhances proliferation of RA synovial cells
2. Enhances IL-6 and IL-8 production
3. Enhances MMP-3 production
4. Reduces TIMP-1 production
PRL producing cells
1. Synovium-infiltrating CD4+ T lymphocytes
2. Synovial cells in the sublining layer
PRLR expressing cells
1. Synovium infiltrating lymphocytes
2. Synovial cells in the sublining layer

emphasize the importance of locally produced PRL by infiltrating T lymphocytes in the induction of aberrant synovial cell functions in patients with RA.

1. INTRODUCTION

In the past few years an increasing number of studies has focused on the role of hormonal modulation of the immune system in the evelopment of autoimmune diseases, such as RA [1–5]. RA is a chronic inflammatory joint disease [6]. The major pathological changes in RA lesions include dysregulated proliferation of synovial cells, intensive lymphocyte infiltration, neoangiogenesis, and cartilage destruction in the affected joints [7]. The majority of these events primarily depend on the excessive production by synovial cells of proinflammatory cytokines and tissue destructive proteolytic enzymes [8]. Anti-tumor necrosis factor (TNF) therapy brings about potent clinical improvement in patients with RA, suggesting the involvement of proinflammatory cytokines in disease manifestation [9].

A dysfunction of the hormonal network in RA has been described, including PRL secretion [10–12]. Indeed, there are numerous reports describing elevated circulating PRL levels and its role in the pathogenesis in patients with RA [13–17].

2. PROLACTIN

PRL is a mammotropic hormone produced by pituitary and extrapituitary cells as different isoforms [18]. The secretion of pituitary PRL is under the control of hypothalamic factors, but is also influenced by factors released by immune cells. The cytokines interleukin (IL)-1, IL-2, and IL-6 stimulate production, while interferon-gamma and endothelin-3 are inhibitory [19]. PRL exerts its effects through binding to specific receptors (PRLR), which exist as three isoforms. PRL regulates reproduction, participates in osmoregulation, and of behavior and has potent immunomodulatory effects [20].

PRL is structurally related to members of the cytokine/hematopoietic growth factor family such as erythropoietin, granulocyte/macrophage-colony stimulating factor, growth hormone, and IL-2 to IL-7. The PRLR is a member of the cytokine/hematopoietic growth factor receptor family. Interaction of PRL with PRLR activates the Jak kinases that phosphorylate latent STAT proteins, resulting in activation of gene transcription [21].

Activated lymphocytes produce PRL [22], and not only lymphocytes but also monocytes in circulation express PRLR [23,24]. Moreover, PRL stimulates B cells to produce immunoglobulin and induces T cell proliferation [25–27]. From our point of view it is quite interesting that PRL counteracts the effects of corticosteroids by enhancing T helper (Th)1 cellular responses. An excessive Th1 cell response has been assigned an important pathogenic role in patients with RA [28,29].

Several investigators have demonstrated clear relationships between circulating PRL concentrations and incidence of developing autoimmune diseases including systemic lupus erythematosus [30], and autoimmune thyroid diseases [31]. As has been described, there are numerous reports describing elevated circulating PRL levels and its pathogenic role in patients with RA [13–17,32]; PRL aggravated collagen-induced arthritis in mice when given during the immunization phase [33,34]. Figueroa et al. studied immunological and clinical effects of PRL suppression in RA patients with active disease [35,36]. They treated the patients for 3 months with BRC, an inhibitor of PRL secretion. They concluded that BRC treatment induced a significant depression of *in vitro* immune function in RA patients and that these changes are related to parameters of disease activity. Similarly, BRC has been beneficial for treating animal models of RA [33,34]. However, there is controversy regarding this issue [37–39]. More recently, Erb et al. reported a patient with RA who had successfully been treated with the prolactin antagonist cabergoline [40]. They suggested that PRL levels should be checked in patients with RA, and that a PRL inhibitor may be beneficial for treating the patients with high serum PRL levels.

Genes encoded in the human leucocyte antigen (HLA) complex, particularly HLA DR4 show only a consistent association with RA [41]. The PRL gene is in close proximity to the HLA region on the short arm of chromosome six. Thus, Brennan et al. investigated linkage disequilibrium between HLA-DRB1 disease susceptibility alleles and microsatellite markers close to the PRL gene in women with RA [42,43]. They found that there may be linkage disequilibrium between HLA-DRB1 alleles and microsatellite marker alleles close to the PRL gene among women with RA. This suggests the possibility of extended haplotypes encoding for HLA-DRB1 susceptibility and high PRL production, which contributes to susceptibility to RA. The next step to do is comparing circulating PRL levels, PRL production by synovium-infiltrating lymphocytes and HLA haplotypes.

As mentioned above clinical observations and experimental animal studies have suggested the importance of circulating PRL in the development of RA. However, its precise role in establishing the local pathological changes of RA has remained to be elucidated. To clarify the role of PRL in the pathologic responses occurring in the affected joints of RA patients, we examined the effects of PRL on RA synovial cell functions, and analyzed the distribution of PRL producing cells and PRLR bearing cells in synovial tissues [44].

3. THE EFFECTS OF PROLACTIN ON RA SYNOVIAL CELL FUNCTIONS

To test whether PRL has a direct effects on RA synovial cell functions, we first measured proliferative responses of RA synovial cells in the presence of PRL. We found that the proliferation was enhanced by the addition of PRL into the cell cultures in a dose-dependent manner. We next studied the effects of PRL on proinflammatory cytokine production by RA synovial cells, including IL-6 and IL-8. RA fibroblast like synovial cells were cultured with various concentrations of PRL for 24 h. Subsequent culture supernatants were recovered. RA synovial cells spontaneously produced substantial amounts of IL-6 and IL-8; treatment of RA synovial cells with PRL enhanced IL-6 and IL-8 production reproducibly.

It has been shown that MMPs secreted by synovial cells are involved in cartilage destruction in RA lesions [45]. The enzymatic activity is abrogated by TIMP, which is produced by the same cells [45]. Thus, imbalance between MMPs and TIMP synthesized in RA synovial cells may be intimately associated with the induction of cartilage destruction. Therefore, we asked whether PRL modulates MMP-3, which is one of the dominant enzymes in RA lesions, and TIMP-1 production by RA fibroblast like synovial cells. We found that PRL enhanced MMP-3 production, and inhibited TIMP-1 production by the RA synovial cells. These results suggest that PRL affects RA synovial cells to promote cartilage destruction in the lesions through the upregulation of overall collagenase activity.

4. PRL PRODUCING AND PRLR BEARING CELLS IN THE SYNOVIUM OF RA PATIENTS

Because we found that PRL enhances several aspects of RA synovial cell function, it is of interest to clarify whether PRL is actually present within the RA joints. To this end, we first examined whether PRL producing cells are present in RA synovial tissues. We found that the vast majority of PRL producing cells are lymphocytes that infiltrate areas of RA synovial tissues. We also found that the infiltrating lymphocytes consisted of a large number of CD4+ cells and relatively small number of CD8+ cells. These results suggest that the majority of PRL producing cells are synovium infiltrating CD4+ T cells. In addition, synovial cells in the sublining layer were consistently stained with anti-PRL mAb, while synovial cells residing in the lining layer of synovium were not stained at all. Furthermore, RA fibroblast like synovial cell lines which were devoid of macrophage like synovial cells were positive for PRL. These results suggest that synovium infiltrating lymphocytes and fibroblast like synovial cells are largely responsible for PRL production in the joints of RA patients.

As a next turn, we have studied the distribution of PRLR in RA synovial tissues. PRLR was expressed on lymphocytes in the lymphocyte infiltrating areas and fibroblast like synovial cells in sublining layer of RA synovium. PRLR was also detected in long-term cultured fibroblast like synovial cell lines of RA patients. We confirmed PRLR expression at the mRNA expression levels by RT-PCR. The results indicate that infiltrating T lymphocytes and fibroblast like synovial cells produce PRL in the RA synovium. In turn, the PRL stimulates fibroblast like synovial cells to produce proinflammatory cytokines and MMPs, leading to the paracrine and autocrine stimulation of synovial cells and to the exacerbation of pathological responses in patients with RA.

5. MECHANISMS OF ACTION OF PRL IN RA SYNOVIAL CELLS

The binding of PRL to PRLR results in intracellular signaling which triggers the translocation of STAT from cytoplasm into the nucleus, as a final afferent biochemical event in lymphocytes [21,46,47]. To examine if this is the case for RA synovial cells, we examined the intracellular localization of STAT-5 in RA synovial cells in the presence or absence of exogenous PRL by immunocytochemical staining. We found that STAT-5 was stained in cytoplasm of RA synovial cells without PRL stimulation. Stimulation of the synovial cells by PRL for one hour induced the translocation of STAT-5 into the nucleus. These results suggest that STAT is involved in the PRL induced enhancement of RA synovial cell functions, through its transcription factor activity.

6. EFFECTS OF BROMOCRIPTINE (BRC) ON RA SYNOVIAL CELL FUNCTIONS

If PRL is really involved in the pathogenesis of RA, BRC, which inhibits the secretion of PRL, could ameliorate the disease. Indeed, it has already been reported that BRC is effective for treating patients with RA. We thus next studied the effects of BRC on the RA synovial cell functions *in vitro*. *In vitro* BRC treatment inhibited PRL mRNA expression and TNF-alpha mRNA expression in primary RA synovial cells including synovium infiltrating T cells. We found that BRC inhibited spontaneous PRL secretion and TNF-alpha production by the primary RA synovial cells in a concentration dependent manner. These data suggest that PRL stimulated macrophage like synovial cells as well as fibroblast like synovial cells, because macrophage like synovial cells produce a large amount of TNF-alpha in the RA synovium.

In vitro BRC treatment of primary RA synovial cells inhibited their proliferation. IL-6 production and tissue destructive MMP activities including MMP-2, 8 and 9 secreted by primary RA synovial cells were similarly inhibited by the BRC. Excessive amounts of exogenous PRL reversed the suppressive effect of BRC *in vitro*. Thus, BRC as a PRL inhibitor, reduces excessive RA synovial cell functions.

7. DISCUSSION

Here, we have focused on the local production of PRL within the affected joints of RA patients and found the existence of PRL producing lymphocytes and PRLR bearing synovial cells in RA synovial tissues. We next examined its effects on RA synovial cells, and found that *in vitro* treatment with PRL induced an enhanced proliferation of and excessive production of proinflammatory cytokines and tissue destructive proteolytic enzymes by RA synovial cells. We also found that PRL inhibited TIMP-1 production by the synovial cells. Thus, total collagenase activity in the joints may be upregulated by PRL [44].

However, several studies have argued against the pathological role of PRL in RA. PRL inhibitors have shown both positive and negative results for treating patients with RA [35,36,38–40]. In spite of disease promoting effects of PRL on collagen induced arthritis during the immunization phase, the same reports revealed that BCR caused exacerbation at a later stage of collagen induced arthritis [33,34]. Presumably, this discrepancy arises from the diverse pharmacological effects of BCR, as an agent that inhibits not only PRL, but also regulates other hormonal factors, which may modulate the disease processes of RA in parallel.

Furthermore, even if the therapeutic protocols are targeted only for PRL production, the effects would be too diverse because of a broad distribution of PRL producing cells and PRLR bearing cells in humans [18,21]. The fact that too many complicated factors are involved in the effects of BCR on RA makes straightforward interpretation of the *in vivo* data difficult. However, our *in vitro* experiments employing purified fibroblast like synovial cells, exogenous PRL and BRC showed beneficial effects of BRC on the pathological process of RA. Thus, BRC, or other types of PRL specific inhibitors may be applicable for treating patients with RA who have high PRL levels in their circulation and/or in their joints. We want to emphasize the possibility that even low levels of RA synovial cell proliferation could be inhibited by treatment with a PRL inhibitor, and this could be beneficial for the patients on long-term follow up.

In conclusion, PRL was spontaneously produced by synovium infiltrating T lymphocytes and fibroblast like synovial cells in RA synovial tissue. This hormone has the potential to induce excessive synovial cell functions including enhanced proliferation, proinflammatory cytokine and MMP production. Our results support the hypothesis that PRL produced locally plays an important role in the pathogenesis of rheumatoid arthritis.

REFERENCES

1. Straub RH, Cutolo M. Involvement of the hypothalamic–pituitary–adrenal/gonadal axis and the peripheral nervous system in rheumatoid arthritis: viewpoint based on a systemic pathogenetic role. Arthritis Rheum 2001;44:493–507.
2. Sakane T, Suzuki N. Neuro-endocrine-immune axis in human rheumatoid arthritis. Arch Immunol Ther Exp (Warsz) 2000;48:417–27.
3. Takeba Y, Suzuki N, Kaneko A, Asai T, Sakane T. Evidence for neural regulation of inflammatory synovial cell functions by secreting calcitonin gene-related peptide and vasoactive intestinal peptide in patients with rheumatoid arthritis. Arthritis Rheum 1999;42:2418–29.
4. Sakane T, Suzuki N. The role of somatostatin in the pathophysiology of rheumatoid arthritis. Clin Exp Rheumatol 1998;16:745–9.
5. Takeba Y, Suzuki N, Takeno M, et al. Modulation of synovial cell function by somatostatin in patients with rheumatoid arthritis. Arthritis Rheum 1997;40:2128–38.
6. Choy EH, Panayi GS. Cytokine pathways and joint inflammation in rheumatoid arthritis. N Engl J Med 2001;344:907–16.
7. Soden M, Rooney M, Cullen A, et al. Immunohistological features in the synovium obtained from clinically uninvolved knee joints of patients with rheumatoid arthritis. Arthritis Rheum 1989;28:287–92.
8. Klimiuk PA, Yang H, Goronzy JJ, Weyand CM. Production of cytokines and metalloproteinases in rheumatoid synovitis is T cell dependent. Clin Immunol 1999;90:65–78.
9. Feldmann M, Maini RN. Anti-TNFalpha therapy of rheumatoid arthritis: What have we learned? Annu Rev Immunol 2001;19:163–196.
10. McMurray RW, Allen SH, Pepmueller PH, Keisler D, Cassidy JT. Elevated serum prolactin levels in children with juvenile rheumatoid arthritis and antinuclear antibody seropositivity. J Rheumatol 1995;22:1577–80.
11. Panayi GS. Hormonal control of rheumatoid inflammation. Br Med Bulletin 1995;51:462–71.
12. Neidhart M. Prolactin in autoimmune diseases. Proc Soc Exp Biol Med

1998;217:408–19.
13. Chikanza IC, Petrou P, Chrousos G, Kingsley G, Panayi G. Excessive and dysregulated secretion of prolactin in rheumatoid arthritis: Immunopathogenetic and therapeutic implications. Br J Rheum 1993;32:445–8.
14. Gagnerault M, Touraine P, Savino W, Kelly PA, Dardenne M. Expression of prolactin receptors in murine lymphoid cells in normal and autoimmune situations. J Immunol 1993;150:5673–81.
15. Folomeev M, Prokaeva T, Nassonova V. Prolactin levels in men with SLE and RA [letter]. J Rheumatol 1993;17:1569–70.
16. Jorgensen C, Sany J. Modulation of the immune responses by the neuro-endocrine axis in rheumatoid arthritis. Clin Exp Rheumatol 1994;12:435–41.
17. Jorgensen C, Maziad H, Bologna C, Sany J. Kinetics of prolactin release in rheumatoid arthritis. Clin Exp Rheumatol 1995;13:705–942.
18. Freeman ME, Kanyicska B, Lerant A, Nagy G. Prolactin: structure, function, and regulation of secretion. Physiol Rev 2000;80:1523–631.
19. Chikanza IC. Prolactin and neuroimmunomodulation: *in vitro* and *in vivo* observations. Ann NY Acad Sci 1999;876:119–30.
20. Goffin V, Kelly PA. The prolactin/growth hormone receptor family: structure/function relationships. J Mammary Gland Biol Neoplasia 1997;2:7–17.
21. Gallego MI, Binart N, Robinson GW, et al. Prolactin, growth hormone, and epidermal growth factor activate Stat5 in different compartments of mammary tissue and exert different and overlapping developmental effects. Dev Biol 2001;229:163–75.
22. Goffin V, Bouchard B, Ormandy CJ, et al. Prolactin: a hormone at the crossroads of neuroimmunoendocrinology. Ann NY Acad Sci 1998;840:498–509.
23. Russell DH, Kibler R, Matrisian L, et al. Prolactin receptors on human T and B lymphocytes: antagonism of prolactin binding by cyclosporin. J Immunol 1985;134:3027–31.
24. Clevenger CV, Medaglia MV. The protein tyrosine kinase p56fyn is associated with prolactin (PRL) receptor and is activated by PRL stimulation of T-lymphocytes. Mol Endocrinol 1994;8:674–81.
25. Reber PM. Prolactin and immunomodulation. Am J Med 1993;95:637–44.
26. Sabharwal P, Glaser R, Lafuse W, et al. Prolactin synthesized and secreted by human peripheral blood mononuclear cells: an autocrine growth factor for lymphoproliferation. Proc Natl Acad Sci USA 1992;89:7713–16.
27. Clevenger CV, Freier DO, Kline JB. Prolactin receptor signal transduction in cells of the immune system. J Endocrinol 1998;157:187–97.
28. Wilder RL. Adrenal and gonadal steroid hormone deficiency in the pathogenesis of rheumatoid arthritis. J Rheumatol Suppl 1996;44:10–2.
29. Matera L, Mori M, Geuna M, Buttiglieri S, Palestro G. Prolactin in autoimmunity and antitumor defence. J Neuroimmunol 2000;109:47–55.
30. Lahita RG. The role of sex hormones in systemic lupus erythematosus. Curr Opin Rheumatol 1999;11:352–6.
31. Ferrari C, Mattei A, Melis GB, et al. Cabergoline: long acting oral treatment of hyperprolactinemic disorders. J Clin Endocrinol Metab 1989;68:1201–6.
32. Kullich WC, Klein G. High levels of macrophage inflammatory protein-1 alpha correlate with prolactin in female patients with active rheumatoid arthritis. Clin Rheumatol 1998;17:263–4.
33. Whyte A, Williams RO. Bromocriptine suppresses postpartum exacerbation of collagen-

induced arthritis. Arthritis Rheum 1988;31:927–8.

34. Mattsson R, Mattsson A, Hansson I, et al. Increased levels of prolactin during, but not after, the immunisation with rat collagen II enhances the course of arthritis in DBA/1 mice. Autoimmunity 1992;11:163–70.
35. Figueroa F, Carrion F, Martinez ME, et al. Effect of bromocriptine in patients with active rheumatoid arthritis. Rev Med Chil 1998;126:33–41.
36. Figueroa FE, Carrion F, Martinez ME, Rivero S, Mamani I. Bromocriptine induces immunological changes related to disease parameters in rheumatoid arthritis [letter]. Br J Rheum 1997;36:1022–3.
37. Jara LJ, Lavalle C, Fraga A, et al. Prolactin, immunoregulation, and autoimmune diseases. Semin Arthritis Rheum 1991;20:273–84.
38. Lavalle C. Prolactin-a hormone with immunoregulatory properties that leads to new therapeutic approaches in rheumatic diseases. J Rheumatol 1992;19:839–41.
39. Dougados M, Duchesne L, Amor B. Bromocriptine and cyclosporin A combination therapy in rheumatoid arthritis. Arthritis Rheum 1988;31:1333–4.
40. Erb N, Pace AV, Delamere JP, Kitas GD. Control of unremitting rheumatoid arthritis by the prolactin antagonist cabergoline. Rheumatology (Oxford) 2001;40:237–9.
41. Weyand CM, Goronzy JJ. Association of MHC and rheumatoid arthritis. HLA polymorphisms in phenotypic variants of rheumatoid arthritis. Arthritis Res 2000;2:212–6.
42. Brennan P, Hajeer A, Ong KR, et al. Related Articles Allelic markers close to prolactin are associated with HLA-DRB1 susceptibility alleles among women with rheumatoid arthritis and systemic lupus erythematosus. Arthritis Rheum 1997;40:1383–6.
43. Brennan P, Ollier B, Worthington J, Hajeer A, Silman A. Are both genetic and reproductive associations with rheumatoid arthritis linked to prolactin? Lancet 1996;348:106–9.
44. Nagafuchi H, Suzuki N, Kaneko A, Asai T, Sakane T. Prolactin locally produced by synovium infiltrating T lymphocytes induces excessive synovial cell functions in patients with rheumatoid arthritis. J Rheumatol 1999;26:1890–900.
45. Fearon U, Reece R, Smith J, Emery P, Veale DJ. Synovial cytokine and growth factor regulation of MMPs/TIMPs: implications for erosions and angiogenesis in early rheumatoid and psoriatic arthritis patients. Ann NY Acad Sci 1999;878:619–21.
46. Kirken RA, Malabarba MG, Xu J, et al. Prolactin stimulates serine/tyrosine phosphorylation and formation of heterocomplexes of multiple Stat5 isoforms in Nb2 lymphocytes. J Biol Chem 1997;272:14098–103.
47. Gouilleux F, Moritz D, Humar M, et al. Prolactin and interleukin-2 receptors in T lymphocytes signal through a MGF-STAT5-like transcription factor. Endocrinol 1995;136:5700–8.

Keyword index